AN INTRODUCTION TO STATISTICAL MECHANICS AND THERMODYNAMICS

T0291139

An Introduction to Statistical Mechanics and Thermodynamics

Second Edition

Robert H. Swendsen

OXFORD
UNIVERSITY PRESS

OXFORD
UNIVERSITY PRESS

Great Clarendon Street, Oxford, OX2 6DP,
United Kingdom

Oxford University Press is a department of the University of Oxford.
It furthers the University's objective of excellence in research, scholarship,
and education by publishing worldwide. Oxford is a registered trade mark of
Oxford University Press in the UK and in certain other countries

© Robert H. Swendsen 2020

The moral rights of the author have been asserted

First Edition published in 2012
Second Edition published in 2020
First published in paperback 2024

Published in the United States of America by Oxford University Press
198 Madison Avenue, New York, NY 10016, United States of America

British Library Cataloguing in Publication Data
Data available

Library of Congress Cataloging in Publication Data
Data available

ISBN 978-0-19-885323-7 (Hbk.)
ISBN 978-0-19-890778-7 (Pbk.)

DOI: 10.1093/oso/9780198853237.001.0001

To the memory of Herbert B. Callen, physicist and mentor,
and to my wife, Roberta L. Klatzky,
without whom this book could never have been written.

Preface to the Second Edition

In preparing to write the preface to the second edition of this book, I realized anew my debt to the work of Herbert B. Callen. He was not only my thesis advisor and my friend, but it was through his teaching and his book on thermodynamics that I first understood the subject in any depth. I take this opportunity once again to acknowledge how much his pedagogy and advice have meant to my work.

The postulational approach to thermodynamics, which is primarily based on his work and that of his thesis advisor, László Tisza, provides a clear basis for the theory. It is not difficult to understand but can seem rather abstract when first encountered as a student – as, indeed, it did to me many years ago. Many professors have told me that they thought that Callen's book was too daunting to give to their students, but that it was the book that they consulted for thermodynamics.

Part I of my book originated as an introduction to Callen's *Thermodynamics* in my teaching. One difficulty that I had found as a student was that Callen's book started off presenting entropy and the postulates of thermodynamics in the first chapter, and temperature as a partial derivative of the entropy in the second chapter. I had only a vague idea at the time of what the entropy was, and its partial derivative with respect to energy was a complete mystery. I have tried to avoid this difficulty in my own teaching of thermodynamics by presenting the students with an explicit calculation of the entropy of a classical ideal gas. All assumptions are stated, and all mathematics is explained. I felt – and my students generally agreed – that they were then ready to understand Callen's postulates.

Part II developed from my notes for teaching from Callen's textbook. I found that while the ideas in Callen's postulates provided a great foundation for thermodynamics, their specific form was less than ideal. For the first edition of this book, I separated them into six new postulates, each of which expressed a separate idea. I also generalized the postulates to include non-homogeneous systems.

I gave an explicit guide to the use of Jacobians in deriving thermodynamic identities, which I have not found anywhere else, but which my students have found to be easy to apply. Callen mentioned Jacobians in his first edition, but not in his second. Similarly, I simplified the derivation of Maxwell relations, with the result that my students have regarded them (correctly) as being easy to derive.

I also gave an explicit derivation of the stability criteria for second partial derivatives with respect to intensive variables because many students had difficulty with them.

Parts III (classical statistical mechanics) and IV (quantum statistical mechanics) used computer calculations extensively. They allowed many calculations to be carried out explicitly. I firmly believe that the future of physics will rely heavily on the computer, and I think that computation is currently being neglected in university curricula.

The second edition has come into being because I have discovered how to clarify the presentation of many of the central concepts, especially in the derivation of the entropy in Part 1. Along the way, I have corrected a significant number of typographical errors.

In Part I, Chapters 4 and 6, I have more clearly distinguished generic variables from variables describing particular systems used in derivations. My previous labeling convention did not cause any problems in the classes I taught, but it has caused confusion with some readers. I have also generalized the derivation of the entropy from treating only two systems at a time to deriving the entropy simultaneously for all systems that might interact.

In the second edition, I have again changed the list of postulates to include the possibility of negative temperatures. Callen had mentioned negative temperatures in his book, but had excluded them in the interest of simplicity.

In Chapter 11, I have expanded the review of the Carnot cycle with two new illustrations. This chapter now also contains a discussion of negative temperatures, and how they affect the analysis of heat engines.

Massieu functions were mentioned by Callen, but not developed. I did the same in the first edition. I have expanded the treatment of Massieu functions in Chapter 12, after realizing that they are much more useful than I had previously thought. They are essential when considering negative temperatures because the corresponding entropy is not monotonic.

The discussion of the Nernst Postulate (Third Law of Thermodynamics) in Chapter 18 includes a discussion of why zero temperature would not be possible to attain if classical mechanics were valid instead of quantum mechanics. In fact, it would be *more* difficult to attain very low temperatures if the Nernst Postulate were not valid.

A new chapter (Chapter 21) has been added to discuss the consequences of including the widths of the energy and particle-number distributions in the calculation of the entropy. It is both a more realistic assumption and gives better expressions for the entropy. These results are based on new work since the publication of the first edition of this book.

In Chapters 28 on Bose-Einstein statistics and 29 on Fermi-Dirac statistics, I've introduced numerical calculations based on work with a former student, Tyson Price. The numerical results show many of the thermal properties of Bose and Fermi gases more clearly and simply than would be possible with analytic calculations alone.

The Index has been thoroughly updated and expanded.

My recommendations for a programming language to use for the computational problems have changed. I still advocate the use of Python, although not VPython. I have found that plots using MatPlotLib are much better, as well as being easier for students (and professors) to program. On the other hand, I have found that students prefer the freedom to use a wide variety of programming languages, and I have never insisted that they use Python.

I would like to thank my colleagues, Markus Deserno and Michael Widom, for their helpful comments based on their own experiences from using my book to teach both undergraduate and graduate courses in thermal physics.

I would also like to thank my former students, William Griffin, Lachlan Lancaster, and Michael Matty, for their contributions to some of the results presented here. I would especially like to thank Michael Matty for his extensive constructive criticism of the text and his contributions to my class. Finally, I would like to thank Karpur Shukla for many useful conversations.

As in the first preface, I would like to thank my wife, Roberta L. Klatzky, for her unwavering support.

Robert H. Swendsen
Pittsburgh, April 2019

Preface to the First Edition

Habe Muth dich deines eigenen Verstandes zu bedienen.
(Have the courage to think for yourself.)

Immanuel Kant, in *Beantwortung der Frage: Was ist Aufklärung?*

The disciplines of statistical mechanics and thermodynamics are very closely related, although their historical roots are separate. The founders of thermodynamics developed their theories without the advantage of contemporary understanding of the atomic structure of matter. Statistical mechanics, which is built on this understanding, makes predictions of system behavior that lead to thermodynamic rules. In other words, statistical mechanics is a conceptual precursor to thermodynamics, although it is a historical latecomer.

Unfortunately, despite their theoretical connection, statistical mechanics and thermodynamics are often taught as separate fields of study. Even worse, thermodynamics is usually taught first, for the dubious reason that it is older than statistical mechanics. All too often the result is that students regard thermodynamics as a set of highly abstract mathematical relationships, the significance of which is not clear.

This book is an effort to rectify the situation. It presents the two complementary aspects of thermal physics as a coherent theory of the properties of matter. My intention is that after working through this text a student will have solid foundations in both statistical mechanics and thermodynamics that will provide direct access to modern research.

Guiding Principles

In writing this book I have been guided by a number of principles, only some of which are shared by other textbooks in statistical mechanics and thermodynamics.

- I have written this book for students, not professors. Many things that experts might take for granted are explained explicitly. Indeed, student contributions have been essential in constructing clear explanations that do not leave out 'obvious' steps that can be puzzling to someone new to this material.
- The goal of the book is to provide the student with conceptual understanding, and the problems are designed in the service of this goal. They are quite challenging, but the challenges are primarily conceptual rather than algebraic or computational.

- I believe that students should have the opportunity to program models themselves and observe how the models behave under different conditions. Therefore, the problems include extensive use of computation.

- The book is intended to be accessible to students at different levels of preparation. I do not make a distinction between teaching the material at the advanced undergraduate and graduate levels, and indeed, I have taught such a course many times using the same approach and much of the same material for both groups. As the mathematics is entirely self-contained, students can master all of the material even if their mathematical preparation has some gaps. Graduate students with previous courses on these topics should be able to use the book with self-study to make up for any gaps in their training.

- After working through this text, a student should be well prepared to continue with more specialized topics in thermodynamics, statistical mechanics, and condensed-matter physics.

Pedagogical Principles

The over-arching goals described above result in some unique features of my approach to the teaching of statistical mechanics and thermodynamics, which I think merit specific mention.

Teaching Statistical Mechanics

- The book begins with *classical* statistical mechanics to postpone the complications of quantum measurement until the basic ideas are established.

- I have defined ensembles in terms of probabilities, in keeping with Boltzmann's vision. In particular, the discussion of statistical mechanics is based on Boltzmann's 1877 definition of entropy. This is not the definition usually found in textbooks, but what he actually wrote. The use of Boltzmann's definition is one of the key features of the book that enables students to obtain a deep understanding of the foundations of both statistical mechanics and thermodynamics.

- A self-contained discussion of probability theory is presented for both discrete and continuous random variables, including all material needed to understand basic statistical mechanics. This material would be superfluous if the physics curriculum were to include a course in probability theory, but unfortunately, that is not usually the case. (A course in statistics would also be very valuable for physics students— but that is another story.)

- Dirac delta functions are used to formulate the theory of continuous random variables, as well as to simplify the derivations of densities of states. This is not the way mathematicians tend to introduce probability densities, but I believe that it is by far the most useful approach for scientists.

- Entropy is presented as a logical consequence of applying probability theory to systems containing a large number of particles, instead of just an equation to be memorized.
- The entropy of the classical ideal gas is derived in detail. This provides an explicit example of an entropy function that exhibits all the properties postulated in thermodynamics. The example is simple enough to give every detail of the derivation of thermodynamic properties from statistical mechanics.
- The book includes an explanation of Gibbs' paradox—which is not really paradoxical when you begin with Boltzmann's 1877 definition of the entropy.
- The apparent contradiction between observed irreversibility and time-reversal-invariant equations of motion is explained. I believe that this fills an important gap in a student's appreciation of how a description of macroscopic phenomena can arise from statistical principles.

Teaching Thermodynamics

- The four fundamental postulates of thermodynamics proposed by Callen have been reformulated. The result is a set of six thermodynamic postulates, sequenced so as to build conceptual understanding.
- Jacobians are used to simplify the derivation of thermodynamic identities.
- The thermodynamic limit is discussed, but the validity of thermodynamics and statistical mechanics does not rely on taking the limit of infinite size. This is important if thermodynamics is to be applied to real systems, but is sometimes neglected in textbooks.
- My treatment includes thermodynamics of non-extensive systems. This allows me to include descriptions of systems with surfaces and systems enclosed in containers.

Organization and Content

The principles I have described above lead me to an organization for the book that is quite different from what has become the norm. As was stated above, while most texts on thermal physics begin with thermodynamics for historical reasons, I think it is far preferable from the perspective of pedagogy to begin with statistical mechanics, including an introduction to those parts of probability theory that are essential to statistical mechanics.

To postpone the conceptual problems associated with quantum measurement, the initial discussion of statistical mechanics in Part I is limited to classical systems. The entropy of the classical ideal gas is derived in detail, with a clear justification for every step. A crucial aspect of the explanation and derivation of the entropy is the use of Boltzmann's 1877 definition, which relates entropy to the probability of a macroscopic state. This definition provides a solid, intuitive understanding of what entropy is all about. It is my experience that after students have seen the derivation of the entropy of the classical ideal gas, they immediately understand the postulates of thermodynamics, since those postulates simply codify properties that they have derived explicitly for a special case.

The treatment of statistical mechanics paves the way to the development of thermodynamics in Part II. While this development is largely based on the classic work by Herbert Callen (who was my thesis advisor), there are significant differences. Perhaps the most important is that I have relied entirely on Jacobians to derive thermodynamic identities. Instead of regarding such derivations with dread—as I did when I first encountered them—my students tend to regard them as straightforward and rather easy. There are also several other changes in emphasis, such as a clarification of the postulates of thermodynamics and the inclusion of non-extensive systems; that is, finite systems that have surfaces or are enclosed in containers.

Part III returns to classical statistical mechanics and develops the general theory directly, instead of using the common roundabout approach of taking the classical limit of quantum statistical mechanics. A chapter is devoted to a discussion of the apparent paradoxes between microscopic reversibility and macroscopic irreversibility.

Part IV presents quantum statistical mechanics. The development begins by considering a probability distribution over all quantum states, instead of the common *ad hoc* restriction to eigenstates. In addition to the basic concepts, it covers black-body radiation, the harmonic crystal, and both Bose and Fermi gases. Because of their practical and theoretical importance, there is a separate chapter on insulators and semiconductors. The final chapter introduces the Ising model of magnetic phase transitions.

The book contains about a hundred multi-part problems that should be considered as part of the text. In keeping with the level of the text, the problems are fairly challenging, and an effort has been made to avoid 'plug and chug' assignments. The challenges in the problems are mainly due to the probing of essential concepts, rather than mathematical complexities. A complete set of solutions to the problems is available from the publisher.

Several of the problems, especially in the chapters on probability, rely on computer simulations to lead students to a deeper understanding. In the past I have suggested that my students use the C++ programming language, but for the last two years I have switched to VPython for its simplicity and the ease with which it generates graphs. An introduction to the basic features of VPython is given in in Appendix A. Most of my students have used VPython, but a significant fraction have chosen to use a different language—usually Java, C, or C++. I have not encountered any difficulties with allowing students to use the programming language of their choice.

Two Semesters or One?

The presentation of the material in this book is based primarily on a two-semester undergraduate course in thermal physics that I have taught several times at Carnegie Mellon University. Since two-semester undergraduate courses in thermal physics are rather unusual, its existence at Carnegie Mellon for several decades might be regarded as surprising. In my opinion, it should be the norm. Although it was quite reasonable to teach two semesters of classical mechanics and one semester of thermodynamics to undergraduates in the nineteenth century—the development of statistical mechanics was just beginning—it is not reasonable in the twenty-first century.

However, even at Carnegie Mellon only the first semester of thermal physics is required. All physics majors take the first semester, and about half continue on to the second semester, accompanied by a few students from other departments. When I teach the course, the first semester covers the first two parts of the book (Chapters 1 through 18), plus an overview of classical canonical ensembles (Chapter 18) and quantum canonical ensembles (Chapter 22). This gives the students an introduction to statistical mechanics and a rather thorough knowledge of thermodynamics, even if they do not take the second semester.

It is also possible to teach a one-semester course in thermal physics from this book using different choices of material. For example:

- If the students have a strong background in probability theory (which is, unfortunately, fairly rare), Chapters 3 and 5 might be skipped to include more material in Parts III and IV.

- If it is decided that students need a broader exposure to statistical mechanics, but that a less detailed study of thermodynamics is sufficient, Chapters 14 through 17 could be skimmed to have time to study selected chapters in Parts III and IV.

- If the students have already had a thermodynamics course (although I do not recommend this course sequence), Part II could be skipped entirely. However, even if this choice is made, students might still find Chapters 9 to 18 useful for review.

One possibility that I do not recommend would be to skip the computational material. I am strongly of the opinion that the undergraduate physics curricula at most universities still contain too little instruction in the computational methods that students will need in their careers.

Acknowledgments

This book was originally intended as a resource for my students in Thermal Physics I (33–341) and Thermal Physics II (33–342) at Carnegie Mellon University. In an important sense, those students turned out to be essential collaborators in its production.

I would like to thank the many students from these courses for their great help in suggesting improvements and correcting errors in the text. All of my students have made important contributions. Even so, I would like to mention explicitly the following students: Michael Alexovich, Dimitry Ayzenberg, Conroy Baltzell, Anthony Bartolotta, Alexandra Beck, David Bemiller, Alonzo Benavides, Sarah Benjamin, John Briguglio, Coleman Broaddus, Matt Buchovecky, Luke Ceurvorst, Jennifer Chu, Kunting Chua, Charles Wesley Cowan, Charles de las Casas, Matthew Daily, Brent Driscoll, Luke Durback, Alexander Edelman, Benjamin Ellison, Danielle Fisher, Emily Gehrels, Yelena Goryunova, Benjamin Greer, Nils Guillermin, Asad Hasan, Aaron Henley, Maxwell

Hutchinson, Andrew Johnson, Agnieszka Kalinowski, Patrick Kane, Kamran Karimi, Joshua Keller, Deena Kim, Andrew Kojzar, Rebecca Krall, Vikram Kulkarni, Avishek Kumar, Anastasia Kurnikova, Thomas Lambert, Grant Lee, Robert Lee, Jonathan Long, Sean Lubner, Alan Ludin, Florence Lui, Christopher Magnollay, Alex Marakov, Natalie Mark, James McGee, Andrew McKinnie, Jonathan Michel, Corey Montella, Javier Novales, Kenji Oman, Justin Perry, Stephen Poniatowicz, Thomas Prag, Alisa Rachubo, Mohit Raghunathan, Peter Ralli, Anthony Rice, Svetlana Romanova, Ariel Rosenburg, Matthew Rowe, Kaitlyn Schwalje, Omar Shams, Gabriella Shepard, Karpur Shukla, Stephen Sigda, Michael Simms, Nicholas Steele, Charles Swanson, Shaun Swanson, Brian Tabata, Likun Tan, Joshua Tepper, Kevin Tian, Eric Turner, Joseph Vukovich, Joshua Watzman, Andrew Wesson, Justin Winokur, Nanfei Yan, Andrew Yeager, Brian Zakrzewski, and Yuriy Zubovski. Some of these students made particularly important contributions, for which I have thanked them personally. My students' encouragement and suggestions have been essential in writing this book.

Yutaro Iiyama and Marilia Cabral Do Rego Barros have both assisted with the grading of Thermal Physics courses, and have made very valuable corrections and suggestions.

The last stages in finishing the manuscript were accomplished while I was a guest at the Institute of Statistical and Biological Physics at the Ludwig-Maximilians-Universität, Munich, Germany. I would like to thank Prof. Dr. Erwin Frey and the other members of the Institute for their gracious hospitality.

Throughout this project, the support and encouragement of my friends and colleagues Harvey Gould and Jan Tobochnik have been greatly appreciated.

I would also like to thank my good friend Lawrence Erlbaum, whose advice and support have made an enormous difference in navigating the process of publishing a book.

Finally, I would like to thank my wife, Roberta (Bobby) Klatzky, whose contributions are beyond count. I could not have written this book without her loving encouragement, sage advice, and relentless honesty.

My thesis advisor, Herbert Callen, first taught me that statistical mechanics and thermodynamics are fascinating subjects. I hope you come to enjoy them as much as I do.

Robert H. Swendsen
Pittsburgh, January 2011

Contents

Part IV   Quantum Statistical Mechanics

    23.1   Basic Quantum Mechanics   286
    23.2   Energy Eigenstates   287
    23.3   Many-Body Systems   290
    23.4   Two Types of Probability   290
    23.5   The Density Matrix   293
    23.6   Uniqueness of the Ensemble   294
    23.7   The Planck Entropy   295
    23.8   The Quantum Microcanonical Ensemble   296

24 Quantum Canonical Ensemble   297

    24.1   Derivation of the QM Canonical Ensemble   297
    24.2   Thermal Averages and the Average Energy   299
    24.3   The Quantum Mechanical Partition Function   299
    24.4   The Quantum Mechanical Entropy   302
    24.5   The Origin of the Third Law of Thermodynamics   303
    24.6   Derivatives of Thermal Averages   305
    24.7   Factorization of the Partition Function   306
    24.8   Special Systems   308
    24.9   Two-Level Systems   309
    24.10 Simple Harmonic Oscillator   311
    24.11 Einstein Model of a Crystal   313
    24.12 Problems   315

25 Black-Body Radiation   322

    25.1   Black Bodies   322
    25.2   Universal Frequency Spectrum   322
    25.3   A Simple Model   323
    25.4   Two Types of Quantization   323
    25.5   Black-Body Energy Spectrum   325
    25.6   Total Energy   328
    25.7   Total Black-Body Radiation   329
    25.8   Significance of Black-Body Radiation   329
    25.9   Problems   330

26 The Harmonic Solid   331

    26.1   Model of a Harmonic Solid   331
    26.2   Normal Modes   332
    26.3   Transformation of the Energy   336
    26.4   The Frequency Spectrum   338

1

Introduction

If, in some cataclysm, all scientific knowledge were to be destroyed, and only one sentence passed on to the next generation of creatures, what statement would contain the most information in the fewest words? I believe it is the atomic hypothesis (or atomic fact, or whatever you wish to call it) that all things are made of atoms—little particles that move around in perpetual motion, attracting each other when they are a little distance apart, but repelling upon being squeezed into one another. In that one sentence you will see an enormous amount of information about the world, if just a little imagination and thinking are applied.

Richard Feynman, in *The Feynman Lectures on Physics*

1.1 Thermal Physics

This book is about the things you encounter in everyday life: the book you are reading, the chair on which you are sitting, the air you are breathing. It is about things that can be hot or cold; hard or soft; solid, liquid, or gas. It is about machines that work for you: automobiles, heaters, refrigerators, air conditioners. It is even about your own body and the stuff of life. The whole subject is sometimes referred to as thermal physics, and it is usually divided into two main topics: thermodynamics and statistical mechanics.

Thermodynamics is the study of everything connected with heat. It provides powerful methods for connecting observable quantities by equations that are not at all obvious, but are nevertheless true for *all* materials. Statistical mechanics is the study of what happens when large numbers of particles interact. It provides a foundation for thermodynamics and the ultimate justification of why thermodynamics works. It goes beyond thermodynamics to reveal deeper connections between molecular behavior and material properties. It also provides a way to calculate the properties of specific objects, instead of just the universal relationships provided by thermodynamics.

The ideas and methods of thermal physics differ from those of other branches of physics. Thermodynamics and statistical mechanics each require their own particular ways of thinking. Studying thermal physics is not about memorizing formulas; it is about gaining a new understanding of the world.

An Introduction to Statistical Mechanics and Thermodynamics. Robert H. Swendsen, Oxford University Press (2020).
© Robert H. Swendsen. DOI: 10.1093/oso/9780198853237.001.0001

1.2   What are the Questions?

Thermal physics seeks quantitative explanations for the properties of macroscopic objects, where the term 'macroscopic' means two things:

1. A macroscopic object is made up of a large number of particles.
2. Macroscopic measurements have limited resolution.

The apparently vague terms 'large number' and 'limited resolution' take on specific meanings with respect to the law of large numbers that we will study in Chapter 3. We will see that the relative statistical uncertainty in quantities like the energy and the number of particles is usually inversely proportional to the square root of the number of particles. If this uncertainty is much smaller than the experimental resolution, it can be neglected. This leads to thermodynamics, which is a description of the macroscopic properties that ignores small statistical fluctuations.

We will assume that we somehow know what the object is made of; that is, what kinds of atoms and molecules it contains, and how many of each. We will generally speak of 'systems' instead of objects. The difference is that a system can consist of any number of objects, and is somehow separated from the rest of the universe. We will concentrate on systems in equilibrium; that is, systems that have been undisturbed long enough for their properties to take on constant values. We shall be more specific about what 'long enough' means in Chapter 22.

In the simplest case—which we will consider in Part I—a system might be completely isolated by 'adiabatic' walls; that is, rigid barriers that let nothing through. We will also assume, at least at first, that we know how much energy is contained in our isolated system.

Given this information, we will ask questions about the properties of the system. We will first study a simple model of a gas, so we will want to know what temperature and pressure are, and how they are related to each other and to the volume. We will want to know how the volume or the pressure will change if we heat the gas. As we investigate more complicated systems, we will see more complex behavior and find more questions to ask about the properties of matter.

As a hint of things to come, we will find that there is a function of the energy E, volume V, and number of particles N that is sufficient to answer *all* questions about the thermal properties of a system. It is called the entropy, and it is denoted by the letter S.

If we can calculate the entropy as a function of E, V, and N, we know everything about the macroscopic properties of the system. For this reason, $S = S(E, V, N)$ is known as a fundamental relation, and it is the focus of Part I.

1.3   History

Atoms exist, combine into molecules, and form every object we encounter in life. Today, this statement is taken for granted. However, in the nineteenth century, and even into the early twentieth century, these were fighting words. The Austrian physicist Ludwig Boltzmann (1844–1906) and a small group of other scientists championed the idea of atoms and molecules as the basis of a fundamental theory of the thermal properties of matter, against the violent objections of another Austrian physicist, Ernst Mach (1838–1916), and many of his colleagues. Boltzmann was right, of course, but the issue had not been settled even at the time of his tragic suicide in 1906. In the tradition of Boltzmann, the intention in this book is to present thermodynamics as a consequence of the molecular nature of matter.

The theory of thermodynamics was developed without benefit of the atomic hypothesis. It began with the seminal work of Sadi Carnot (French physicist, 1792–1832), who initiated the scientific discussion of heat engines. The First Law of Thermodynamics was discovered by James Prescott Joule (English physicist, 1818–1889), when he established that heat was a form of energy and measured the mechanical equivalent of heat. The Second Law of Thermodynamics was first stated in 1850 by Rudolf Clausius (German physicist, 1822–1888), who in 1865 also invented the related concept of entropy. The Second Law can be expressed by the statement that the entropy of an isolated system can increase, but not decrease.

Entropy was a mystery to nineteenth-century scientists. Clausius had given entropy an experimental definition that allowed it to be calculated, but its meaning was puzzling. Like energy, it could not be destroyed, but unlike energy, it could be created. It was essential to the calculation of the efficiency of heat engines (machines that turn the energy in hot objects into mechanical work), but it did not seem to be related to any other physical laws.

The reason why scientists working in the middle of the nineteenth century found entropy so mysterious is that few of them thought in terms of atoms or molecules. Even with molecular theory, explaining the entropy required brilliant insight; without molecular theory, there was no hope.

Serious progress in understanding the properties of matter and the origins of thermodynamic laws on the basis of the molecular nature of matter began in the 1860s with the work of Boltzmann and the American physicist J. Willard Gibbs (1839–1903).

Gibbs worked from a formal starting point, postulating that observable quantities could be calculated from an average over many replicas of an object in different microscopic configurations, and then working out what the equations would have to be. His work is very beautiful (to a theoretical physicist), although somewhat formal. However, it left certain questions unresolved—most notably, what has come to be called 'Gibbs' paradox'. This issue concerned a discontinuous change in the entropy when differences in the properties of particles in a mixture were imagined to disappear continuously. Gibbs himself did not regard this as a paradox. The issues involved are still a matter of debate in the twenty-first century.

Boltzmann devoted most of his career to establishing the molecular theory of matter and deriving the consequences of the existence of atoms and molecules. One of his great achievements was his 1877 definition of entropy—a definition which provides a physical interpretation of the entropy and the foundation of statistical mechanics. Unfortunately, although Boltzmann's paper is often cited, it is rarely read. This had led to frequent misinterpretations of what he meant. At least part of the problem is that he wrote in a fairly dense nineteenth-century German, and an English translation has only appeared recently.[1] We will discuss what he actually wrote in some detail.

Part I of this book is devoted to developing an intuitive understanding of the concept of entropy, based on Boltzmann's definition. We will present an explicit, detailed derivation of the entropy for a simple model to provide insight into its significance. Later parts of the book will develop more sophisticated tools for investigating thermodynamics and statistical mechanics, but they are all based on Boltzmann's definition of the entropy.

1.4   Basic Concepts and Assumptions

This book is concerned with the macroscopic behavior of systems containing many particles. This is a broad topic since everything we encounter in the world around us contains enormous numbers of atoms and molecules. Even in relatively small objects, there are 10^{20} or more atoms.

Atoms and molecules are not the only examples of large numbers of particles in a system. Colloids, for example, can consist of 10^{12} or more microscopic particles suspended in a liquid. The large number of particles in a typical colloid means that they are also well described by statistical mechanics.

Another aspect of macroscopic experiments is that they have limited resolution. We will see in Part I that the fluctuations in quantities like the density of a system are approximately inversely proportional to the square root of the number of particles in the system. If there are 10^{20} molecules, this gives a relative uncertainty of about 10^{-10} for the average density. Because it is rare to measure the density to better than one part in 10^6, these tiny fluctuations are not seen in macroscopic measurements. Indeed, an important reason for Mach's objection to Boltzmann's molecular hypothesis was that no direct measurement of atoms had ever been carried out in the nineteenth century.

Besides the limited accuracy, it is rare for more than a million quantities to be measured in an experiment, and usually only a handful of measurements are made. Since it would take about $6N$ quantities to specify the microscopic state of N atoms, thermodynamic measurements provide relatively little information about the microscopic state.

Due to our lack of detailed knowledge of the microscopic state of an object, we need to use probability theory—discussed in Chapters 3 and 5—to make further progress. However, we do not know the probabilities of the various microscopic states either. This means that we have to make assumptions about the probabilities. We will make

---

[1] Sharp, K. and Matschinsky, F. Translation of Ludwig Boltzmann's 'Paper On the Relationship between the Second Fundamental Theorem of the Mechanical Theory of Heat and Probability Calculations Regarding the Conditions for Thermal Equilibrium', (Sitzungberichte der Kaiserlichen Akademie der Wissenschaften. Mathematisch-Naturwissen Classe). Abt. II, LXXVI 1877, pp. 373–435 (Wien. Ber. 1877, 76: 373–435). Reprinted in Wiss. Abhandlungen, Vol. II, reprint 42, pp. 164–223, Barth, Leipzig, 1909. *Entropy* 2015, 17: 1971–2009.

the simplest assumptions that are consistent with the physical constraints (number of particles, total energy, and so on): we will assume that everything we do not know is equally probable.

Based on our assumptions about the probability distribution, we will calculate the macroscopic properties of the system and compare our predictions with experimental data. We will find that our predictions are correct. This is comforting. However, we must realize that it does not necessarily mean that our assumptions were correct. In fact, we will see in Chapter 22 that many different assumptions would also lead to the same predictions. This is not a flaw in the theory but merely a fact of life. Recognizing this fact helps a great deal in resolving apparent paradoxes, as we will see.

The predictions we make based on assumptions about the probabilities of microscopic states lead to the postulates of thermodynamics. These postulates, in turn, are sufficient to derive the very powerful formalism of thermodynamics, as we will examine in Part II.

These same assumptions about the probabilities of microscopic states also lead to the even more powerful formalism of statistical mechanics, which we will investigate in the Parts III and IV of the book.

1.4.1   State Functions

It has long been known that when most macroscopic systems are left by themselves for a long period of time, their measurable properties stop changing and become time-independent. Simple systems, like a container of gas, evolve into macroscopic states that are well described by a small number of variables. For a simple gas, the energy, volume, and number of particles might be sufficient. These quantities, along with other quantities that we will discuss, are known as 'state functions'.

The molecules in a macroscopic system are in constant motion so that the microscopic state is constantly changing. This fact leads to a fundamental question. How is it that the macroscopic state can be time-independent with precisely defined properties? The answer to this question should become clear in Part I of this book.

1.4.2   Irreversibility

A second fundamental question is this. Even if there exist macroscopic equilibrium states that are independent of time, how can a system evolve toward such a state but not away from it? How can a system that obeys time-reversal-invariant laws of motion show irreversible behavior? This question has been the subject of much debate, at least since Johann Josef Loschmidt's (Austrian physicist, 1821–1895) formulation of the 'reversibility paradox' in 1876. We will present a resolution of the paradox in Chapter 22.

1.4.3   Entropy

The Second Law of Thermodynamics states that there exists a state function called the 'entropy' that is maximized in equilibrium. Boltzmann's 1877 definition of the entropy provides an account of what this means. A major purpose of the calculation of the entropy of a classical ideal gas in Part I is to obtain an intuitive understanding of the significance of Boltzmann's entropy.

1.5 Plan of the Book

This book presents thermal physics as a consequence of the molecular nature of matter. The book is divided into four parts, in order to provide a systematic development of the ideas for someone coming to the subject for the first time, as well as for someone who knows the basic material but would like to review a particular topic.

Part I: Entropy The entropy will be introduced by an explicit calculation for the classical ideal gas. The ideal gas is a simple model with macroscopic properties that can be calculated exactly. This allows us to derive the entropy in closed form without any hidden assumptions. Since the entropy of the classical ideal gas exhibits most of the thermodynamic properties of more general systems, it will serve as an introduction to the otherwise rather abstract postulates of thermodynamics. In the last two chapters, Chapters 7 and 8 of Part I, the formal expression for the entropy of a classical gas with interacting particles is obtained, along with general expressions for the temperature, pressure, and chemical potential, which establish the foundations of classical statistical mechanics.

Part II: Thermodynamics The formal postulates of thermodynamics are introduced, based on the properties of the entropy derived in Part I. Our treatment follows the vision of László Tisza (Hungarian physicist, 1907–2009) and Herbert B. Callen (American physicist, 1919–1993). Although the original development of thermodynamics by nineteenth-century physicists was a brilliant achievement, their arguments were somewhat convoluted because they did not understand the microscopic molecular basis of the laws they had discovered. Deriving the equations of thermodynamics from postulates is much easier than following the historical path. The full power of thermodynamics can be developed in a straightforward manner and the structure of the theory made transparent.

Part III: Classical statistical mechanics Here we return to classical statistical mechanics to discuss more powerful methods of calculation. In particular, we introduce the canonical ensemble, which describes the behavior of a system in contact with a heat bath at constant temperature. The canonical ensemble provides a very powerful approach to solving most problems in classical statistical mechanics. We also introduce the grand canonical ensemble, which describes a system that can exchange particles with a large system at a fixed chemical potential. This ensemble will prove to be particularly important when we encounter it again in Chapters 27 through 30 of Part IV, where we discuss quantum systems of indistinguishable particles. Statistical mechanics can derive results that go beyond those of thermodynamics. We discuss and resolve the apparent conflict between time-reversal-invariant microscopic equations with the obvious existence of irreversible behavior in the macroscopic world. Part III will introduce molecular dynamics and Monte Carlo computer simulations to demonstrate some of the modern methods for obtaining information about many-particle systems. We will also refine the calculation of the entropy to include the non-zero width of the energy and particle number distributions.

Part IV: Quantum statistical mechanics In the last part of the book, we develop quantum statistical mechanics, which is necessary for the understanding of the properties of most real systems. After two introductory chapters on basic ideas we will devote chapters to black-body radiation and lattice vibrations. There is a chapter on the general theory of indistinguishable particles, followed by individual chapters on the properties of bosons and fermions. Since the application of the theory of fermions to the properties of insulators and semiconductors has both theoretical and practical importance, this topic has a chapter of its own. The last chapter provides an introduction to the Ising model of ferromagnetism as an example of the theory of phase transitions.

Part I

Entropy

2

The Classical Ideal Gas

The laws of thermodynamics, as empirically determined, express the approximate and probable behavior of systems of a great number of particles, or, more precisely, they express the laws of mechanics of such systems as they appear to beings who have not the fineness of perception to enable them to appreciate quantities of the order of magnitude of those which relate to single particles, and who cannot repeat their experiments often enough to obtain any but the most probable results.

J. Willard Gibbs

The purpose of Part I of this book is to provide an intuitive understanding of the entropy, based on calculations for a simple model. The model chosen is the classical ideal gas, for which the entropy can be calculated explicitly and completely without any approximations or hidden assumptions.

The treatment is entirely in terms of the theory of classical mechanics. No quantum mechanical concepts are used. All the ideas will follow directly from the work of Ludwig Boltzmann (1844–1906), and especially his 1877 paper on the Second Law of Thermodynamics. We will use more modern mathematical methods than he did to derive the entropy of the classical ideal gas, but we will make no assumptions with which he was not familiar.

In Chapters 7 and 8 the formal expression for the entropy will be extended to classical systems with interacting particles. Although the expression we obtain can rarely be evaluated exactly, the formal structure will be sufficient to provide a basis for the development of thermodynamics in Part II. The same formal structure will also lead to more powerful methods of calculation for statistical mechanics in Parts III and IV.

2.1 Ideal Gas

What distinguishes an 'ideal' gas from a 'real' gas is the absence of interactions between the particles. Although an ideal gas might seem to be an unrealistic model, its properties are experimentally accessible by studying real gases at low densities. Since even the molecules in the air you are breathing are separated by an average distance of about ten times their diameter, nearly ideal gases are easy to find.

The most important feature that is missing from a classical ideal gas is that it does not exhibit any phase transitions. Other than that, its properties are qualitatively the same

An Introduction to Statistical Mechanics and Thermodynamics. Robert H. Swendsen, Oxford University Press (2020).
© Robert H. Swendsen. DOI: 10.1093/oso/9780198853237.001.0001

as those of real gases, which makes it valuable for developing intuition about statistical mechanics and thermodynamics.

The great advantage of the ideal gas model is that all of its properties can be calculated exactly, and nothing is obscured by mathematical complexity.

2.2 Phase Space of a Classical Gas

Our model of a classical gas consists of N particles contained in some specified volume. Each particle has a well-defined position and momentum. The positions of all particles can be represented as a point in configuration space—an abstract $3N$-dimensional space, with axes for every coordinate of every particle. These coordinates can be given in various forms:

$$\begin{aligned}
q &= \{\vec{r}_i | i = 1, \ldots, N\} \\
&= \{x_i, y_i, z_i | i = 1, \ldots, N\} \\
&= \{q_j | j = 1, \ldots, 3N\}.
\end{aligned} \tag{2.1}$$

The momenta of all particles can be represented as a point in momentum space—an abstract $3N$-dimensional space, with axes for every component of the momentum of every particle

$$\begin{aligned}
p &= \{\vec{p}_i | i = 1, \ldots, N\} \\
&= \{p_{x,i}, p_{y,i}, p_{z,i} | i = 1, \ldots, N\} \\
&= \{p_j | j = 1, \ldots, 3N\}.
\end{aligned} \tag{2.2}$$

The kinetic energy of the i-th particle is given by the usual expression, $|\vec{p}_i|^2/2m$, and the total kinetic energy is just the sum of the kinetic energies of all particles,

$$E = \sum_{i=1}^{N} \frac{|\vec{p}_i|^2}{2m}. \tag{2.3}$$

The complete microscopic state of the system can be described by a point in phase space—an abstract $6N$-dimensional space with axes for every coordinate and every momentum component for all N particles. Phase space is the union of configuration space and momentum space, $\{p, q\}$:

$$\{p, q\} = \{q_j, p_j | j = 1, \ldots, 3N\}. \tag{2.4}$$

Since, by definition, there are no interactions between the particles in an ideal gas, the potential energy of the system is zero.

2.3   Distinguishability

Particles will be regarded as distinguishable, in keeping with classical concepts. To be specific, particles are distinguishable when the exchange of two particles results in a different microscopic state. In classical mechanics, this is equivalent to saying that every point in phase space represents a different microscopic state. Distinguishability does not necessarily mean that the particles have different properties; classically, particles were always regarded as distinguishable because their trajectories could, at least in a thought experiment, be followed and the identity of individual particles determined.

On the other hand, it will be important to remember that experiments on macroscopic systems are always assumed to have limited resolution. In both statistical mechanics and thermodynamics, we are concerned with measurements that do not have the sensitivity to resolve the positions or identities of individual atoms.

2.4   Probability Theory

Because of the importance of probability theory in statistical mechanics, two chapters are devoted to the topic. The chapters discuss the basic principles of the probability theory of discrete random variables (Chapter 3) and continuous random variables (Chapter 5).

The mathematical treatment of probability theory has been separated from the physical application for several reasons: (1) it provides an easy reference for the mathematics, (2) it makes the derivation of the entropy more compact, and (3) it is unobtrusive for those readers who are already entirely familiar with probability theory.

If the reader is already familiar with probability theory, Chapters 3 and 5 might be skipped. However, since the methods for the transformation of random variables presented differ from those usually found in mathematics textbooks, these chapters might still be of some interest. It should be noted that we will need the transformation methods using Dirac delta functions, which are rarely found in mathematical texts on probability theory.

The chapters on probability theory are placed just before the chapters in which the material is first needed to calculate contributions to the entropy. Chapter 3 provides the methods needed to calculate the contributions of the positions in Chapter 4, and Chapter 5 provides the methods required to calculate the contributions of the momenta in Chapter 6.

To apply probability theory to the calculation of the properties of the classical ideal gas—or any other model, for that matter—we will have to make assumptions about the probability distribution of the positions and momenta of 10^{20} or more particles. Our basic strategy will be to make the simplest assumptions consistent with what we know about the system and then calculate the consequences.

Another way of describing our strategy is that we are making a virtue of our ignorance of the microscopic states and assume that everything we don't know is equally likely. How this plays out in practice is the subject of the rest of the book.

2.5   Boltzmann's Definition of the Entropy

In 1877, after a few less successful attempts, Boltzmann defined the entropy in terms of the probability of a macroscopic state. His explanation of the Second Law of Thermodynamics was that isolated systems naturally develop from less probable macroscopic states to more probable macroscopic states. Although Boltzmann's earlier efforts to prove this with his famous H-theorem were problematic and highly controversial, his fundamental insight is essentially correct.

In his 1877 paper, Boltzmann also specified that the entropy should be defined in terms of a composite system; that is, a system composed of two or more subsystems with some sort of constraint between them. An example of such a composite system would be a volume of gas divided into two smaller volumes (or subsystems) by a partition. The partition acts as a constraint by restricting the number of particles in each subsystem to be constant. The removal of the partition would then allow the system to develop from a less probable macroscopic state to a more probable macroscopic state for the distribution of particle positions. The final state, after the composite system had come to equilibrium, would correspond to the most probable macroscopic state. According to the Second Law of Thermodynamics, the thermodynamic entropy should also be maximized in the equilibrium state. The comparison of these two properties of the equilibrium state led Boltzmann to associate the entropy with the probability of a macroscopic state, or more precisely with the logarithm of the probability.

In the following chapters, we will make a direct application of Boltzmann's definition to the calculation of the entropy of the classical ideal gas, to within additive and multiplicative constants that we will determine later.

2.6 $S = k \log W$

Boltzmann's achievement has been honored with the inscription of the equation $S = k \log W$ on his tombstone. The symbol S denotes the entropy. The symbol W denotes the German word *Wahrscheinlichkeit*, which means 'probability'. The natural logarithm is intended, although 'log' is written instead of the more modern 'ln'. Curiously enough, Boltzmann never wrote this equation. It does reflect his ideas, except for the important use of composite systems in his definition. The equation was first written in this form by the German physicist Max Planck (1858–1947) in 1900. The constant k, also written as k_B, is known as the Boltzmann constant.

The symbol W has often been misinterpreted to mean a volume in phase space, which has caused a considerable amount of trouble. This misinterpretation is so common that many scientists are under the impression that Boltzmann defined the entropy as the logarithm of a volume in phase space. Going back to the original meaning of W and Boltzmann's 1877 definition (with composite systems) eliminates much of the confusion about the statistical interpretation of entropy.

The main differences between Boltzmann's treatment of entropy and the one in this book lie in the use of modern mathematical methods and the explicit treatment of the dependence of the entropy on the number of particles.

2.7   Independence of Positions and Momenta

In the derivation of the properties of the classical ideal gas, we will assume that the positions and momenta of the particles are independent. We will present a more formal definition of independence in the Chapter 3, but the idea is that knowing the position of a particle tells us nothing about its momentum, and knowing its momentum tells us nothing about its position. As demonstrated at the beginning of Chapter 4, the independence of the positions and momenta means that their contributions to the total entropy can be calculated separately and simply added to produce the final answer.

2.8   Road Map for Part I

The analysis of the entropy in Part I has been divided into chapters to make it easier to keep track of the different aspects of the derivation.

The concepts and equations of discrete probability theory are developed in Chapter 3, just before they are needed in Chapter 4 to calculate the contributions of the positions to the entropy.

Probability theory for continuous random variables is discussed in Chapter 5, just before its application to the calculation of the contributions of the momenta to the entropy in Chapter 6.

Chapter 7 generalizes the entropy to systems with interacting particles.

Chapter 8 completes the foundations of classical statistical mechanics by relating the partial derivatives of the entropy to the temperature, pressure, and chemical potential.

The following flowchart is intended to illustrate the organization of Part I. The derivation of the entropy of the classical ideal gas follows the arrows down the right-hand side of the flowchart.

Flow chart for Part I

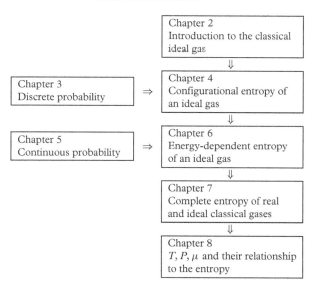

3

Discrete Probability Theory

The initial state will usually be a very improbable one. The system will run from this state to ever more probable states until it reaches the most probable one, which is the state of thermal equilibrium. If we apply this to the second law, we can identify the quantity that is usually called the entropy with the probability of that state.

Ludwig Boltzmann, 1877

3.1   What is Probability?

The definition of probability is sufficiently problematic to have occasioned something akin to religious wars in academic departments of statistics. There are two basic factions: frequentists and Bayesians.

To understand why the definition of probability can be a problem, consider an experiment with N trials, each of which can result in a success or failure by some criterion. If N_s be the number of successes, then N_s/N is called the frequency of success for those trials.

Frequentists define the probability by the asymptotic frequency of success in the limit of an infinite number of trials,

$$p = \lim_{N \to \infty} \frac{N_s}{N}. \tag{3.1}$$

This definition looks precise and objective. Indeed, it is found as *the* definition in many physics texts. The major problem is that humans have a finite amount of time available, so that by the frequentist definition we can never determine the probability of anything. Bayesian probability provides a solution to this quandary.

The Bayesian view of probability is based on the work of Thomas Bayes (English mathematician and Presbyterian minister, 1702–1761). Bayesians define probability as a description of a person's knowledge of the outcome of an experiment, based on whatever evidence is at that person's disposal.

One great advantage of the Bayesian definition of probability is that it gives a clear meaning to a statement such as: 'The mass of a proton is $1.672621637(83) \times 10^{-27} kg$', where the '(83)' is called the uncertainty in the last digit. What does '(83)' mean? Certainly the mass of a proton is an unambiguous number that does not take on

An Introduction to Statistical Mechanics and Thermodynamics. Robert H. Swendsen, Oxford University Press (2020).
© Robert H. Swendsen. DOI: 10.1093/oso/9780198853237.001.0001

different values for different experiments. Nevertheless, the '(83)' does have meaning as an expression of our uncertainty as to the exact value of a proton's mass.

Bayesian probability is accepted by most statisticians. However, it is in disrepute among some physicists because they regard it as subjective, in the sense that it describes what an individual knows, rather than being absolute truth. However, none of us has access to absolute truth. Bayesian statistics provide an appropriate way to describe the knowledge that an individual does have, based on the experimental results that he or she has access to.

In my opinion, the only reasonable form of objectivity that we can demand is that two observers with the same information will come to the same conclusions, and Bayesian statistics fulfills this requirement.[1]

The word 'probability' is used somewhat differently in statistical mechanics. I shall call it 'model probability'. It is most closely related to the frequentist meaning in that the specific values of model probabilities are assumed to be known exactly, as if we had an infinite number of trials. Of course, we really have only finite number of (imprecise) measurements, and as we shall see, we need to fix the probabilities of an enormous number of random variables (in excess of 10^{20}), so that we have no direct access to an experimental confirmation of our assumptions.

The model probability is only assumed to be valid for an equilibrium state. Indeed, all calculations will be for equilibrium, except in Chapter 22, where we discuss irreversibility. Even though we speak of thermodynamics, we limit our discussions to transitions between equilibrium states, primarily because the transitions themselves are too difficult to calculate.

The assumed form of the model probability is usually based on the plausible argument that events that are essentially alike have equal probabilities in equilibrium.

Another common argument uses the property of ergodicity, which is an additional assumption. A system is ergodic if, starting from an arbitrary point in phase space, the trajectory of the system will pass arbitrarily close to every allowable point in phase space. This trajectory is necessarily a one-dimensional subset of a $6N$-dimensional space. Regions in phase space that are visited equally frequently are then assigned equal probabilities. The argument to ergodicity is very popular, but it has certain weaknesses. For one thing, ergodicity by itself does not determine which regions will be visited equally frequently. Next, ergodicity has only been proven for a very limited number of systems (not the ideal gas). Such an argument also has the difficulty that it requires an enormous amount of time for a trajectory to spread through phase space—of the same order of magnitude as the Poincaré recurrence time discussed in Chapter 22.

A somewhat better argument, although still not proven for general systems, starts with an initial probability distribution that is spread over a non-zero region in phase space, which is a much more realistic assumption. If the time development of such a distribution

---

[1] This interpretation is sometimes called 'objective Bayesian probability' to distinguish it from 'subjective Bayesian probability'. Some workers do not regard the 'prior' in the Bayesian formalism as describing the information available to an experimenter before the current experiment, but rather as being an arbitrary choice. With this interpretation, Bayesian probability is subjective—and useless for physics.

leads to a unique prediction for observable quantities, that is used to justify the model probability. It has the advantage of converging *much* faster, as shown in Chapter 22 for the classical ideal gas.[2]

We can then use our assumed model probability to calculate the predicted outcome of an experiment. If measurements agree with our predictions, we can say that our model is consistent with the experiment. This is not the same thing as saying that our model probabilities are correct, and we will see in Chapter 22 that many different distributions can lead to the same predictions. Nevertheless, agreement with experiment is always comforting.

Statistical mechanics is based on simple assumptions, expressed as model probabilities, that lead to a wide variety of predictions in excellent agreement with experiments. How this is done is the main subject of this book.

3.2   Discrete Random Variables and Probabilities

We begin by defining a set of elementary events

$$A = \left\{ a_j | j = 1, N_A \right\} \tag{3.2}$$

and assigning a probability $P(a_j)$ to each event. The combination of random events and their probabilities is called a 'random variable'. If the number of elementary events is finite or countable, it is called a 'discrete random variable'.

The probabilities must satisfy the conditions that

$$0 \leq P(a_j) \leq 1 \tag{3.3}$$

for all a_j. An impossible event has probability zero, and a certain event has probability 1.

Random events can be anything: heads/tails, red/black, and so on. If the random events are all numbers, the random variable is called a 'random number'.

Elementary events are defined to be exclusive—one, and only one elementary event can occur at each trial—so that the probabilities must also satisfy the normalization condition

$$\sum_{j=1}^{N_A} P(a_j) = 1. \tag{3.4}$$

---

[2] In general, the probability distribution after an irreversible process is not exactly the same as the model probability, although it gives identical predictions for the probabilities of macroscopic variables. This supports the use of model probabilities for equilibrium calculations.

To simplify notation, we will often write this equation as

$$\sum_a P(a) = 1,\tag{3.5}$$

suppressing explicit mention of the number of elementary events.

3.3   Probability Theory for Multiple Random Variables

If more than one thing can happen at each trial we can describe the situation with two or more sets of random events. For example, both an event from A and an event from

$$B = \{b_k | k = 1, N_B\}\tag{3.6}$$

might occur. We can then define a joint probability $P(a_j, b_k)$—or more simply, $P(a, b)$—which must satisfy

$$0 \le P(a, b) \le 1\tag{3.7}$$

and

$$\sum_a \sum_b P(a, b) = 1.\tag{3.8}$$

3.3.1   Marginal and Conditional Probability

Naturally, if we have $P(a, b)$ we can retrieve the information for either A or B alone. The marginal probability of A is defined by

$$P_A(a) = \sum_b P(a, b)\tag{3.9}$$

with a similar expression for $P_B(b)$. A nice feature of marginal probabilities is that they automatically satisfy the positivity and normalization criteria in eqs. (3.3) and (3.4).

The name marginal probability comes from the practice of calculating it in the margins of a table of probabilities. Table 3.1 shows an example of two random variables that each take on two values.

Conditional probability is the probability of an event a, given that event b has occurred. It is denoted by $P(a|b)$, and is related to the full probability of both A and B occurring by the equations:

$$P(a, b) = P(a|b)P_B(b)\tag{3.10}$$
$$= P(b|a)P_A(a).\tag{3.11}$$

Table 3.1 *Example of a table of probabilities for independent random variables: The events of A (labeled '3' and '4') are listed down the left column, and events of B (labeled '1' and '2') across the top. The values of the probabilities are given in the four center squares. The margins contain the marginal probabilities, as defined in eq. (3.9).*

$A\backslash B$	1	2	$P_A(a)$
3	1/2	1/4	3/4
4	1/6	1/12	1/4
$P_B(b)$	2/3	1/3	1

If $P_B(b) \neq 0$, the conditional probability $P(a|b)$ can be written as

$$P(a|b) = \frac{P(a,b)}{P_B(b)}. \tag{3.12}$$

Combining eqs. (3.10) and eqs. (3.11) gives us Bayes' theorem

$$P(a|b) = \frac{P(b|a)P_A(a)}{P_B(b)}. \tag{3.13}$$

We will discuss some of the consequences of Bayes' theorem in Section 5.5.

3.3.2 Independent Variables

A particularly interesting case occurs when the probability distribution can be written as a product:

$$P(a,b) = P_A(a)P_B(b). \tag{3.14}$$

When eq. (3.14) is true, the two random variables are said to be independent because, if $P_B(b) \neq 0$, the conditional probability $P(a|b)$ is then independent of b,

$$P(a|b) = \frac{P(a,b)}{P_B(b)} = \frac{P_A(a)P_B(b)}{P_B(b)} = P_A(a), \tag{3.15}$$

and $P(b|a)$ is independent of a.

Table 3.1 gives an example of independent random variables, and Table 3.2 gives an example of random variables that are not independent.

Table **3.2** *An example of a table of probabilities for random variables that are not independent. The table is arranged in the same way as Table 3.1.*

$C\backslash D$	1	2	$P_C(c)$
3	0.5	0.25	0.75
4	0.1	0.15	0.25
$P_D(d)$	0.6	0.4	1

3.3.3   Pairwise Independence and Mutual Independence

If we have more than two random variables, the set of random variables might satisfy two kinds of independence: pairwise or mutual.

Pairwise independence means that the marginal distribution of any *pair* of random variables can be written as the product of the marginal distributions of the individual random variables.

Mutual independence means that the marginal distribution of any *subset* of random variables can be written as the product of the marginal distributions of the individual random variables.

It is obvious that mutual independence implies pairwise independence. Whether the converse is true is the subject of a problem in this chapter.

3.4   Random Numbers and Functions of Random Variables

Given an arbitrary random variable $A = \{a_j | j = 1, \cdots N_A\}$, we can define a numerical function on the set of elementary events, $F = \{F(a_j) | 1, \cdots, N_A\}$. The set of random numbers F, together with their probabilities, is then also a random number.

If we introduce the Kronecker delta function

$$\delta_{x,y} = \begin{cases} 1 & x = y \\ 0 & x \neq y \end{cases} \tag{3.16}$$

then we can write the probability distribution of F compactly,

$$P_F(f) = \sum_a \delta_{f,F(a)} P_A(a). \tag{3.17}$$

As a simple illustration, consider a random variable that takes on the three values -1, 0, and 1, with probabilities $P_A(-1) = 0.2$, $P_A(0) = 0.3$, and $P_A(1) = 0.5$. Define a function $F(a) = |a|$, so that F takes on the values 0 and 1. The probabilities $P_F(f)$ are found from eq. (3.17),

$$P_F(0) = \sum_{a=-1}^{1} \delta_{0,F(a)} P_A(a)$$
$$= \delta_{0,F(-1)} P_A(-1) + \delta_{0,F(0)} P_A(0) + \delta_{0,F(1)} P_A(1)$$
$$= 0 + P(0) + 0 = 0.3 \tag{3.18}$$

$$P_F(1) = \sum_{a=-1}^{1} \delta_{1,F(a)} P_A(a)$$
$$= \delta_{1,F(-1)} P_A(-1) + \delta_{1,F(0)} P_A(0) + \delta_{1,F(1)} P_A(1)$$
$$= P_A(-1) + 0 + P_A(1) = 0.2 + 0.5 = 0.7. \tag{3.19}$$

We can also use the Kronecker delta to express arbitrary combinations of random numbers to form new compound random numbers. For example, if X and Y are random numbers, and $G(x,y)$ is an arbitrary function, we can define a new random variable, Z. The probability distribution of Z is given by a sum over all combinations of the events of X and Y, with the Kronecker delta picking out the ones that correspond to particular events of Z

$$P_Z(z) = \sum_x \sum_y \delta_{z,G(x,y)} P(x,y). \tag{3.20}$$

A warning is necessary because the limits on the sums in eq. (3.20) have been suppressed. The only difficult thing about actually doing the sums is in keeping track of those limits. Since being able to keep track of limits will also be important when we get to continuous distributions, we will illustrate the technique with the simple example of rolling two dice and asking for the probability distribution of their sum.

Example: Probability of the Sum of Two Dice

We will assume that the dice are honest, which they tend to be in physics problems, if not in the real world. Let $X = \{x | x = 1, 2, 3, 4, 5, 6\}$ be the random number representing the outcome of the first die, with Y a corresponding random number for the outcome of the second die. The sum $S = X + Y$. The values taken on by S range from 2 to 12. Since all elementary events are equally likely,

$$P(x,y) = P_X(x) P_Y(x) = \left(\frac{1}{6}\right)\left(\frac{1}{6}\right) = \frac{1}{36}, \tag{3.21}$$

and eq. (3.20) then becomes

$$P(s) = \frac{1}{36} \sum_{x=1}^{6} \sum_{y=1}^{6} \delta_{s,x+y}. \tag{3.22}$$

Do the sum over y first. Its value depends on whether $s = x + y$ for some value of y, or equivalently, whether $s - x$ is in the interval $[1,6]$

$$\sum_{y=1}^{6} \delta_{s,x+y} = \begin{cases} 1 & 1 \le s - x \le 6. \\ 0 & \text{otherwise} \end{cases} \tag{3.23}$$

This places two conditions on the remaining sum over x. Only those values of x for which both $x \le s - 1$ and $x \ge s - 6$ contribute to the final answer. These limits are in addition to the limits of $x \le 6$ and $x \ge 1$ that are already explicit in the sum on X. Since all four of these inequalities must be satisfied, we must take the more restrictive of the inequalities in each case. Which inequality is the more restrictive depends on the value of s, as indicated in Table 3.3.

For $s \le 7$, the lower bound on x is 1 and the upper bound is $s - 1$. For $s \ge 7$, the lower bound on x is $s - 6$ and the upper bound is 6. The sums can then be evaluated explicitly,

$$P(s) = \begin{cases} \displaystyle\sum_{x=1}^{s-1} \frac{1}{36} = \frac{s-1}{36} & s \le 7 \\[4mm] \displaystyle\sum_{x=s-6}^{6} \frac{1}{36} = \frac{13-s}{36} & s \ge 7. \end{cases} \tag{3.24}$$

Even though the probability distribution of both X and Y are uniform, the probability distribution of S is not. This is an important result. Although we will be generally assuming a uniform underlying probability distribution for most microscopic quantities, we will see that the distribution of macroscopic observables will be very sharply peaked.

Table 3.3 *Determining the limits for the second sum when evaluating eq. (3.22).*

	lower limit	upper limit
Limits on sum:	$x \ge 1$	$x \le 6$
From the Kronecker delta:	$x \ge s-6$	$x \le s-1$
More restrictive if $s \le 7$	$x \ge 1$	$x \le s-1$
More restrictive if $s \ge 7$	$x \ge s-6$	$x \le 6$

It must be admitted that there are easier ways of numerically computing the probability distribution of the sum of two dice, especially if a numerical result is required. Eq. (3.20) is particularly easy to evaluate in a computer program, since the Kronecker delta just corresponds to an 'IF' statement. However, we will have ample opportunity to apply the method just described to problems for which it is the simplest approach. One reason why the example in this section was chosen is that the correct answer is easy to recognize, which makes the method more transparent.

3.5 Mean, Variance, and Standard Deviation

The mean, or average, of a function $F(A)$, defined on the random variable A, is given by

$$\langle F \rangle \equiv \sum_a F(a) P_A(a). \tag{3.25}$$

Similarly, the n-th moment of F can be defined as

$$\langle F^n \rangle \equiv \sum_a F(a)^n P_A(a). \tag{3.26}$$

The n-th central moment is defined by subtracting the mean before taking the n-th power in the definition

$$\langle (F - \langle F \rangle)^n \rangle \equiv \sum_a (F(a) - \langle F \rangle)^n P_A(a). \tag{3.27}$$

The second central moment plays an important role in statistical analysis and is called the variance, σ^2

$$\sigma^2 = \langle (F - \langle F \rangle)^2 \rangle = \sum_a (F(a) - \langle F \rangle)^2 P_A(a). \tag{3.28}$$

It can also be written as

$$\sigma^2 = \langle (F - \langle F \rangle)^2 \rangle = \langle F^2 \rangle - \langle F \rangle^2. \tag{3.29}$$

The square root of the variance is called the standard deviation,

$$\sigma \equiv \sqrt{\langle F^2 \rangle - \langle F \rangle^2}. \tag{3.30}$$

The standard deviation is frequently used as a measure of the width of a probability distribution.

3.6 Correlation Function

Suppose we have two random numbers, A and B, and their joint probability distribution, $P(A,B)$, along with functions $F(A)$ and $G(B)$ defined on the random variables. F and G are random numbers, and we can ask questions about how they are related. In particular, we can define a correlation function, f_{FG},

$$f_{FG} = \langle FG \rangle - \langle F \rangle \langle G \rangle. \tag{3.31}$$

If F and G are independent random numbers, we would expect the correlation function to vanish, which it does:

$$
\begin{aligned}
f_{FG} &= \sum_a \sum_b F(a)G(b)P(a,b) \\
&\quad - \sum_a F(a)P_A(a) \sum_b G(b)P_B(b) \\
&= \sum_a \sum_b F(a)G(b)P_A(a)P_B(b) \\
&\quad - \sum_a F(a)P_A(a) \sum_b G(b)P_B(b) \\
&= 0.
\end{aligned}
\tag{3.32}
$$

3.7 Sets of Independent Random Numbers

Given a set of random numbers $\{F_j | j = 1, \cdots, N\}$, we are often interested in a new random number formed by taking the sum

$$S = \sum_{j=1}^{N} F_j. \tag{3.33}$$

We can easily calculate the mean of the random number S, which is just the sum of the means of the individual random numbers

$$\langle S \rangle = \left\langle \sum_{j=1}^{N} F_j \right\rangle = \sum_{j=1}^{N} \langle F_j \rangle. \tag{3.34}$$

If the random numbers are pairwise independent, we can also calculate the variance and the standard deviation,

$$\sigma_S^2 \equiv \langle S^2 \rangle - \langle S \rangle^2 \tag{3.35}$$

$$= \sum_{j=1}^{N} \sum_{k=1}^{N} \langle F_j F_k \rangle - \sum_{j=1}^{N} \langle F_j \rangle \sum_{k=1}^{N} \langle F_k \rangle$$

$$= \sum_{j=1}^{N} \sum_{k=1(k \neq j)}^{N} \langle F_j \rangle \langle F_k \rangle + \sum_{j=1}^{N} \langle F_j^2 \rangle - \sum_{j=1}^{N} \langle F_j \rangle \sum_{k=1}^{N} \langle F_k \rangle$$

$$= \sum_{j=1}^{N} \left(\langle F_j^2 \rangle - \langle F_j \rangle^2 \right)$$

$$= \sum_{j=1}^{N} \sigma_j^2.$$

In this derivation, σ_j^2 denotes the variance of the j-th random number. We see that the variance of the sum of a set of *pairwise independent* random numbers is just the sum of the variances.

If the random numbers F_j all have the same mean and variance, these equations simplify further. If $\langle F_j \rangle = \langle F \rangle$ for all j, then

$$\langle S \rangle = \sum_{j=1}^{N} \langle F_j \rangle = N \langle F \rangle. \tag{3.36}$$

Similarly, if $\sigma_j^2 = \sigma^2$ for all j, then

$$\sigma_S^2 = \sum_{j=1}^{N} \sigma_j^2 = N \sigma^2. \tag{3.37}$$

Note that the standard deviation of S grows as the square root of the number of variables

$$\sigma_S = \sigma \sqrt{N}. \tag{3.38}$$

On the other hand, the relative standard deviation decreases with the square root of the number of variables

$$\frac{\sigma_S}{\langle S \rangle} = \frac{\sigma \sqrt{N}}{N \langle F \rangle} = \frac{\sigma}{\langle F \rangle \sqrt{N}}. \tag{3.39}$$

It might be argued that this is the most important result of probability theory for statistical mechanics. For many applications in statistical mechanics, the appropriate value of N is 10^{20} or higher, so that the relative uncertainties for macroscopic quantities are generally of the order of 10^{-10} or smaller. This is far smaller than most experimental errors, leading to predictions with no measurable uncertainty.

3.8   Binomial Distribution

A particularly important case is that of N independent, identically distributed random numbers, $\{F_j\}$, that can each take on the value 1 with probability p and 0 with probability $1 - p$. An example would be N flips of a coin, where the coin might be fair ($p = 0.5$) or biased ($p \neq 0.5$).

The mean and variance of each random number are easily seen to be $\langle F \rangle = p$ and $\sigma^2 = p(1 - p)$. The mean and variance of the sum

$$S = \sum_{j=1}^{N} F_j \tag{3.40}$$

are then

$$\langle S \rangle = pN \tag{3.41}$$

and

$$\sigma_S^2 = p(1 - p)N. \tag{3.42}$$

The standard deviation is

$$\sigma_S = \sqrt{p(1 - p)N} \tag{3.43}$$

and the relative standard deviation is

$$\frac{\sigma_S}{\langle S \rangle} = \frac{\sqrt{p(1 - p)N}}{pN} = \sqrt{\frac{1 - p}{pN}}. \tag{3.44}$$

3.8.1   Derivation of the Binomial Distribution

We can go further and calculate the explicit probability distribution $P(S)$ of the sum of random numbers. This result will be extremely important in the analysis of the classical ideal gas.

The probability of a specific subset of n random variables taking on the value 1, while the remaining $N - n$ random variables that take on the value 0 is easily seen to be

$$p^n(1-p)^{N-n}. \tag{3.45}$$

To complete the calculation, we only need to determine the number of permutations of the random variables with the given numbers of ones and zeros. This is the same as the number of ways by which N distinct objects can be put into two boxes, such that n of them are in the first box and $N - n$ are in the second.

To calculate the number of permutations, first consider the simpler problem of calculating the number of ways in which N distinct objects can be ordered. Since any of the N objects can be first, $N - 1$ can be second, and so on, there are a total of $N! = N(N-1)(N-2)\cdots 2 \cdot 1$ permutations.

For our problem of putting objects into two boxes, the order of the objects in each box does not matter. Therefore, we must divide by $n!$ for over counting in the first box and by $(N-n)!$ for over counting in the second box. The final number of permutations is known as the binomial coefficient and has its own standard symbol

$$\binom{N}{n} = \frac{N!}{n!(N-n)!}. \tag{3.46}$$

Multiplying by the probability given in eq. (3.45) gives us the binomial distribution

$$P(n|N) = \frac{N!}{n!(N-n)!}p^n(1-p)^{N-n} = \binom{N}{n}p^n(1-p)^{N-n}. \tag{3.47}$$

The binomial distribution acquires its name from the binomial theorem, which states that for any numbers p and q,

$$(p+q)^N = \sum_{n=0}^{N} \frac{N!}{n!(N-n)!}p^n q^{N-n} = \sum_{n=0}^{N} \binom{N}{n}p^n q^{N-n}. \tag{3.48}$$

Setting $q = 1 - p$ proves that the binomial distribution in eq. (3.47) is normalized.

3.8.2   Useful Identities for the Binomial Coefficients

Although the evaluation of the binomial coefficients appears to be straightforward, $N!$ grows so rapidly as a function of N that the numbers are too large for a direct application of the definition. For example, a popular spreadsheet overflows at $N = 171$, and my hand calculator cannot even handle $N = 70$.

On the other hand, the binomial coefficients themselves do not grow very rapidly with N. The following identities, which follow directly from eq. (3.46), allow us to calculate the binomial coefficients for moderately large values of n and N without numerical difficulties,

$$\binom{N}{0} = \binom{N}{N} = 1 \tag{3.49}$$

$$\binom{N-1}{n} + \binom{N-1}{n-1} = \binom{N}{n} \tag{3.50}$$

$$\binom{N}{n+1} = \frac{N-n}{n+1}\binom{N}{n}. \tag{3.51}$$

The proofs of these identities will be left as exercises.

3.9 Gaussian Approximation to the Binomial Distribution

For a fixed value of p and large values of N, the binomial distribution can be approximated by a Gaussian function. This is known as the central limit theorem. We will not prove it, but we will show how to determine the appropriate parameters in the Gaussian approximation.

Consider a general Gaussian function,

$$g(x) = \frac{1}{\sqrt{2\pi\sigma^2}} \exp\left[-\frac{(x-x_o)^2}{2\sigma^2}\right]. \tag{3.52}$$

The mean and the mode (the location of the maximum) coincide

$$\langle x \rangle = x_o = x_{max}. \tag{3.53}$$

The variance is given by the second central moment

$$\begin{aligned}
\langle (x - x_o)^2 \rangle &= \frac{1}{\sqrt{2\pi\sigma^2}} \int_{-\infty}^{\infty} (x - x_o)^2 \exp\left[-\frac{(x-x_o)^2}{2\sigma^2}\right] dx \\
&= \frac{1}{\sqrt{2\pi\sigma^2}} \int_{-\infty}^{\infty} y^2 \exp\left[-\frac{y^2}{2\sigma^2}\right] dy \\
&= \frac{1}{\sqrt{2\pi\sigma^2}} \frac{1}{2} \sqrt{\pi} [2\sigma^2]^{3/2} \\
&= \frac{1}{\sqrt{2\pi\sigma^2}} \sqrt{\pi} \sqrt{2\sigma^2} \sigma^2 \\
&= \sigma^2.
\end{aligned} \tag{3.54}$$

It is now easy to find a Gaussian approximation to the binomial distribution, since both the mean and the variance are known exactly from eqs. (3.41) and (3.42),

$$\langle n \rangle = pN \tag{3.55}$$

and

$$\sigma^2 = p(1-p)N. \tag{3.56}$$

A Gaussian function with this mean and variance gives a good approximation to the binomial distribution for sufficiently large n and N

$$P(n|N) \approx \frac{1}{\sqrt{2\pi p(1-p)N}} \exp\left[-\frac{(n-pN)^2}{2p(1-p)N}\right]. \tag{3.57}$$

How large n and N must be for eq. (3.57) to be a good approximation will be left as a numerical exercise.

3.10   A Digression on Gaussian Integrals

To derive a good approximation for $N!$ that is practical and accurate for values of N as large as 10^{20}, we need to develop another mathematical tool: Gaussian integrals.

Gaussian integrals are so important in statistical mechanics that they are well worth a slight detour. Although the formulas derived in this section can be found wherever fine integrals are sold, a student of statistical mechanics should be able to evaluate Gaussian integrals without relying on outside assistance.

The first step in evaluating the integral

$$G = \int_{-\infty}^{\infty} e^{-ax^2} dx \tag{3.58}$$

will later prove to be very useful in non-Gaussian integrals later in the book because we sometimes only need the dependence of the integral on the parameter a. Make the integral in eq. (3.58) dimensionless by the transformation $y = x\sqrt{a}$

$$G = \frac{1}{\sqrt{a}} \int_{-\infty}^{\infty} e^{-y^2} dy. \tag{3.59}$$

Note that the dependence on the parameter a appears as a simple factor in front of the integral.

To evaluate the dimensionless integral, we square it and recast the product as a two-dimensional integral

$$G^2 = \frac{1}{\sqrt{a}} \int_{-\infty}^{\infty} e^{-x^2} \, dx \frac{1}{\sqrt{a}} \int_{-\infty}^{\infty} e^{-y^2} \, dy \qquad (3.60)$$

$$= \frac{1}{a} \int_{-\infty}^{\infty} \int_{-\infty}^{\infty} e^{-(x^2+y^2)} \, dx \, dy$$

$$= \frac{1}{a} \int_{0}^{\infty} e^{-r^2} 2\pi r \, dr$$

$$= \frac{\pi}{a} \left[-e^{-r^2} \right]_{0}^{\infty} = \frac{\pi}{a}.$$

The value of a Gaussian integral is therefore:

$$\int_{-\infty}^{\infty} e^{-ax^2} \, dx = \sqrt{\frac{\pi}{a}}. \qquad (3.61)$$

3.11   Stirling's Approximation for *N!*

As mentioned previously, a difficulty in using the binomial distribution is that $N!$ becomes enormously large when N is even moderately large. For $N = 25$, $N! \approx 1.6 \times 10^{25}$, and we need to consider values of N of 10^{20} and higher! We would also like to differentiate and integrate expressions for the probability distribution, which is inconvenient in the product representation.

The problem is solved by Stirling's approximation, which is valid for large numbers—exactly the case in which we are interested. We will discuss various levels of Stirling's approximation, beginning with the simplest.

3.11.1   The Simplest Version of Stirling's Approximation

Consider approximating $\ln N!$ by an integral.

$$\ln N! = \ln \left(\prod_{n=1}^{N} n \right) = \sum_{n=1}^{N} \ln(n) \approx \int_{1}^{N} \ln(x) \, dx = N \ln N - N + 1. \qquad (3.62)$$

This is equivalent to the approximation

$$N! \approx N^N \exp(1 - N). \qquad (3.63)$$

3.11.2   A Better Version of Stirling's Approximation

A better approximation can be obtained from an exact integral representation of $N!$,

$$N! = \int_{0}^{\infty} e^{-x} x^N \, dx. \qquad (3.64)$$

The correctness of eq. (3.64) can be shown by induction. It is clearly true for $N = 0$, since $0! = 1 = \int_0^\infty e^{-x} dx$. If it is true for a value N, then using integration by parts we can then prove that it is true for $N+1$,

$$(N+1)! = \int_0^\infty e^{-x} x^{N+1} dx \tag{3.65}$$

$$= \left[-e^{-x} x^{N+1} \right]_0^\infty - \int_0^\infty \left(-e^{-x}(N+1) x^N \right) dx$$

$$= (N+1)N!.$$

The integral in eq. (3.64) can be approximated by noting that the integrand is sharply peaked for large values of N. This suggests using the method of steepest descent, which involves approximating the integrand by a Gaussian function of the form

$$g(x) = A \exp\left[-\frac{(x-x_0)^2}{2\sigma^2} \right], \tag{3.66}$$

where A, x_0, and σ are constants that must be evaluated.

We can find the location of the maximum of the integrand in eq. (3.64) by setting the first derivative of its logarithm equal to zero. For the Gaussian function, this gives

$$\frac{d}{dx} \ln g(x) = \frac{d}{dx}\left[\ln A - \frac{(x-x_0)^2}{2\sigma^2} \right] = -\frac{(x-x_0)}{\sigma^2} = 0, \tag{3.67}$$

or $x = x_0$.

Comparing eq. (3.67) to the first derivative of the logarithm of the integrand in eq. (3.64) equal to zero, we find the location x_0 of the maximum

$$0 = \frac{d}{dx}[-x + N\ln x] = -1 + \frac{N}{x}, \tag{3.68}$$

or $x = x_0 = N$. The value of the amplitude is then just the integrand in eq. (3.64), $e^{-x} x^N$ evaluated at $x = x_0 = N$, or $A = e^{-N} N^N$.

The second derivative of the logarithm of a Gaussian function tells us the value of the variance

$$\frac{d^2}{dx^2} \ln g(x) = \frac{d}{dx}\left[-\frac{(x-x_0)}{\sigma^2} \right] = -\frac{1}{\sigma^2}. \tag{3.69}$$

When using this method of evaluating the variance of a Gaussian approximation, the second derivative of the logarithm of the function being approximated should be evaluated at x_0.

To find the value of σ^2, take the second derivative of the logarithm of the integrand in eq. (3.64) and evaluate it at x_0,

Table 3.4 *Comparison of different levels of Stirling's approximation with exact results for N! 'Stirling (simple)' refers to eq. (3.63), 'Stirling (improved)' to eq. (3.72), and 'Gosper' to eq. (3.73).*

N	N!	Stirling (simple)	error	Stirling (improved)	error	Stirling (Gosper)	error
1	1	1	0%	0.922	−7.79%	0.996	−0.398%
2	2	1.47	−26%	1.919	−4.05%	1.997	−0.132%
3	6	3.65	−39%	5.836	−2.73%	5.996	−0.064%
4	24	12.75	−47%	23.506	−2.06%	23.991	−0.038%
5	120	57.24	−52%	118.019	−1.65%	119.970	−0.025%
10	3628800	1234098	−66%	3598694	−0.83%	3628559	−0.007%
20	2.43×10^{18}	5.87×10^{17}	−76%	2.43×10^{18}	−0.42%	2.43×10^{18}	−0.002%

$$-\frac{1}{\sigma^2} = \frac{d^2}{dx^2}[-x + N\ln x] = \frac{d}{dx}[-1 + \frac{N}{x}] = -\frac{N}{x^2}, \quad (3.70)$$

or

$$\sigma^2 = \frac{x_0^2}{N} = N. \quad (3.71)$$

We can now use the formula for a Gaussian integral derived in Section 3.10,

$$N! = \int_0^\infty e^{-x} x^N dx \approx \int_0^\infty e^{-N} N^N \exp\left[-\frac{(x-N)^2}{2N}\right] dx \approx e^{-N} N^N \sqrt{2\pi N}. \quad (3.72)$$

As can be seen in Table 3.4, eq. (3.72) is a considerable improvement over eq. (3.63).

The procedure used in this section to approximate a sharply peaked function by a Gaussian is extremely useful in statistical mechanics because a great many functions encountered are of this form.

3.11.3 Gosper's Approximation

Finally, we mention a very interesting variant of Stirling's approximation due to Gosper,[3]

$$N! \approx e^{-N} N^N \sqrt{\left(2N + \frac{1}{3}\right)\pi}. \quad (3.73)$$

---

[3] Cited in http://mathworld.wolfram.com/StirlingsApproximation.html

Table 3.4 shows that Gosper's approximation is extremely good, even for very small values of N.

3.11.4 Using Stirling's Approximation

In applications to statistical mechanics, we are most often interested in the logarithm of the probability distribution, and consequently in $\ln N!$. This has the curious consequence that the simplest form of Stirling's approximation is by far the most useful, even though Table 3.4 shows that its accuracy for $N!$ is very poor for large N.

From Gosper's approximation, we have a very accurate equation for $\ln N!$

$$\ln N! \approx N \ln N - N + \frac{1}{2} \ln\left[\left(2N + \frac{1}{3}\right)\pi\right]. \tag{3.74}$$

In statistical mechanics we are most interested in large values of N, ranging from perhaps 10^{12} to 10^{24}. The relative error in using the simplest version of Stirling's approximation, $\ln N! \approx N \ln N - N$, is of the order of $1/N$, which is completely negligible even if N is as 'small' as 10^{12}. For $N = 10^{24}$, $\ln N! \approx N \ln N - N$ is one of the most accurate approximations in physics.

3.12 Binomial Distribution with Stirling's Approximation

Since the last term in eq. (3.74) is negligible for large values of N, the most common approximation is to keep only the first two terms. If we use this form of Stirling's formula, we can write the binomial distribution in eq. (3.47) in an approximate but highly accurate form,

$$P(n|N) = \frac{N!}{n!(N-n)!} p^n (1-p)^{N-n} \tag{3.75}$$

$$\ln P(n|N) \approx N \ln N - n \ln n - (N-n)\ln(N-n) \tag{3.76}$$
$$+ n \ln p + (N-n)\ln(1-p).$$

Note that the contributions from the second term in Stirling's approximation in eq. (3.74) cancel in eq. (3.76).

Using Stirling's approximation to the binomial distribution turns out to have a number of pleasant features. We know that the binomial distribution will be peaked and that its relative width will be small. We can use Stirling's approximation to find the location of the peak by treating n as a continuous variable and setting the derivative of the logarithm of the probability distribution with respect to n in eq. (3.47) equal to zero,

$$\frac{\partial}{\partial n} \ln P(n|N) = 0 \tag{3.77}$$

or

$$0 = \frac{\partial}{\partial n} [\ln N! - \ln n! - \ln(N-n)! + n\ln p + N - n\ln(1-p)]$$

$$= \frac{\partial}{\partial n} [N\ln N - N - n\ln n - n - (N-n)\ln(N-n) - (N-n)$$

$$+ n\ln p + N - n\ln(1-p)]$$

$$= -\ln n + \ln(N-n) + \ln p - \ln(1-p). \tag{3.78}$$

This equation determines the location n_o of the maximum probability from the equation:

$$\frac{n_o}{N - n_o} = \frac{p}{1-p} \tag{3.79}$$

which has the solution

$$n_o = pN. \tag{3.80}$$

This means that the location of the peak, within the simplest of Stirling's approximations, gives the *exact* value of $\langle n \rangle = pN$.

3.13   Multinomial Distribution

An important generalization of the binomial distribution is the multinomial distribution. Instead of considering N_T particles divided among just two boxes, we divide them among $M \geq 2$ boxes, labelled $j = 1, \ldots, M$.

The assignment of each particle is assumed to be independent of the assignment of every other particle. If the probability of a given particle being assigned to box j is p_j, the probability of a specific subset of N_j particles being assigned to box j is

$$\prod_{j=1}^{M} p_j^{N_j}, \tag{3.81}$$

which is the generalization of eq. (3.45). This must be multiplied by the number of ways of choosing the set of N_j particles from N_T total particles. The argument runs parallel to that in Subsection 3.8.1. There are $N_T!$ total permutations of the N_T particles. Divide them up into M boxes of N_j particles, where $\sum_j N_j = N_T$. The order of the particles in each box is irrelevant, so we must divide by a factor of $N_j!$ for each box j. The resultant multinomial probability distribution is

$$P(\{N_j\}) = \left(\frac{N_T!}{\prod_{j=1}^{M} N_j!}\right) \prod_{j=1}^{M} p_j^{N_j}. \tag{3.82}$$

For application to the ideal gas, we will assume that the probability of finding a particular particle in the volume V_j of box j is proportional to V_j. If the total volume is $V_T = \sum_j V_j$, we have $p_j = V_j/V_T$. Inserting this into eq. (3.82) it becomes

$$P(\{N_j\}) = \left(\frac{N_T!}{\prod_{j=1}^{M} N_j!}\right) \prod_{j=1}^{M} \left(\frac{V_j}{V_T}\right)^{N_j} = \frac{N_T!}{V_T^{N_T}} \prod_{j=1}^{M} \left(\frac{V_j^{N_j}}{N_j!}\right)^{N_j}. \tag{3.83}$$

This equation will be very important in the next chapter. Indeed, it is very important to all of classical statistical mechanics.

3.14   Problems

Sample program to simulate rolling a single die

The sample program uses Python 2.

The computer program supplied first asks you how many times you want to roll a single die. It then chooses a random integer between 1 and 6, records its value, and repeats the process until it has chosen the number of random integers you specified. It then prints out the histogram (the number of times each integer occurred), tells you how many seconds the program took, and quits.

Histograms will play a significant role in the questions in this and other chapters. They are simply a record of the number of times each possible outcome is realized during a particular experiment or simulation. Histograms are a compact way of displaying information, as opposed to a list of all results generated. An estimate of the probability of each outcome is found by simply dividing the entries in the histogram by the number of trials.

The histogram is printed out in two formats to illustrate the possibilities. The time required to run the program is calculated and it will prove very useful. Since both 'trials' and 'sides' are integers, the division 'trials/sides' truncates the result and produces an integer, as is common in many computer languages. If Python 3 had been used, this division would have produced a floating point number.

Two special Python commands are used in this program: numpy.zeros(sides,int) and random.random(). The command numpy.zeros(sides,int) is in the 'module' numpy, which is imported in the first line of the program. It produces a vector of integer zeros. The command random.random() is in the 'module' random, which is imported in the second line of the program.

Text to be printed is enclosed in either single or double quotes, but they must match.

Take a few minutes and play with this program. Change any command and see what happens. One of the nice things about programming is that you can't break the computer.

When you finish coding a program, run it several times exploring various regions of parameters. It takes very little of your time, it gives you a good overview of how the parameters affect the probability distributions, and it is the aspect of programming that is the most fun.

The following sample Python code is for the program OneDie_PARTIAL.py.

```python
import numpy
import random
from time import clock
t0 = clock ()

trials = 100
print "_Number_of_trials _=_", trials

sides = 6
histogram = numpy. zeros (sides ,int)
print   histogram

j =0
while j < trials :
    r = int (random .random () * sides )
    histogram [r] =   histogram [r] + 1
    j=j+1
print   histogram

s=0
while s < sides :
    print   s, histogram [s], histogram [s]  —  trials / sides
    s=s+1

t1 = clock ()
print "\n_Program_time _=_" , '%_10.4f' % (t1 — t0) , "_seconds"
```

....

PROBLEM 3.1
Rolling a die with a computer

For this problem, either modify the Python program (OneDie_PARTIAL.py) supplied, or write your own from scratch, using whatever computer language you prefer.

1. Write a program (or modify the one given) to calculate and print out the histogram of the number of times each side occurs, the deviation of this number from one-sixth of the number of trials, the frequency with which each side occurs, and the deviation of this from one sixth. Hand in a copy of the code you used.
2. Show a typical print-out of your program.
3. Run the program for various numbers of random integers, starting with a small number, say 10, and increasing to a substantially larger number. The only upper limit is that the program should not take more than about one second to run. (Your time is valuable.)

THIS IS A COMPUTATIONAL PROBLEM.
ANSWERS MUST BE ACCOMPANIED BY DATA.
Please hand in hard copies for all data to which you refer in your answers.

4. As the number of trials increases, does the magnitude (absolute value) of the differences between the number of times a given side occurs and one-sixth of the number of trials increase or decrease?
(Hint: This is not the same question as the next one.)

5. As you increase the number of trials, does the ratio of the number of times each side occurs to the total number of trials come closer to $1/6$?

PROBLEM 3.2
Mutual independence

We defined the concept of pairwise independence in this chapter. There is a related concept called *mutual independence*. Consider the set of random variables

$$\{A_j | j = 1, \ldots, N\}.$$

They are said to be mutually independent if for any subset of $\{A_j\}$ containing n of these random variables, the marginal distribution satisfies the condition

$$P_{i,j,\ldots,n}(a_i, a_j, a_k, \ldots, a_n) = P_i(a_i)P_j(a_j)P_k(a_k)\cdots P_n(a_n).$$

Obviously, mutual independence implies pairwise independence. The question is whether pairwise independence implies mutual independence.

Provide a proof or a counter example.

PROBLEM 3.3
A die with an arbitrary number of sides

Suppose we have a die with S sides, where S is an integer, but not necessarily equal to 6. The set of possible outcomes is then $\{n | n = 1, \ldots, S\}$ (or $\{n | n = 0, \ldots, S-1\}$, your choice). Assume that all sides of the die are equally probable, so that $P(n) = 1/S$.

Since this is partly a computational problem, be sure to support your answers with data from your computer simulations.

1. What are the theoretical values of the mean, variance, and standard deviation as functions of S? The answers should be in closed form, rather than expressed as a sum. (Hint: It might be helpful to review the formulas for the sums of integers and squares of integers.)

2. Modify the program you used for an earlier problem to simulate rolling a die with S sides, and print out the mean, variance, and standard deviation for a number of trials. Have the program print out the theoretical predictions in each case, as well as the deviations from theory to facilitate comparisons.

3. Run your program for two very different values of S. Are the results for the mean, variance, and standard deviation consistent with your predictions?
(Do not use such long runs that the program takes more than about a second to run. It would be a waste of your time to wait for the program to finish.)

4. Experiment with different numbers of trials. How many trials do you need to obtain a *rough* estimate for the values of the mean and standard deviation? How many trials do you need to obtain an error of less than 1%? Do you need the same number of trials to obtain 1% accuracy for the mean and standard deviation?

PROBLEM 3.4
Independence and correlation functions

We have shown that if the random variables A and B were independent and $F(A)$ and $G(B)$ were numerical functions defined on A and B, then

$$\langle F(A)G(B)\rangle = \langle F(A)\rangle\langle G(B)\rangle.$$

Suppose have two random *numbers*, X and Y, and we know that:

$$\langle XY\rangle = \langle X\rangle\langle Y\rangle.$$

Does that imply that X and Y are independent?
Provide a proof or counter-example.

PROBLEM 3.5
Some probability calculations (the Chevalier de Méré's problem)

1. Suppose we roll an honest die ten times. What is the probability of *not* finding a '3' on any of the rolls.
2. Calculate the following probabilities and decide which is greater.

The probability of finding at least one '6' on four rolls is greater than the probability of finding at least one 'double 6' on 24 rolls of two dice.

   Historical note:

This is a famous problem, which is attributed to the French writer Antoine Gombaud (1607–1684), who called himself the Chevalier de Méré (although according to Wikipedia he was not a nobleman). He was an avid but not terribly successful gambler, and he enlisted his friend Blaise Pascal (French mathematician, 1623–1662) to calculate the correct odds to bet on the dice rolls, described previously. Pascal's solution was one of the early triumphs of probability theory.

PROBLEM 3.6
Generalized dice

1. Modify your computer program to simulate the roll of N dice. Your program should let the dice have any number of sides, but the same number of sides for each die. The number of dice and the number of sides should be read in at the start of the program.
   One trial will consist of N rolls. Your program should calculate the sum of the N numbers that occur during each trial. It should also compare the results for the mean, variance, and standard deviation of that sum with the theoretical predictions.

2. Test the calculations that we have carried out for the mean, variance, and standard deviation of the sum of the numbers on the dice. In each case, obtain data for a couple of different run-lengths.
Investigate cases (a) and (b).

   (a) Two dice, each with ten sides.
   (b) Ten dice, each with twenty sides.

3. Use your program to investigate the width of the distribution for various numbers of two-sided dice. Does the width of the distribution increase or decrease with increasing numbers of dice? Do your results agree with the theory?

PROBLEM 3.7
Mismatched dice

We derived the probability distribution for the sum of two dice using Kronecker deltas, where each die had S sides.

For this problem, calculate the probability distribution for the sum of two dice using Kronecker deltas, when one die has four sides, and the other has six sides.

PROBLEM 3.8
Computer simulations of mismatched dice

1. Write a computer program to compute the probability distribution for the sum of two dice when each die has an arbitrary number of sides.
Run your program for dice of four and six sides.
2. Modify the computer program you wrote for the previous problem to compute the probability distribution for the sum of *three* dice when each die has an arbitrary number of sides.
Run your program once for all dice having six sides, and once for any combination you think is interesting.

PROBLEM 3.9
Sums of 0s and 1s

The following questions are directly relevant to the distribution of ideal-gas particles between two boxes of volume V_A and V_B. If $V = V_A + V_B$, then $p = V_A/V$ and $1 - p = V_B/V$.

1. Prove the following identity for binomial coefficients:

$$\binom{N}{n+1} = \frac{N-n}{n+1}\binom{N}{n}.$$

2. Modify a new copy of your program to simulate the addition of an arbitrary number of independent random numbers, $\{n_j | j = 1, \ldots, N\}$, each of which can take on the value 1 with probability $P(1) = p$ and 0 with probability $P(0) = 1 - p$.

Have the program print the histogram of values generated. (To save paper, print only the non-zero parts of the histograms.)

Include a calculation of the theoretical probability from the binomial distribution, using the identity you proved at the beginning of this assignment. Have your program calculate the mean, variance, and standard deviation, and compare them to the theoretical values that we have calculated.

Have the program plot the theoretical prediction for the histogram on the same graph as your histogram. Plot the deviations of the simulation data from the theoretical predictions on the second graph. (A discussion of how to make plots in VPython is included in Section 4 of the Appendix.)

3. Run your program for the following cases, using a reasonable number of trials. Comment on the agreement between the theory and your computer experiment.

   a. $N = 10, p = 0.5$
   b. $N = 30, p = 0.85$
   c. $N = 150, p = 0.03$

PROBLEM 3.10
Sums of 0s and 1s revisited: The Gaussian approximation

1. Modify the program you used for the simulation of sums of 0s and 1s to include a calculation of a Gaussian approximation to the binomial distribution.

   Keep the curve in the first graph that shows the theoretical binomial probabilities, but add a curve representing the Gaussian approximation in a contrasting color.

   Modify the second graph in your program to plot the difference between the full binomial approximation and the Gaussian approximation.

2. Run simulations for various values of the probability p and various numbers of 'dice', with a sufficient number of trials to obtain reasonable accuracy. (Remember not to run it so long that it wastes your time.)

   Comment on the accuracy (or lack thereof) of the Gaussian approximation.

3. Run your program for the following cases, using a reasonable number of trials. Comment on the agreement between the theory and your computer experiment.

   1. $N = 10, p = 0.5$
   2. $N = 30, p = 0.85$
   3. $N = 150, p = 0.03$

PROBLEM 3.11
The Poisson distribution

We have derived the binomial distribution for the probability of finding n particles in a subvolume out of a total of N particles in the full system. We assumed that the probabilities for each particle were independent and equal to $p = V_A/V$.

When you have a very large number N of particles with a very small probability p, you can simplify this expression in the limit that $p \to 0$ and $N \to \infty$, with the product fixed at $pN = \mu$. The answer is the Poisson distribution:

$$P_\mu(n) = \frac{1}{n!} \mu^n \exp(-\mu).$$

Derive the Poisson distribution from the binomial distribution.

(Note that Stirling's formula is an approximation, and should not be used in a derivation or a proof.)

Calculate the mean, variance, and standard deviation for the Poisson distribution as functions of μ.

PROBLEM 3.12
Numerical evaluation of the Poisson distribution

In an earlier problem you derived the Poisson distribution,

$$P_\mu(n) = \frac{1}{n!} \mu^n \exp(-\mu).$$

1. Modify (a copy of) your program to read in the values of μ and N and calculate the value of the probability p. Include an extra column in your output that gives the theoretical probability based on the Poisson distribution.

   (Note: It might be quite important to suppress rows in which the histogram is zero, otherwise, the print-out could get out of hand for large values of N.)

2. Run your program for various values of μ and N. How large does N have to be for the agreement to be good?

4

The Classical Ideal Gas: Configurational Entropy

The single, all-encompassing problem of thermodynamics is the determination of the equilibrium state that eventually results after the removal of internal constraints in a closed, composite system.

Herbert B. Callen

This chapter begins the derivation of the entropy of the classical ideal gas, as outlined in Chapter 2. Our goal is to derive the entropy function for each member of a large set of isolated thermodynamic systems, with the total entropy given by the sum over the individual entropies. The essential property of these entropy functions is that when two or more systems are allowed to exchange particles, the sum of the total entropy is a maximum at the equilibrium value of the quantity which can be exchanged. We will use this property as a basis for the definition of entropy, following the ideas of Boltzmann. The exchange of energy or volume will be considered in Chapter 6.

The first step is to separate the calculation of the entropy into two parts: One for the contributions of the positions of the particles, and one for the contributions of their momenta. In an ideal gas, these are assumed to be independent, although this is not true for all interacting particles. As we will see, the total entropy is then just the sum of the two contributions.

The contributions of the probability distribution for the positions of the particles, which we will call the configurational entropy, will be calculated in this chapter. After probabilities are generalized to continuous variables in Chapter 5, the contributions of the momenta to the entropy will be given in Chapter 6.

4.1 Separation of Entropy into Two Parts

The assumption that the positions and momenta of the particles in an ideal gas are independent allows us to consider each separately.

As we saw in Section 3.3, the independence of the positions and momenta means that their joint probability can be expressed as a product of functions of q and p alone,

An Introduction to Statistical Mechanics and Thermodynamics. Robert H. Swendsen, Oxford University Press (2020). © Robert H. Swendsen. DOI: 10.1093/oso/9780198853237.001.0001

$$P(q,p) = P_q(q)P_p(p).$$ (4.1)

According to Boltzmann's 1877 definition, the entropy *of a composite system* is proportional to the logarithm of the probability of the extensive variables to within additive and multiplicative constants. Since eq. (4.1) shows that the probability distribution in phase space can be expressed as a product, the total entropy will be expressed as a sum of the contributions of the positions and the momenta.

The probability distribution in configuration space, $P_q(q)$, depends only on the volume V and the number of particles, N. Consequently, the configurational entropy, S_q, depends only on V and N; that is, $S_q = S_q(V,N)$.

The probability distribution in momentum space, $P_p(p)$, depends only on the total energy, E, and the number of particles, N. Consequently, the energy-dependent contribution to the entropy from the momenta, S_p, depends only on E and N; that is, $S_p = S_p(E,N)$.

The total entropy of the ideal gas is given by the sum of the configurational and the energy-dependent terms:

$$S_{total}(E, V, N) = S_q(V,N) + S_p(E,N).$$ (4.2)

The thermodynamic quantities E, V, and N are referred to as 'extensive' parameters (or observables, or variables) because they measure the amount, or extent, of something. They are to be contrasted with 'intensive' parameters, such as temperature or pressure (to be defined later), which do not automatically become bigger for bigger systems.

4.2   Probability Distribution of Particles

Before presenting the solution for a large number of systems in Section 4.8, we will first restrict the discussion to the distribution of particles between two systems, and see how this leads to an expression for the configurational entropy.

Let us consider a composite system consisting of two boxes (or subsystems) containing a total of $N_{T,jk}$ distinguishable, non-interacting particles. We will label the boxes j and k, with volumes V_j and V_k. The total volume of the two boxes is $V_{T,jk} = V_j + V_k$. The number of particles in subsystem j is N_j, with $N_k = N_{T,jk} - N_j$ being the number of particles in subsystem k.

We can either constrain the number of particles in each box to be fixed, or allow the numbers to fluctuate by removing the wall that separates the boxes. The total number of particles $N_{T,jk}$ is constant in either case. We will first consider the probability distribution for the number of particles in each subsystem when the constraint (impenetrable wall) is released (by removing the wall or making a hole in it). We will consider the case of isolated systems in Section 4.3.

In keeping with our intention of making the simplest reasonable assumptions about the probability distributions of the configurations, we will assume that the positions of the particles are not only independent of the momenta, but are also mutually independent of each other. The probability density $P_q(q)$ can then be written as a product,

$$P_q(q) = P_{N_{T,jk}}(\{\vec{r}_i\}) = \prod_{i=1}^{N_{T,jk}} P_i(\vec{r}_i), \tag{4.3}$$

where $P_i(\vec{r}_i)$ is the probability distribution of the single particle i. If we further assume that a given particle is equally likely to be anywhere in the composite system, the probability of it being in subsystem j is $V_j/V_{T,jk}$.[1]

Remember that this is an assumption that could, in principle, be tested with repeated experiments. It is not necessarily true. However, we are strongly prejudiced in this matter. If we were to carry out repeated experiments and find that a particular particle was almost always in subsystem j, we would probably conclude that there is something about the system that breaks the symmetry.

The assumption that everything we do not know is equally probable—subject to the constraints—is the simplest assumption we can make. It is the starting point for all statistical mechanics. Fortunately, it provides extremely good predictions.

If the $N_{T,jk}$ particles are free to go back and forth between the two subsystems, the probability distribution of N_j is given by the binomial distribution, eq. (3.47), as discussed in Section 3.8:

$$P(N_j|N_{T,jk}) = \frac{N_{T,jk}!}{N_j!(N_{T,jk}-N_j)!} \left(\frac{V_j}{V_{T,jk}}\right)^{N_j} \left(1 - \left(\frac{V_j}{V_{T,jk}}\right)\right)^{N_{T,jk}-N_j}$$

$$= \binom{N_{T,jk}}{N_j} \left(\frac{V_j}{V_{T,jk}}\right)^{N_j} \left(1 - \left(\frac{V_j}{V_{T,jk}}\right)\right)^{N_{T,jk}-N_j}. \tag{4.4}$$

To emphasize the equal standing of the two subsystems, it is often useful to write this equation in the symmetric form

$$P(N_j, N_k) = \frac{N_{T,jk}!}{N_j!N_k!} \left(\frac{V_j}{V_{T,jk}}\right)^{N_j} \left(\frac{V_k}{V_{T,jk}}\right)^{N_k}, \tag{4.5}$$

with the implicit constraint that $N_j + N_k = N_{T,jk}$.

4.3 Distribution of Particles between Two Isolated Systems that were Previously in Equilibrium

If the two subsystems are first in equilibrium with each other and then separated by inserting a partition or closing the hole, the probability of finding N_j particles in system j and N_k particles in system k is exactly the same as if the subsystems remained in

---

[1] This can be shown more formally after the discussion of continuous random variables in Section 5.2 of Chapter 5.

equilibrium with each other. It is still given by eq. (4.5). This probability is defined for all values of N_j and $N_k = N_{T,jk} - N_j$ in two subsystems that are isolated from each other.

Therefore, eq. (4.4) gives the probabilities of the variables N_j and N_k, even if the two systems, j and k, are isolated from each other. Since the initial state of the composite system in this example is specified by N_j and N_k, eq. (4.4) gives the probability of the initial state that Boltzmann was referring to in his 1877 paper, which was quoted at the beginning of Chapter 3.

4.4   Consequences of the Binomial Distribution

As shown in Section 3.7, the average value of N_j from the binomial distribution is

$$\langle N_j \rangle = N_{T,jk} \left(\frac{V_j}{V_{T,jk}} \right). \tag{4.6}$$

By symmetry, we also have

$$\langle N_k \rangle = N_{T,jk} \left(\frac{V_k}{V_{T,jk}} \right), \tag{4.7}$$

so that

$$\frac{\langle N_j \rangle}{V_j} = \frac{\langle N_k \rangle}{V_k} = \frac{N_{T,jk}}{V_{T,jk}}. \tag{4.8}$$

The width of the probability distribution for N_j is given by the standard deviation, as given in eq. (3.43),

$$\delta N_j = \left[N \left(\frac{V_j}{V_{T,jk}} \right) \left(1 - \frac{V_j}{V_{T,jk}} \right) \right]^{1/2} \tag{4.9}$$

$$= \left[N \left(\frac{V_j}{V_{T,jk}} \right) \left(\frac{V_k}{V_{T,jk}} \right) \right]^{1/2}$$

$$= \left[\langle N_j \rangle \left(\frac{V_k}{V_{T,jk}} \right) \right]^{1/2}.$$

In general, the mode, or location of the maximum of the probability distribution, differs from the location of the average value $\langle N_j \rangle$ by something of order $1/\langle N_j \rangle$. If $\langle N_j \rangle \approx 10^{20}$, this error is unmeasurable. By a fortunate quirk in the mathematics, the average value has exactly the same value as the mode of *approximate* probability density using Stirling's approximation. This turns out to be very convenient, although not necessary.

4.5   Actual Number versus Average Number

It is important to make a clear distinction between the actual number of particles N_j at any given time and the average number of particles $\langle N_j \rangle$.

The actual number of particles, N_j, is a property of the system. It is an integer, and it fluctuates with time if the system can exchange particles with another system.

The average number of particles, $\langle N_j \rangle$, is part of a description of the system and not a property of the system itself. It is not an integer, and is time-independent in equilibrium.

The magnitude of the fluctuations of the actual number of particles is given by the standard deviation, δN_j, from eq. (4.9), which is of the order of $\sqrt{\langle N_j \rangle}$. If there are about 10^{20} particles in subsystem j, the actual number of particles N_j will fluctuate around the value $\langle N_j \rangle$ by about $\delta N_j \approx 10^{10}$ particles. The numerical difference between N_j and $\langle N_j \rangle$ is very big—and it becomes even bigger for bigger systems!

Given the large difference between N_j and $\langle N_j \rangle$, why is $\langle N_j \rangle$ at all useful? The answer lies in the fact that macroscopic measurements do not count individual molecules. The typical method used to measure the number of molecules is to weigh the sample and divide by the weight of a molecule. The weight is measured experimentally with some relative error—usually between 1% and one part in 10^5. Consequently, using $\langle N_j \rangle$ as a description of the system is good whenever the relative width of the probability distribution, $\delta N_j / \langle N_j \rangle$, is small.

From eq. (4.6), the relative width of the probability distribution is given by

$$\frac{\delta N_j}{\langle N_j \rangle} = \frac{1}{\langle N_j \rangle} \left[\langle N_j \rangle \left(\frac{V_k}{V_{T,jk}} \right) \right]^{1/2} = \sqrt{\frac{1}{\langle N_j \rangle}} \sqrt{\frac{V_k}{V_{T,jk}}}. \tag{4.10}$$

The relative width is proportional to $1/\sqrt{\langle N_j \rangle}$, which becomes very small for a macroscopic system. For 10^{20} particles, the relative uncertainty in the probability distribution is about 10^{-10}, which is much smaller than the sensitivity of thermodynamic experiments.

In the nineteenth century, as thermodynamics was being developed, the atomic hypothesis was far from being accepted by all scientists. Thermodynamics was formulated in terms of the mass of a sample, rather than the number of molecules—which in any case, many scientists did not believe in. Fluctuations were not seen experimentally, so scientists did not make a distinction between the average mass in a subsystem and the actual mass.

Maintaining the distinction between N_j and $\langle N_j \rangle$ is made more difficult by the common tendency to use a notation that obscures the difference. In thermodynamics, it would be more exact to use $\langle N_j \rangle$ and similar expressions for the energy and other observable quantities. However, it would be tiresome to continually include the brackets in all equations, and the brackets are invariably dropped. We will also follow this practice in the chapters on thermodynamics, although we will maintain the distinction in discussing statistical mechanics.

Fortunately, it is fairly easy to remember the distinction between the actual energy or number of particles in a subsystem and the average values, once the distinction has been recognized.

4.6 The 'Thermodynamic Limit'

It is sometimes said that thermodynamics is only valid in the limit of infinite system size. This is implied by the standard terminology, which defines the 'thermodynamic limit' as the limit of infinite size while holding the ratio of the number of particles to the volume fixed.

An obvious difficulty with such a point of view is that we only carry out experiments on finite systems, which would imply that thermodynamics could never apply to the real world.

Another difficulty with restricting thermodynamics to infinite systems is that the finite-size effects due to containers, surfaces, interfaces, and phase transitions are lost.

The point of view taken in this book is that thermodynamics is valid in the real world when the uncertainties due to statistical fluctuations are much smaller than the experimental errors in measured quantities. For macroscopic systems containing 10^{20} or more molecules, this means that the statistical uncertainties are of the order 10^{-10} or less, which is several orders of magnitude smaller than typical experimental errors. Even if we consider colloids with about 10^{12} particles, the statistical uncertainties are about 10^{-6}, which is still smaller than most measurements.

Taking the limit of infinite system size can be a very useful mathematical approximation, especially when studying phase transitions, but it is not essential to either understand or apply thermodynamics.

4.7 Probability and Entropy

We have seen that the equilibrium value of N_j is determined by the maximum of the probability distribution (or mode) given by

$$P(N_j, N_k) = \frac{N_{T,jk}!}{N_j! N_k!} \left(\frac{V_j}{V_{T,jk}} \right)^{N_j} \left(\frac{V_k}{V_{T,jk}} \right)^{N_k}, \tag{4.11}$$

with the constraint that $N_j + N_k = N_{T,jk}$. We can show the dependence of eq. (4.11) on the composite nature of the total system by introducing a new function,

$$\Omega_q(N, V) = \frac{V^N}{N!}, \tag{4.12}$$

where V and N are generic variables that stand for the relevant volume and particle number (V_j, V_k or $V_{T,jk}$), and (N_j, N_k or $N_{T,jk}$). This allows us to rewrite eq. (4.11) as

$$P(N_j, N_k) = \frac{\Omega_q(N_j, V_j)\,\Omega_q(N_k, V_k)}{\Omega_q(N_{T,jk}, V_{T,jk})}.$$ (4.13)

At this point, we are ready to make an extremely important observation. Since the logarithm is a monotonic function of its argument, the maximum of $P(N_j, N_k)$ and the maximum of $\ln[P(N_j, N_k)]$ occur at the same values of N_j and N_k; they both represent the mode of the probability distribution. Using $\ln[P(N_j, N_k)]$ turns out to be much more convenient than using $P(N_j, N_k)$, partly because we can divide it naturally into three distinct terms:

$$\ln[P(N_j, N_k)] = \ln\left[\frac{V_j^{N_j}}{N_j!}\right] + \ln\left[\frac{V_k^{N_k}}{N_k!}\right] - \ln\left[\frac{V_{T,jk}^{N_{T,jk}}}{N_{T,jk}!}\right]$$ (4.14)

$$= \ln\Omega_q(N_j, V_j) + \ln\Omega_q(N_k, V_k) - \ln\Omega_q(N_{T,jk}, V_{T,jk}).$$

The first term on the right (in both forms of this equation) depends only on the variables for subsystem j, the second depends only on the variables for subsystem k, and the third term depends only on the variables for the total composite system. Note that since $N_{T,jk}$ and $V_{T,jk}$ are assumed to be constants, $\ln\Omega_q(N_{T,jk}, V_{T,jk})$ is also a constant.

It will be convenient to define a function

$$S_q(N, V) \equiv k\ln\left(\frac{V^N}{N!}\right) + kXN$$ (4.15)

$$\equiv k\ln\Omega_q(N, V) + kXN,$$ (4.16)

where N and V are again generic variables, and k and X are both (at this point) arbitrary constants. The maximum of the function

$$S_{q,jk}(N_j, V_j, N_k, V_k) = k\ln[P(N_j, N_k)] + S_q(N_{T,jk}, V_{T,jk})$$ (4.17)
$$= S_q(N_j, V_j) + S_q(N_k, V_k),$$

with the usual constraint that $N_j + N_k = N_{T,jk}$, then gives us the location of the mode of the distribution of the number of particles, which is an excellent approximation to $\langle N_j \rangle$.

We have seen in Section 3.8 that the width of the probability distribution is proportional to $\sqrt{\langle N_j \rangle}$. Since the probability distribution is normalized, the value of its peak must be proportional to $1/\sqrt{\langle N_j \rangle}$. This can also be seen from the gaussian approximation in eq. (3.57), with $pN \to \langle N_j \rangle$, and $1 - p \to V_k/V_{T,jk}$. At the equilibrium values of N_j and N_k, this gives

$$\ln P\left(N_j, N_k\right)\big|_{equil} \approx -\frac{1}{2}\ln\left(2\pi \langle N_j\rangle (V_k/V_{T,jk})\right). \tag{4.18}$$

Since the function $S_q\left(N_{T,jk}, V_{T,jk}\right)$ is of order $N_{T,jk}$, and $N_{T,jk} > \langle N_j\rangle >> \ln\langle N_j\rangle$, the term $k\ln\left[P\left(N_j, N_k\right)\right]$ in eq. (4.17) is completely negligible at the equilibrium values of N_j and N_k. Therefore, *in equilibrium*, we have

$$S_{q,jk}\left(N_j, V_j, N_k, V_k\right) = S_q\left(N_j, V_j\right) + S_q\left(N_k, V_k\right) = S_q\left(N_{T,jk}, V_{T,jk}\right). \tag{4.19}$$

Note that in eq. (4.19) we have used the numbers N_j and N_k instead of the averages $\langle N_j\rangle$ and $\langle N_k\rangle$. This represents the approximation that the width of the particle-number distribution vanishes. It corresponds to the microcanonical ensemble, which assumes that width of the energy distribution vanishes. It should be remembered that they are both approximations, although very good ones. We will discuss their consequences in some detail in Chapter 21.

We can identify $S_{q,jk}\left(N_j, V_j, N_k, V_k\right)$ as the part of the entropy of the composite system that is associated with the configurations; that is, with the positions of the particles. We will call $S_{q,jk}\left(N_j, V_j, N_k, V_k\right)$ the total configurational entropy of the composite system. The functions $S_q\left(N_j, V_j\right)$ and $S_q\left(N_k, V_k\right)$ are called the configurational entropies of subsystems j and k.

We have therefore found functions of the variables of each subsystem, such that the maximum of their sum yields the location of the equilibrium values

$$S_{q,jk}\left(N_j, V_j, N_k, V_k\right) = S_q\left(N_j, V_j\right) + S_q\left(N_k, V_k\right) \tag{4.20}$$

subject to the usual constraint that $N_j + N_k = N_{T,jk}$.

The fact that the contributions to the configurational entropy from each subsystem are simply added to find the configurational entropy of the composite system is a very important property. It is traditionally known, reasonably enough, as 'additivity'. On the other hand, the property might also be called 'separability', since we first derived the entropy of the composite system and were then able to separate the expression into the sum of individual contributions from each subsystem.

The entropy of a composite system has the property of being additive whether or not the subsystems are in equilibrium with each other.

The function that we have identified as the configurational entropy follows Boltzmann's definition (see Section 2.5) in being the logarithm of the probability *of a composite system* (within additive and multiplicative constants). The fact that it is maximized at equilibrium (takes on the value of its mode) agrees with Boltzmann's intuition. This turns out to be the most important property of the entropy in thermodynamics, as we will see in Chapter 9.

Boltzmann's definition of the entropy always produces a function that takes on its maximum, or mode, in equilibrium. Anticipating the discussion of the laws of thermodynamics in Part II, this property is equivalent to the Second Law of Thermodynamics, which we have therefore derived from statistical principles. (There is still the matter of irreversibility, which will be addressed in Chapter 22.)

4.8   The Generalization to $M \geq 2$ Systems

In thermodynamics, we are usually interested in the interactions between more than just two (sub)systems. We will need to consider the effects of putting two systems into equilibrium with each other, separating them, and then letting one of the systems equilibrate with a different system. We will also need to consider equilibrium conditions involving three or more systems.

The equations we have derived are easily generalized to an arbitrarily large number of systems using the results for the multinomial probability distribution in eq. (3.82), as derived in Section 3.13. In fact, the systems can be taken to include all systems in the world that might possibly be brought into contact with each other.

If we have $M \geq 2$ systems, whether exchanging particles or isolated from each other, the probability distribution for $\{N_j | j = 1, \ldots, M\}$ is

$$P(\{N_j | j = 1, \ldots, M\}) = \left(\frac{N_T!}{\prod_{j=1}^M N_j!} \right) \prod_{k=1}^M \left(\frac{V_k}{V_T} \right)^{N_k}. \tag{4.21}$$

In this equation, $N_T = \sum_{j=1}^M N_j$ and $V_T = \sum_{j=1}^M V_j$ are constants. The individual system volumes, V_j are also regarded as constants for now. Eq. (4.21) was derived in Chapter 3 as eq. (3.83).

Using the definition of $\Omega_q(N, V)$ in eq. (4.12), the multinomial probability can be written as

$$P(\{N_j, V_j | j = 1, \ldots, M\}) = \frac{\prod_{k=1}^M \Omega_Q(N_k, V_k)}{\Omega_Q(N_T, V_T)}. \tag{4.22}$$

This equation is a generalization of eq. (4.13). It is valid whether some or all of the individual systems are in equilibrium with each other, or if all M systems are isolated from each other. If every system is isolated from every other system, eq. (4.22) is the probability of the initial state $\{N_j, V_j | j = 1, \ldots, M\}$.

Finally, using the definition of entropy in eq. (4.15), we can write

$$S_q(\{N_j, V_j | j = 1, \ldots, M\}) = \sum_{k=1}^M S_q(N_k, V_k) - S_q(N_T, V_T) + C, \tag{4.23}$$

where C is an arbitrary constant. N_T, V_T, and C are constants, whose values have no physical consequences.

Eq. (4.23) shows that the entropy of an individual system, j, $S_q(N_j, V_j)$, is exactly the same as derived from considering just two systems, and, in fact, the expression for $S_q(N_j, V_j)$, is independent of the total number of systems. The validity of this equation is not dependent on whether the systems are isolated or in equilibrium.

The fact that the expression for the entropy of system j contains a factor of $1/N_j!$, is due to the use of interacting systems in the definition of entropy, following Boltzmann and the form of the multinomial coefficient. This derivation has not been universally accepted, and a wide variety of other explanations have been proposed, some claiming that it is impossible to understand the factor of $1/N_j!$ without invoking quantum mechanics. The controversies have run for over 140 years without any sign of slowing down. I recommend the literature on Gibbs' paradox to any student interested in pursuing the matter.

4.9   An Analytic Approximation for the Configurational Entropy

The expression for the configurational entropy becomes much more practical if we introduce Stirling's approximation from Section 3.11. The entropy can then be differentiated and integrated with standard methods of calculus, as well as being much easier to deal with numerically.

Because we are interested in large numbers of particles, only the simplest version of Stirling's approximation is needed: $\ln N! \approx N \ln N - N$. The relative magnitude of the correction terms is only of the order of $\ln N/N$, which is completely negligible. It is rare to find an approximation quite this good, and we should enjoy it thoroughly.

Using Stirling's approximation, our expression for the configurational entropy becomes

$$S_q(N, V) \approx kN \left[\ln \left(\frac{V}{N} \right) + X \right], \tag{4.24}$$

where N and V are generic variables that take the values N_T and V_T, as well as $\{N_k, V_k | k = 1, \ldots, M\}$. X and k are still arbitrary constants. In Chapter 8 we will discuss further constraints that give them specific values.

Eq. (4.24) is our final result for the configurational entropy of the classical ideal gas, under the assumption that the number of particles is known exactly. This definition will be supplemented in Chapter 6 with contributions from the momentum terms in the Hamiltonian. But first, we need to develop mathematical methods to deal with continuous random variables, which we will do in Chapter 5.

4.10 Problems

....

PROBLEM 4.1
Using the entropy to find the equilibrium equations

This object of this problem is to find the equilibrium conditions for ideal-gas systems that are allowed to exchange particles. Use Stirling's approximation and eqs. (4.15), (6.40), and (4.24).

1. Consider two ideal-gas systems with volumes V_j and V_k. They are initially isolated and contain N_j and N_k, respectively. After they are brought together and allowed to come to equilibrium, find the equation that expresses the condition for the equilibrium values, N_j^* and N_k^*, and solve that equation.
2. Now consider three ideal-gas systems with volumes V_j, V_k, and V_3. The systems are initially isolated and contain N_j, N_k, and N_3 particles. Now let them all exchange particles and come to equilibrium. Find two equations to express the equilibrium conditions, and solve them to find the equilibrium values N_j^*, N_k^*, and N_3^*.
3. Again consider the same three ideal-gas systems with volumes V_j, V_k, and V_3. The systems are again initially isolated and contain N_j, N_k, and N_3 particles.
   First, bring systems j and k together and let them come to equilibrium. Find the equilibrium values N_j^* and N_k^*.
   Now separate systems j and k. Bring systems k and 3 together and let them come to equilibrium. Find the number of particles in each of the three systems.

....

5

Continuous Random Numbers

It may be that the race is not always to the swift nor the battle to the strong—but that's the way to bet.

Damon Runyon

In Chapter 3 we discussed the basics of probability theory for discrete events. However, the components of the momenta of the particles in an ideal gas are continuous variables. This requires an extension of probability theory to deal with continuous random numbers, which we will develop in this chapter.

5.1 Continuous Dice and Probability Densities

To illustrate the essential features of continuous random numbers we will use a simple generalization of throwing a die. While a real die has six distinct sides, our continuous die can take on any real value, x, between 0 and 6.

Assume that our continuous die is 'honest' in the sense that all real numbers between 0 and 6 are equally probable. Since there is an infinite number of possibilities, the probability of any particular number being found is $p = 1/\infty = 0$.

On the other hand, if we ask for the probability of the continuous random number x being in some interval, the probability can be non-zero. If all values of x are equally probable, we would assume that the probability of finding x in the interval $[a, b]$ is proportional to the length of that interval,

$$P([a,b]) = A \int_a^b dx = A(b-a) \tag{5.1}$$

where A is a normalization constant. Since the probability of x being somewhere in the interval $[0, 6]$ must be 1,

$$P([0,1]) = A \int_0^6 dx = 6A = 1, \tag{5.2}$$

we have $A = 1/6$.

An Introduction to Statistical Mechanics and Thermodynamics. Robert H. Swendsen, Oxford University Press (2020).
© Robert H. Swendsen. DOI: 10.1093/oso/9780198853237.001.0001

This leads us to define a 'probability density function' or simply a 'probability density', $P(x) = 1/6$, to provide us with an easy way of computing the probability of finding x in the interval $[a, b]$,

$$P([a,b]) = \int_a^b P(x)dx = \int_a^b \frac{1}{6}dx = \frac{b-a}{6}. \tag{5.3}$$

I apologize for using P to denote both probabilities and probability densities, even though they are very different things. This confusing notation certainly does not help keep the distinction between the two concepts in mind. On the other hand, it is usually easy to tell which concept is intended. Since much of the literature assumes that the reader can figure it out, this is probably a good place to become accustomed to doing so.

5.2 Probability Densities

We can use the idea of a probability density to extend consideration to cases in which all values of a continuous random number are not equally likely. The only change is that $P(x)$ is then no longer a constant. The probability of finding x in an interval is given by the integral over that interval,

$$P([a,b]) = \int_a^b P(x)dx. \tag{5.4}$$

Generalization to multiple dimensions is simply a matter of defining the probability density of a multidimensional function. If there are two continuous random numbers, x and y, the probability density is $P(x,y)$.

There is one feature of probability densities that might not be clear from our example of continuous dice. In general, although probabilities are dimensionless, probability densities have units. For example, if the probability of finding a single classical particle is uniform in a box of volume V, the probability density within the volume is

$$P(x, y, z) = \frac{1}{V}, \tag{5.5}$$

and it has units of $[m^{-3}]$. When the probability density is integrated over a sub-volume of the box, the result is a dimensionless probability, as expected.

Because probability densities have units, there is no upper limit on their values. Their values must, of course, be positive. However, unlike probabilities, they do not have to be less than 1. In fact, it is even possible for a probability density to diverge at one or more points, as long as the integral over all values is 1.

Like probabilities, probability densities must be normalized. If Ω indicates the entire range over which the probability density is defined, then

$$\int_\Omega P(x)dx = 1 \tag{5.6}$$

with obvious extensions to multi-dimensional, continuous random numbers.

Marginal probabilities are defined in analogy to the definition for discrete numbers

$$P_x(x) = \int_{-\infty}^{\infty} P(x,y)dy. \tag{5.7}$$

Conditional probabilities can also be defined if $P_x(x) \neq 0$. The conditional probability $P(y|x)$ can be written as

$$P(y|x) = \frac{P(x,y)}{P_x(x)}. \tag{5.8}$$

In general, we have

$$\begin{aligned} P(x,y) &= P(x|y)P_y(y) \\ &= P(y|x)P_x(x). \end{aligned} \tag{5.9}$$

Bayes' theorem for continuous variables then takes the form

$$P(y|x) = \frac{P(x|y)P_y(y)}{P_x(x)}. \tag{5.10}$$

Independence is also defined in analogy to discrete probability theory. Two continuous random numbers are said to be independent if

$$P(x,y) = P_x(x)P_y(y). \tag{5.11}$$

Using eq. (5.9), we can see that if x and y are independent random numbers and $P_y(y) \neq 0$, the conditional probability $P(x|y)$ is independent of y:

$$P(x|y) = \frac{P(x,y)}{P_y(y)} = \frac{P_x(x)P_y(y)}{P_y(y)} = P_x(x). \tag{5.12}$$

The average of any function $F(x)$ can be calculated by integrating over the probability density

$$\langle F(x) \rangle = \int_{-\infty}^{\infty} F(x)P(x)dx. \tag{5.13}$$

Averages, moments, and central moments are all defined as expected from discrete probability theory.

Mean:

$$\langle x \rangle = \int_{-\infty}^{\infty} x P(x) dx \tag{5.14}$$

n-th moment:

$$\langle x^n \rangle = \int_{-\infty}^{\infty} x^n P(x) dx \tag{5.15}$$

n-th central moment:

$$\langle (x - \langle x \rangle)^n \rangle = \int_{-\infty}^{\infty} (x - \langle x \rangle)^n P(x) dx. \tag{5.16}$$

As for discrete probabilities, the variance is defined as the second central moment, and the standard deviation is the square root of the variance.

A word of warning:

Many books refer to continuous random numbers as if they were discrete. They will refer to the probability of finding a random number in a given interval as being proportional to the 'number of points' in that interval. This is entirely incorrect and can cause considerable confusion. The number of points in an interval is infinite. It is even the same infinity as the number of points in any other interval, in the sense that the points in any two intervals can be mapped one-to-one onto each other. This error seems to have arisen in the nineteenth century, before physicists were entirely comfortable working with continuous random numbers. Although we are now in the twenty-first century, the error seems to be quite persistent. I can only hope that the next generation of physicists will finally overcome it.

5.3   Dirac Delta Functions

In Chapter 3 we discussed finding the probability of the sum of two dice using Kronecker delta functions. In the next section we will carry out the analogous calculation for two continuous dice. Although the problem of two continuous dice is highly artificial, it is useful because its mathematical structure will often be encountered in statistical mechanics.

For these calculations we will introduce the Dirac delta function, named for its inventor, Paul Adrien Maurice Dirac (English physicist, 1902–1984). The Dirac delta function is an extension of the idea of a Kronecker delta to continuous functions. This approach is not usually found in textbooks on probability theory—which is unfortunate, because Dirac delta functions make calculations much easier.

Dirac delta functions are widely used in quantum mechanics, so most physics students will have encountered them before reading this book. However, since we will need more properties of the delta function than are usually covered in courses in quantum mechanics, we will present a self-contained discussion in the next section.

5.3.1 Definition of Delta Functions

To provide a simple definition of a Dirac delta function, we will first consider an ordinary function that is non-zero only in the neighborhood of the origin,[1]

$$
\delta_\epsilon(x) \equiv
\begin{cases}
0 & x < -\epsilon \\
\dfrac{1}{2\epsilon} & -\epsilon \le x \le \epsilon. \\
0 & x > \epsilon
\end{cases}
$$

Clearly, this function is normalized,

$$
\int_{-\infty}^{\infty} \delta_\epsilon(x)\,dx = 1. \tag{5.17}
$$

In fact, since $\delta_\epsilon(x)$ is only non-zero close to the origin,

$$
\int_{a}^{b} \delta_\epsilon(x)\,dx = 1 \tag{5.18}
$$

as long as $a < -\epsilon$ and $b > \epsilon$.

It is also obvious that the function is symmetric,

$$
\delta_\epsilon(x) = \delta_\epsilon(-x). \tag{5.19}
$$

If we consider the function $\delta_\epsilon(cx)$, where c is a constant, the symmetry of the function means that the sign of c is irrelevant,

$$
\delta_\epsilon(cx) = \delta_\epsilon(|c|x). \tag{5.20}
$$

Note that the width of the function $\delta_\epsilon(cx)$ is a factor of $1/|c|$ times that of $\delta_\epsilon(x)$, while the height of the function remains the same ($1/2\epsilon$).

---

[1] An alternative way of defining the Dirac delta function uses Gaussian functions, which has advantages when discussing derivatives. The approach used in this section has been chosen for its simplicity.

5.3.2   Integrals over Delta Functions

We can integrate over $\delta_\epsilon(cx)$ by defining a new variable $y = cx$,

$$\int_a^b \delta_\epsilon(cx)dx = \int_{a/|c|}^{b/|c|} \delta_\epsilon(y)dy/|c| = \frac{1}{|c|}. \tag{5.21}$$

When the argument of the delta function has a zero at some value of $x \neq 0$, the integral is unaffected as long as the limits of the integral include the zero of the argument of the delta function. As long as $a < d/c - \epsilon$ and $b > d/c + \epsilon$, we have

$$\int_a^b \delta_\epsilon(cx - d)dx = \int_{a/|c|}^{b/|c|} \delta_\epsilon(y)dy/|c| = \frac{1}{|c|}. \tag{5.22}$$

The Dirac delta function, $\delta(x)$, is defined as the limit of $\delta_\epsilon(x)$ as ϵ goes to zero.

$$\delta(x) \equiv \lim_{\epsilon \to 0} \delta_\epsilon(x). \tag{5.23}$$

The Dirac delta function is zero for all $x \neq 0$ and infinite for $x = 0$, so that calling it a function greatly annoys most mathematicians. For this reason, mathematics books on probability theory rarely mention the delta function. Nevertheless, if you are willing to put up with a mathematician's disapproval, the Dirac delta function can make solving problems much easier.

The integral of the Dirac delta function is also unity when the limits of integration include the location of the zero of the argument of the delta function,

$$\int_a^b \delta(x)dx = 1 \tag{5.24}$$

as long as $a < 0$ and $b > 0$, and zero otherwise.

Similarly, when the argument of the delta function is $(cx - d)$, the value of the integral is

$$\int_a^b \delta(cx - d)dx = \frac{1}{|c|} \tag{5.25}$$

as long as $a < d/c$ and $b > d/c$, and zero otherwise.

5.3.3   Integral of $f(x)$ times a Delta Function

The Dirac delta function can be used to pick out the value of a continuous function at a particular point, just as the Kronecker delta does for a discrete function. This is one of its most useful properties.

Consider a function $f(x)$ that is analytic in some region that includes the point x_o. Then we can expand the function as a power series with a non-zero radius of convergence

$$f(x) = \sum_{j=0}^{\infty} \frac{1}{j!} f^{(j)}(x_o)(x - x_o)^j, \tag{5.26}$$

where $f^{(j)}(x_o)$ is the j-th derivative of $f(x)$, evaluated at $x = x_o$,

$$f^{(j)}(x_o) = \frac{d^j}{dx^j} f(x) \bigg|_{x=x_0}. \tag{5.27}$$

Consider the integral over the product of $f(x)$ and $\delta_\epsilon(x - x_o)$

$$\int_{-\infty}^{\infty} f(x)\delta_\epsilon(x - x_o)dx = \frac{1}{2\epsilon} \int_{x_o-\epsilon}^{x_o+\epsilon} f(x)dx$$

$$= \sum_{j=0}^{\infty} \frac{1}{j!} \frac{1}{2\epsilon} \int_{x_o-\epsilon}^{x_o+\epsilon} f^{(j)}(x_o)(x - x_o)^j dx$$

$$= \sum_{j=0}^{\infty} \frac{1}{j!} \frac{1}{2\epsilon} f^{(j)}(x_o) \frac{1}{j+1} \left[\epsilon^{j+1} - (-\epsilon)^{j+1} \right]. \tag{5.28}$$

All terms with odd j vanish. Keeping only the even terms and defining $n = j/2$ for even values of j, we can write the integral as

$$\int_{-\infty}^{\infty} f(x)\delta_\epsilon(x - x_o)dx = \sum_{n=0}^{\infty} \frac{1}{(2n+1)!} f^{(2n)}(x_o)\epsilon^{2n}$$

$$= f(x_o) + \sum_{n=1}^{\infty} \frac{1}{(2n+1)!} f^{(2n)}(x_o)\epsilon^{2n}. \tag{5.29}$$

When we take the limit of $\epsilon \to 0$, the function, $\delta_\epsilon(\cdot)$ becomes a Dirac delta function, and the right-hand side of the equation is just $f(x_o)$,

$$\int_{-\infty}^{\infty} f(x)\delta(x - x_o)dx = f(x_o). \tag{5.30}$$

We can generalize eq. (5.30) to include more general arguments of the delta function. First, if the argument is of the form $c(x - x_o)$, the result is divided by $|c|$,

$$\int_{-\infty}^{\infty} f(x)\delta(c(x - x_o)) dx = \frac{f(x_o)}{|c|}. \tag{5.31}$$

We can further generalize this equation to allow the delta function to have an arbitrary argument, $g(x)$. If $g(x)$ has zeros at the points $\{x_j | j = 1, \ldots, n\}$, and $g'(x)$ is the derivative of $g(x)$, the integral becomes

$$\int_{-\infty}^{\infty} f(x)\delta(g(x))\, dx = \sum_{j=1}^{n} \frac{f(x_j)}{|g'(x_j)|}. \tag{5.32}$$

This can be proved by expanding $g(x)$ about each of its zeros.

Common errors in working with Dirac delta functions:

- Not including all zeros of $g(x)$ inside the region of integration.
- Including zeros of $g(x)$ that are outside the region of integration.
- Not using the absolute value of $g'(x)$ in the denominator.

5.4   Transformations of Continuous Random Variables

We are often interested in functions of random variables, which are themselves random variables. As an example, consider the following problem.

Given two continuous random variables, x and y, along with their joint probability density, $P(x,y)$, we wish to find the probability density of a new random variable, s, that is some function of the original random variables, $s = f(x,y)$. The formal solution can be written in terms of a Dirac delta function,

$$P(s) = \int_{-\infty}^{\infty} \int_{-\infty}^{\infty} P(x,y)\delta(s - f(x,y))dx\,dy. \tag{5.33}$$

As was the case for the corresponding discrete random variables, the probability density of s is automatically normalized,

$$\int_{-\infty}^{\infty} P(s)ds = \int_{-\infty}^{\infty} \int_{-\infty}^{\infty} \int_{-\infty}^{\infty} P(x,y)\delta(s - f(x,y))ds\,dx\,dy \tag{5.34}$$

$$= \int_{-\infty}^{\infty} \int_{-\infty}^{\infty} P(x,y)\,dx\,dy = 1.$$

The last equality is due to the normalization of $P(x,y)$.

5.4.1   Sum of Two Random Numbers

To see how eq. (5.33) works in practice, consider the sum of two uniformly distributed random numbers. We will use the example of two continuous dice, in which the two random numbers x and y each take on values in the interval $[0,6]$, and the probability density is uniform,

$$P(x,y) = 1/36. \tag{5.35}$$

This probability density is consistent with the normalization condition in eq. (5.6).
   The probability density for $s = x + y$ is then

$$P(s) = \int_0^6 \int_0^6 P(x,y)\delta(s - (x+y))dx\,dy$$

$$= \frac{1}{36} \int_0^6 \int_0^6 \delta(s - (x+y))dx\,dy. \tag{5.36}$$

Following the same procedure as in Section 3.4, we first carry out the integral over y.

$$\int_0^6 \delta(s - (x+y))dy = \begin{cases} 1 & 0 \le s - x \le 6 \\ 0 & \text{otherwise} \end{cases}. \tag{5.37}$$

   In direct analogy to the procedure for calculating the sum of two discrete dice in Section 3.4, we have two conditions on the remaining integral over x. Only those values of x for which both $x < s$ and $x > s - 6$ contribute to the final answer. These limits are in addition to the limits of $x < 6$ and $x > 0$ that are already explicit in the integral over x. Since all four of these inequalities must be satisfied, we must take the more restrictive of the inequalities in each case. Which inequality is the more restrictive depends on the value of s. Determining the proper limits on the integral over x is illustrated in Table 5.1, which should be compared to Table 3.3 for the corresponding discrete case.
   For $s \le 6$, the lower bound on x is 0 and the upper bound is s. For $s \ge 6$, the lower bound on x is $s - 6$ and the upper bound is 6. The sums can then be evaluated explicitly,

Table 5.1 *Determining the limits for the second integral when evaluating eq. (5.36).*

	lower limit	upper limit
From limits on integral:	$x \ge 0$	$x \le 6$
From delta function:	$x \ge s - 6$	$x \le s$
More restrictive if $s \le 6$	$x \ge 0$	$x \le s$
More restrictive if $s \ge 6$	$x \ge s - 6$	$x \le 6$

$$P(s) = \begin{cases} \int_0^s \dfrac{1}{36} = \dfrac{s}{36} & s \le 6 \\[4mm] \int_{s-6}^6 \dfrac{1}{36} = \dfrac{12-s}{36} & s \ge 6 \end{cases}.$$ (5.38)

5.5 Bayes' Theorem

Bayes' theorem, eq. (3.13), was derived in Section 3.3.1,

$$P(A|B) = \frac{P(B|A)P_A(A)}{P_B(B)},$$ (5.39)

and given in eq. (5.10) for continuous variables as

$$P(y|x) = \frac{P(x|y)P_y(y)}{P_x(x)}.$$ (5.40)

If we accept the Bayesian definition of probability in Section 3.1, we can apply eq. (5.39) to the determination of theoretical parameters from experiment.

Let X denote the data from an experiment, while θ denotes the theoretical parameter(s) we wish to determine. Using standard methods, we can calculate the conditional probability $P(X|\theta)$ of observing a particular set of data if we know the parameters. However, that is not what we want. We have the results of the experiment and we want to calculate the parameters conditional on that data.

Bayes' theorem gives us the information we want about the theoretical parameters after the experiment. It provides the information on a particular set of measurements X, as the conditional probability for θ in the following equation:

$$P(\theta|X) = \frac{P(X|\theta)P_\theta(\theta)}{P_X(X)}.$$ (5.41)

In Bayesian terminology, $P_\theta(\theta)$ is called the 'prior' and represents whatever knowledge we had of the parameters before we carried out the experiment. The conditional probability $P(X|\theta)$—*viewed as a function of θ*—is called the 'likelihood'. The conditional probability $P(\theta|X)$ is known as the 'posterior', and represents our knowledge of the parameters after we have obtained data from an experiment.

For example, suppose we want to determine the value of the asymptotic frequency f (the frequentist's probability) for a particular random event. Assume we have done an experiment with a binary outcome of success or failure, and we obtained $n = 341,557$ successes in $N = 10^6$ trials. The experimental data, denoted previously by X, is n

in this case. We would immediately guess that $f \approx 0.341557$, but how close to that value is it?

Our theoretical parameter, denoted by θ in eq. (5.41), is the value of f. Suppose we knew nothing of the value of f before the experiment, except that $0 \leq f \leq 1$. A reasonable description of our knowledge (or lack of it) would be to say that our prior is a uniform constant for values between zero and 1. From the normalization condition, the constant must be 1,

$$P_f(f) = \begin{cases} 0 & < 0 \\ 1 & 0 \leq f \leq 1. \\ 0 & f > 1 \end{cases} \tag{5.42}$$

The likelihood is given by the binomial distribution

$$P_N(n|f) = \frac{N!}{n!(N-n)!} f^n (1-f)^{N-n}. \tag{5.43}$$

As is usual in Bayesian calculations, we will ignore $P_{N,n}(n)$; since it does not depend on f, it simply plays the role of a normalization constant. Putting eqs. (5.41), (5.42), and (5.43) together and ignoring multiplicative constants, we have the probability density for f,

$$P_N(f|n) \propto f^n (1-f)^{N-n}. \tag{5.44}$$

For large n and N, this is a sharply peaked function. The maximum can be found by setting the derivative of its logarithm equal to zero:

$$\frac{d}{df} P_N(f|n) = \frac{d}{df} [n \ln f + (N-n) \ln(1-f) + \text{constants}] = \frac{n}{f} - \frac{N-n}{1-f} = 0. \tag{5.45}$$

The maximum is therefore located at

$$f_{max} = \frac{n}{N} \tag{5.46}$$

as expected, since we had n successes in N trials.

The variance can be found by approximating the distribution by a Gaussian, finding the second derivative, and evaluating it at f_{max},

$$\frac{d^2}{df^2} [n \ln f + (N-n) \ln(1-f)] = \frac{d}{df} \left[\frac{n}{f} - \frac{N-n}{1-f} \right] \tag{5.47}$$

$$= -\frac{n}{f^2} - \frac{N-n}{(1-f)^2}. \tag{5.48}$$

Evaluating this derivative at $f = f_{max} = n/N$ The variance is then given by

$$-\frac{1}{\sigma^2} = -\frac{N}{f_{max}} - \frac{N}{1 - f_{max}} = -\frac{N}{f_{max}(1 - f_{max})} \tag{5.49}$$

or

$$\sigma^2 = \frac{f_{max}(1 - f_{max})}{N}. \tag{5.50}$$

The standard deviation is then

$$\sigma = \sqrt{\frac{f_{max}(1 - f_{max})}{N}} \tag{5.51}$$

so that the uncertainty in the value of f is proportional to $1/\sqrt{N}$. In the case of our example with $f \approx 0.341557$ and $N = 10^6$, the standard deviation is ≈ 0.00047.

5.6 Problems

....

PROBLEM 5.1
Probability densities

1. Given a continuous random number x, with the probability density

$$P(x) = A\exp(-2x)$$

for all $x > 0$, find the value of A and the probability that $x > 1$.
2. It is claimed that the function

$$P(y) = \frac{B}{y}$$

where B is a constant, is a valid probability density for $0 < y < 1$.
Do you agree? If so, what is the value of B?
3. It is claimed that the function

$$P(y) = \sqrt{\frac{C}{y}}$$

where C is a constant, is a valid probability distribution for $0 < y < 1$.
Do you agree? If so, what is the value of C?
4. The probability distribution $P(x,y)$ has the form

$$P(x,y) = Dxy$$

for the two random numbers x and y in the region $x > 0, y > 0$, and $x+y < 1$, where D is a constant, and $P(x,y) = 0$ outside this range.

(a) What is the value of the constant D?
(b) What is the probability that $x < 1/2$?
(c) Are x and y independent?

PROBLEM 5.2
How well can we measure the asymptotic frequency?

I have run a number of computer simulations in which a 'success' occurred with probability p. It would be no fun if I told you the value of p, but I will tell you the results of the simulations.

Trials	Successes
10^2	69
10^3	641
10^4	6353
10^5	63738
10^6	637102
10^7	6366524

1. Using Bayesian statistics, we showed that we can represent our knowledge of the value of p by a probability distribution (or probability density) $P(p|n)$, which is the conditional probability density for p, given that an experiment of N trials produced n successes. $P(p|n)$ is found using Bayes' theorem. Note that $P(n|p)$, the conditional probability of finding n successes in N trials when the probability is p, is just the binomial distribution that we have been studying.
   Find the location of the maximum of this probability density by taking the first derivative of $P(p|n)$ and setting

$$\frac{\partial}{\partial p} \ln P(p|n) = 0.$$

   Find the general expression for the width of $P(p|n)$ by taking the second derivative. The procedure is essentially the same as the one we used for calculating the Gaussian approximation to the integrand in deriving Stirling's approximation.
2. Using the result you obtained above, what can you learn about the value of p from each of these trials? Is the information from the various trials consistent? (Suggestion: A spreadsheet can be very helpful for calculations in this problem.)

PROBLEM 5.3
Continuous probability distributions

Determine the normalization constant and calculate the mean, variance, and standard deviation for the following random variables:

1. For $x \geq 0$

$$P(x) = A\exp(-ax).$$

2. For $x \geq 1$

$$P(x) = Bx^{-3}.$$

PROBLEM 5.4
Integrals over delta functions

Evaluate the following integrals in closed form:

1.

$$\int_{-\infty}^{\infty} x^4 \delta(x^2 - y^2)dx$$

2.

$$\int_{-1}^{\infty} \exp(-x)\delta(\sin(x)))dx.$$

3. The probability density $P(x,y)$ has the form

$$P(x,y) = 24xy$$

for the two random numbers x and y in the range $x > 0$, $y > 0$, and $x+y < 1$, and $P(x,y) = 0$ outside this range.
What is the probability distribution for the following new random variable,

$$z = x+y?$$

PROBLEM 5.5
Transforming random variables

Consider the two random variables x and y. Their joint probability density is given by

$$P(x,y) = A\exp\left[-x^2 - 2y\right]$$

where A is a constant, $x \geq 0$, $y \geq 0$, and $y \leq 1$.

1. Evaluate the constant A.
2. Are x and y independent? Justify your answer.
3. If we define a third random number by the equation

$$z = x^2 + 2y$$

what is the probability density $P(z)$?

PROBLEM 5.6
Maxwell–Boltzmann distribution

In the near future we will derive the Maxwell–Boltzmann distribution for the velocities (or the momenta) of gas particles. The probability density for the x-component of the velocity is

$$P(v_x) = A \exp\left[-\beta \frac{1}{2} m v_x^2\right].$$

1. What is the value of the normalization constant A?
2. What is the probability density for the velocity $\vec{v} = (v_x, v_y, v_z)$?
3. What is the probability distribution (probability density) of the speed (magnitude of the velocity)?

PROBLEM 5.7
Two one-dimensional, ideal-gas particles

Consider two ideal-gas particles with masses m_A and m_B, confined to a one-dimensional box of length L. Assume that the particles are in thermal equilibrium with each other, and that the total kinetic energy is $E = E_A + E_B$. Use the usual assumption that the probability is uniform in phase space, subject to the constraints.

1. Calculate the probability distribution $P(E_A)$ for the energy of one of the particles.
2. Calculate the average energy of the particle, $\langle E_A \rangle$.
3. Find the most probable value of the energy E_A (the location of the maximum of the probability density).

PROBLEM 5.8
Energy distribution of a free particle

Suppose we have an ideal gas in a finite cubic box with sides of length L. The system is in contact with a thermal reservoir at temperature T. We have determined the momentum distribution of a single particle to be:

$$P(\vec{p}) = X \exp\left[-\beta \frac{|\vec{p}|^2}{2m}\right].$$

Find $P(E)$, where E is the energy of a single particle.

PROBLEM 5.9
Particle in a gravitational field

An ideal gas particle in a gravitational field has a probability distribution in momentum $\vec{p}$ and height z of the form

$$P(\vec{p}, z) = X \exp\left[-\beta \frac{|\vec{p}|^2}{2m} - \beta mgz\right]$$

for $0 \leq z < \infty$ and all values of momentum.

1. Evaluate the constant X.
2. Calculate the average value of the height: $\langle z \rangle$.
3. Calculate the probability distribution of the total energy of the particle

$$E = \frac{|\vec{p}|^2}{2m} + mgz.$$

6

The Classical Ideal Gas: Energy Dependence of Entropy

You should call it entropy, for two reasons. In the first place your uncertainty function has been used in statistical mechanics under that name, so it already has a name. In the second place, and more important, no one really knows what entropy really is, so in a debate you will always have the advantage.

John von Neumann, suggesting to Claude Shannon a name for his new uncertainty function, as quoted in *Scientific American*, (1971) 225(3): 180.

This chapter provides the second half of the derivation of the entropy of the classical ideal gas, as outlined in Chapter 2 and begun in Chapter 4. In the present chapter we will calculate the probability distribution for the energy of each subsystem. The logarithm of that probability then gives the energy-dependent contribution to the entropy of the classical ideal gas. The total entropy is just the sum of the configurational entropy and the energy-dependent terms, as discussed in Section 4.1.

6.1 Distribution for the Energy between Two Subsystems

We again consider the composite system that consists of the two subsystems discussed in Chapter 2, and we will extend the discussion to $M \geq 2$ systems in Section 6.6. Subsystem j contains N_j particles and subsystem k contains N_k particles, with a total of $N_{T,jk} = N_j + N_k$ particles in the composite system. Since we are dealing with classical, non-interacting particles, the energy of each subsystem is given by

$$E_\alpha = \sum_{i=1}^{N_\alpha} \frac{|\vec{p}_{\alpha,i}|^2}{2m},$$

(6.1)

where $\alpha = j$ or k and m is the mass of a single particle. The momentum of the i-th particle in subsystem α is $\vec{p}_{\alpha,i}$.

An Introduction to Statistical Mechanics and Thermodynamics. Robert H. Swendsen, Oxford University Press (2020).
© Robert H. Swendsen. DOI: 10.1093/oso/9780198853237.001.0001

The composite system is perfectly isolated from the rest of the universe, and the total energy $E_{T,jk} = E_j + E_k$ is fixed. We are interested in the case in which the two subsystems can be brought into thermal contact to enable the subsystems to exchange energy, so that the composite system can come to thermal equilibrium.

A partition that is impervious to particles, but allows energy to be exchanged between subsystems, is called a 'diathermal' wall, in contrast to an adiabatic wall that prevents the exchange of either particles or energy. We wish to calculate the probability distribution of the energy between two subsystems separated by a diathermal wall.

To proceed with the entropy calculation, we must make an assumption about the probability distribution in momentum space. The simplest assumption is that the momentum distribution is constant, subject to the constraints on the energy. This is again the model probability that we introduced in Chapter 4.

With the assumption of uniform probability density in momentum space, we can calculate the probability distribution for the energy from that of the momenta, using the methods from Section 5.4. We use Dirac delta functions to select states for which system j has energy E_j and system k has energy E_k. Conservation of energy in the composite system of course requires that $E_k = E_{T,jk} - E_j$, but this form of writing the probability distribution will be useful in Section 6.2 when we separate the contributions of the two subsystems.

Since the notation can become unwieldy at this point, we use a compact notation for the integrals

$$\int_{-\infty}^{\infty} \cdots \int_{-\infty}^{\infty} (\cdots) dp_{\alpha,1}^3 \cdots dp_{\alpha,N_j}^3 \equiv \int_{-\infty}^{\infty} (\cdots) dp_\alpha, \tag{6.2}$$

and write:

$$P(E_j, E_k) = \frac{\int_{-\infty}^{\infty} \delta\left(E_j - \sum_{i=1}^{N_j} \frac{|\vec{p}_{j,i}|^2}{2m}\right) dp_j \int_{-\infty}^{\infty} \delta\left(E_k - \sum_{\ell=1}^{N_k} \frac{|\vec{p}_{k,\ell}|^2}{2m}\right) dp_k}{\int_{-\infty}^{\infty} \delta\left(E_{T,jk} - \sum_{i=1}^{N_{T,jk}} \frac{|\vec{p}_i|^2}{2m}\right) dp}. \tag{6.3}$$

The integral over dp in the denominator goes over the momenta of all particles in systems j and k.

The denominator of eq. (6.3) is chosen to normalize the probability distribution,

$$\int_0^{\infty} P(E_j, E_{T,jk} - E_j) \, dE_j = 1. \tag{6.4}$$

By defining a function

$$\Omega_p\left(E_\alpha, N_\alpha\right) = \int_{-\infty}^{\infty} \delta\left(E_\alpha - \sum_{i=1}^{N_\alpha} \frac{|\vec{p}_{\alpha,i}|^2}{2m}\right) dp_\alpha, \tag{6.5}$$

we can write eq. (6.3) in an even more compact form that highlights its similarity to eq. (4.13)

$$P\left(E_j, E_k\right) = \frac{\Omega_p\left(E_j, N_j\right)\Omega_p\left(E_k, N_k\right)}{\Omega_p\left(E_{T,jk}, N_{T,jk}\right)}. \tag{6.6}$$

6.2 Evaluation of Ω_p

To complete the derivation of $P\left(E_j, E_k\right)$, we need to evaluate the function $\Omega_p\left(E, N\right)$ in eq. (6.5), where E and N are generic variables that stand for E_j, E_k, or $E_{T,jk}$ and N_j , N_k, or $N_{T,jk}$. Fortunately, this can be done exactly by taking advantage of the symmetry of the integrand to transform the $3N$-dimensional integral to a one-dimensional integral, and then relating the integral to a known function.

6.2.1 Exploiting the Spherical Symmetry of the Integrand

The delta function in eq. (6.5) makes the integrand vanish everywhere in momentum space, except on a sphere in momentum space defined by energy conservation

$$2mE = \sum_{i=1}^{N} |\vec{p}_i|^2. \tag{6.7}$$

The radius of this $3N$-dimensional sphere is clearly $\sqrt{2mE}$.

Since the area of an n-dimensional sphere is given by $A = S_n r^{n-1}$, where S_n is a constant that depends on the dimension, n, the expression for the function can be immediately transformed from a $3N$-dimensional integral to a one-dimensional integral,

$$\Omega_p\left(E, N\right) = \int_{-\infty}^{\infty} \cdots \int_{-\infty}^{\infty} \delta\left(E - \sum_{i=1}^{N} \frac{|\vec{p}_i|^2}{2m}\right) dp_j \cdots dp_{3N}$$

$$= \int_{0}^{\infty} S_n p^{3N-1} \delta\left(E - \frac{p^2}{2m}\right) dp. \tag{6.8}$$

The new integration variable is just the radial distance in momentum space,

$$p^2 = \sum_{i=1}^{N} |\vec{p}_i|^2. \tag{6.9}$$

To evaluate the integral in eq. (6.8), transform the variable of integration by defining $x = p^2/2m$, which implies that $p^2 = 2mx$ and $p\,dp = m\,dx$. Inserting this in the expression for $\Omega_p(E,N)$, we can evaluate the integral

$$\Omega_p(E,N) = S_n \int_0^{\infty} (2mx)^{(3N-1)/2} \delta(E-x)(2mx)^{-1/2} m\,dx$$

$$= S_n m(2mE)^{3N/2-1}. \tag{6.10}$$

The only thing remaining is to evaluate S_n, which is the surface area of an n-dimensional sphere of unit radius.

6.2.2   The Surface Area of an n-Dimensional Sphere

The volume of a sphere in n dimensions is given by $C_n r^n$, where C_n is a constant. By differentiation, the area of the sphere is $nC_n r^{n-1}$, so that $S_n = nC_n$.

To evaluate C_n we can use a trick involving Gaussian integrals. We begin by taking the n-th power of both sides of eq. (3.61):

$$\left[\int_{-\infty}^{\infty} e^{-x^2} dx \right]^n = \left[\sqrt{\pi} \right]^n = \pi^{n/2}$$

$$= \int_{-\infty}^{\infty} \cdots \int_{-\infty}^{\infty} \exp\left(-\sum_{j=1}^{n} x_j^2 \right) dx_1\, dx_2, \cdots, dx_n$$

$$= \int_0^{\infty} \exp\left(-r^2 \right) S_n r^{n-1} dr$$

$$= nC_n \int_0^{\infty} \exp\left(-r^2 \right) r^{n-1} dr. \tag{6.11}$$

We can transform the final integral in eq. (6.11) to a more convenient form by changing the integration variable to $t = r^2$,

$$\pi^{n/2} = nC_n \int_0^{\infty} \exp\left(-r^2 \right) r^{n-1} dr$$

$$= \frac{1}{2} nC_n \int_0^\infty \exp\left(-r^2\right) r^{n-2} 2r \, dr$$

$$= \frac{1}{2} nC_n \int_0^\infty e^{-t} t^{n/2-1} \, dt. \tag{6.12}$$

The integral in eq. (6.12) is well known; it is another representation of the factorial function

$$m! = \int_0^\infty e^{-t} t^m \, dt. \tag{6.13}$$

We can see why this is so by induction. Begin with $m = 0$,

$$\int_0^\infty e^{-t} t^0 \, dt = \int_0^\infty e^{-t} \, dt = 1 = 0!. \tag{6.14}$$

If eq. (6.13) is valid for m, then we can use integration by parts to prove that it is valid for $(m+1)$,

$$\int_0^\infty e^{-t} t^{m+1} \, dt = \left[e^{-t}(m+1)t^m \right]_0^\infty - (m+1) \int_0^\infty (-e^{-t}) t^m \, dt$$

$$= (m+1) \int_0^\infty e^{-t} t^m \, dt$$

$$= (m+1)m!$$

$$= (m+1)!. \tag{6.15}$$

This confirms the validity of eq. (6.13) for all positive integers.

A convenient consequence of eq. (6.13) is that it provides an analytic continuation of $m!$ to non-integer values. It is traditional to call this extension of the concept of factorials a Gamma (Γ) function,

$$\Gamma(m+1) = m! = \int_0^\infty e^{-t} t^m \, dt. \tag{6.16}$$

For many purposes, $\Gamma(\cdot)$ is a very useful notation. However, the shift between m and $m+1$ can be a mental hazard. I will stay with the factorial notation, even for non-integer values.

Returning to eq. (6.12), we can now complete the derivation of C_n,

$$\pi^{n/2} = \frac{1}{2} n C_n \int_0^\infty e^{-t} t^{n/2-1} dt$$

$$= C_n \left(\frac{n}{2}\right)(n/2 - 1)!$$

$$= (n/2)! C_n. \tag{6.17}$$

This gives us values for C_n and $S_n = nC_n$:

$$C_n = \frac{\pi^{n/2}}{(n/2)!} \tag{6.18}$$

$$S_n = n \frac{\pi^{n/2}}{(n/2)!}. \tag{6.19}$$

6.2.3 Exact Expression for Ω_p

Inserting eq. (6.19) into the expression for $\Omega_p(E,N)$ in eq. (6.10) we find

$$\Omega_p(E,N) = \frac{3N\pi^{3N/2}}{(3N/2)!} m(2mE)^{3N/2-1}. \tag{6.20}$$

An excellent approximation to the logarithm of $\Omega_p(E,N)$ can be found using Stirling's approximation. The errors are of order $\ln N / N$,

$$\ln \Omega_p(E,N) \approx N \left[\frac{3}{2} \ln\left(\frac{E}{N}\right) + X\right]. \tag{6.21}$$

The constant X in eq. (6.21) can be calculated, but will not be needed at this point.

6.3 Distribution of Energy between Two Isolated Subsystems that were Previously in Equilibrium

As was the case for the probability distribution of particles, the probability distribution of the energy is exactly the same in equilibrium and for isolated systems that were previously in equilibrium. The act of separating the two systems has no effect on the probability distribution of the energy. Analogous to Subsection 4.3, when the two systems are allowed to exchange energy, the probability distribution for E_j is given by the product of the two factors.

6.4 Probability Distribution for Large N

Putting the functional form of Ω_p into eq. (6.6)—and ignoring constants for the time being—we find the energy-dependence of the probability distribution,

$$P\left(E_j, E_{T,jk} - E_j\right) = \frac{\Omega_p\left(E_j, N_j\right)\Omega_p\left(E_{T,jk} - E_j, N_k\right)}{\Omega_p\left(E_{T,jk}, N_{T,jk}\right)}$$

$$\propto \left(E_j\right)^{3N_j/2-1}\left(E_{T,jk} - E_j\right)^{3N_k/2-1}. \tag{6.22}$$

We are primarily interested in finding the average energy, $\langle E_j\rangle$. The energy E_j has a very narrow probability distribution, and the measured energy will almost always be equal to $\langle E_j\rangle$ within experimental error. We want to find the width of the distribution, δE_j, primarily to confirm that the distribution is narrow. As before, note that the neglect of the width of the distribution is a extremely good approximation, but an approximation nevertheless. We will return to this point in Chapter 21.

The average, $\langle E_j\rangle$, is most easily found by approximating it by the mode of the probability distribution, which is the location of the maximum probability. The mode is found by taking the derivative of the logarithm of the energy distribution and setting it equal to zero,

$$\frac{\partial}{\partial E_j}\ln P\left(E_j, E_{T,jk} - E_j\right) = \left(\frac{3N_j}{2} - 1\right)\frac{1}{E_j} - \left(\frac{3N_k}{2} - 1\right)\frac{1}{E_{T,jk} - E_j} = 0, \tag{6.23}$$

Solving this equation gives us the location of the maximum of the probability distribution, which is located at $\langle E_j\rangle$ to very high accuracy,

$$\langle E_j\rangle = E_{j,\mathrm{max}} = \left(\frac{3N_j - 2}{3N_{T,jk} - 4}\right)E_{T,jk}. \tag{6.24}$$

For a large number of particles, this becomes

$$E_{j,\mathrm{max}} = \left(\frac{N_j}{N_{T,jk}}\right)E_{T,jk}, \tag{6.25}$$

or

$$\frac{E_{j,\mathrm{max}}}{N_j} = \frac{E_{T,jk}}{N_{T,jk}} = \frac{\langle E_j\rangle}{N_j}, \tag{6.26}$$

with a relative error of the order of $N_{T,jk}^{-1}$. When $N_{T,jk} = 10^{20}$, or even $N_{T,jk} = 10^{12}$, this is certainly an excellent approximation.

Note that the energy per particle is the same in each subsystem. This is a special case of the equipartition theorem, which we will discuss in more detail when we come to the canonical ensemble in Chapter 19.

For large N_j, the probability distribution is very sharply peaked. It can be approximated by a Gaussian, and we can use the same methods as we did in Section 3.9 to determine the appropriate parameters. In particular, we can calculate the width of the probability distribution by taking the second derivative of the logarithm, which gives us the negative reciprocal of the variance:

$$\frac{\partial^2}{\partial E_j^2} \ln P\left(E_j, E_{T,jk} - E_j\right) = -\left(\frac{3N_j}{2} - 1\right)\frac{1}{E_j^2} - \left(\frac{3N_k}{2} - 1\right)\frac{1}{(E_{T,jk} - E_j)^2}. \quad (6.27)$$

Evaluating this at the maximum of the function, $E_{j,\max} = E_{T,jk}N_j/N_{T,jk}$, and assuming that we are dealing with large numbers of particles so that we can approximate $3N_j/2 - 1 \approx 3N_j/2$, this becomes

$$\frac{\partial^2}{\partial E_j^2} \ln P\left(E_j, E_{T,jk} - E_j\right)|_{E_j = E_{T,jk}N_j/N_{T,jk}}$$

$$= -\left(\frac{3N_j}{2}\right)\frac{N_{T,jk}^2}{E_{T,jk}^2 N_j^2} - \left(\frac{3N_k}{2}\right)\frac{N_{T,jk}^2}{E_{T,jk}^2 N_k^2}$$

$$= -\frac{3N_{T,jk}^2}{2E_{T,jk}^2}\left(\frac{1}{N_j} + \frac{1}{N_k}\right)$$

$$= -\frac{3N_j^2}{2E_j^2}\left(\frac{N_{T,jk}}{N_j N_k}\right)$$

$$= -\sigma_{E_j}^2. \quad (6.28)$$

The variance of E_j is then:

$$\sigma_{E_j}^2 = \frac{2\langle E_j\rangle^2}{3N_j^2}\left(\frac{N_j N_k}{N_{T,jk}}\right) = \frac{\langle E_j\rangle^2}{N_{T,jk}}\left(\frac{2N_k}{3N_j}\right). \quad (6.29)$$

The width of the energy distribution is then given by the standard deviation $\delta E_j = \sigma_{E_j}$,

$$\sigma_{E_j} = \langle E_j\rangle\sqrt{\frac{1}{N_{T,jk}}\left(\frac{2N_k}{3N_j}\right)}^{1/2}$$

$$= \langle E_j \rangle \sqrt{\frac{1}{N_j} \left(\frac{2N_k}{3N_{T,jk}} \right)^{1/2}}. \tag{6.30}$$

From eq. (6.30), we see that since $E_j \propto N_j$, the width of the probability distribution increases with $\sqrt{N_j}$, while the relative width,

$$\frac{\delta E_j}{\langle E_j \rangle} = \frac{\sigma_{E_j}}{\langle E_j \rangle} = \sqrt{\frac{1}{N_j} \left(\frac{2N_k}{3N_{T,jk}} \right)^{1/2}} \tag{6.31}$$

decreases with $\sqrt{N_j}$.

The behavior of the width and relative width is analogous to the $\sqrt{N_j}$ behavior we saw in eq. (4.9) for the width of the probability distribution for N_j. As the size of the system increases, the width of the probability distribution for E_j increases, but the relative width decreases. For a system with 10^{20} particles, the typical relative deviation of E_j from it average value is of the order of 10^{-10}, which is a very small number.

This increase of the standard deviation δE_j with the size of the system has the same significance discussed in Section 4.5 for δN_j. The average value of the energy is a description of the subsystem that we find useful; it is not the same as the true energy of the subsystem at any given time.

For both the energy and the number of particles in a subsystem, we have found that the relative width of the probability distribution is very narrow. As long as the width of the probability distribution is narrower than the uncertainty in the experimental measurements, we can regard the predictions of probability theory (statistical mechanics) as giving effectively deterministic values.

It is important to note both that it is not necessary to take the 'thermodynamic limit' (infinite size) for thermodynamics to be valid, and that it is necessary for the system to be large enough so that the relative statistical fluctuations of measured quantities is negligible in comparison with the experimental error.

6.5   The Logarithm of the Probability Distribution and the Energy-Dependent Terms in the Entropy

In direct analogy to the derivation of eq. (4.17) for the configurational contributions in Chapter 4, we can find the contributions from the momentum degrees of freedom to the entropy (the energy-dependent term) by taking the logarithm of eq. (6.6).

In analogy to eq. (4.15), we can define a function to describe the contributions to the entropy of the classical ideal gas from the momentum degrees of freedom,

$$S_{p,\alpha} = k \ln \Omega_p \left(E_\alpha, N_\alpha \right) \tag{6.32}$$

where α refers to one of the subsystems, j or k, or the total system when the subscript T, jk is used. Taking the logarithm of eq. (6.22), the energy-dependent contributions to the entropy of the composite system are seen to be:

$$
\begin{aligned}
S_{p,tot}\left(E_j, N_j, E_k, N_k\right) &\equiv k \ln P\left(E_j, N_j, E_k, N_k\right) + S_p\left(E_{T,jk}, N_{T,jk}\right) \\
&= S_p\left(E_j, N_j\right) + S_p\left(E_k, N_k\right).
\end{aligned}
\tag{6.33}
$$

$S_{p,tot}\left(E_j, N_j, E_k, N_k\right)$ is the energy-dependent contribution to the total entropy of the composite system, while $S_p\left(E_j, N_j\right)$ and $S_p\left(E_k, N_k\right)$ as the corresponding energy-dependent contributions to the entropies of the subsystems. Eq. (6.33) shows that the energy-dependent contributions to the entropy of an ideal gas are additive, as the configurational terms were shown to be in Section 4.7.

As was the case for the configurational entropy, the height of the probability distribution for E_j is proportional to $\sqrt{N_j}$ (or $\sqrt{N_k}$), so that in equilibrium

$$
k \ln P\left(E_j, N_j, E_k, N_k\right) \propto \ln N_j \ll S_p\left(E_{T,jk}, N_{T,jk}\right) \propto N_{T,jk}
\tag{6.34}
$$

and the energy-dependent contributions to the entropy of the composite system in equilibrium is given by $S_p\left(E_{T,jk}, N_{T,jk}\right)$,

$$
S_{p,tot}\left(E_j, N_j, E_k, N_k\right) = S_p\left(E_j, N_j\right) + S_p\left(E_k, N_k\right) = S_p\left(E_{T,jk}, N_{T,jk}\right).
\tag{6.35}
$$

If E and N are generic variables standing for N_j or N_k, we can see from eq. (6.21), that the energy contributions of the entropy for large N are given by

$$
S_p(E, N) = k \ln \Omega_p(E, N) \approx kN\left[\frac{3}{2}\ln\left(\frac{E}{N}\right) + X\right]
\tag{6.36}
$$

where the errors are of order $1/N$, and X is a constant that we will not need to evaluate at this point.

6.6 The Generalization to M ≥ 2 systems

As in Section 4.8, the equation for the entropy can be generalized to an arbitrary number $M \geq 2$ systems. Eq. (6.6) simply becomes a product of M factors, one for each system,

$$
P(\{E_k, N_k | k = 1, \ldots, M\}) = \frac{\sum_{k=1}^{M} \Omega_p\left(E_k, N_k\right)}{\Omega_p\left(E_T, N_T\right)},
\tag{6.37}
$$

where $\Omega_p\left(E_T, N_T\right)$ is a normalization constant. In this equation, $E_T = \sum_{k=1}^{M} E_k$ and $N_T = \sum_{k=1}^{M} N_k$ are constants. Eq. (6.37) is analogous to eq. (4.22), which was derived from the discrete probability distribution for the particle distribution.

Finally, in analogy to Chapter 4, we can define the energy contributions to the entropy as

$$S_p(E_k, N_k) = \ln \Omega_p(E_k, N_k) \tag{6.38}$$

so that

$$S_p(\{E_k, N_k | k = 1, \ldots, M\}) = \sum_{k=1}^{M} S_p(E_k, N_k) - S_p(E_T, N_T) + C' \tag{6.39}$$

where E_T, N_T, and C' are constants, whose values have no physical consequences.

Eq. (6.39) shows that contributions to the entropy of an individual system, $S_p(E_k, N_k)$, are exactly the same as derived from considering just two systems, and, in fact, are independent of the total number of systems. The validity of this equation is not dependent on whether the systems are isolated or in equilibrium with each other.

7

Classical Gases: Ideal and Otherwise

All models are wrong, but some are useful.

George Box, British statistician (1919–2013)

In Chapter 2 we introduced Boltzmann's 1877 definition of the entropy in terms of the logarithm of a probability for a composite system. We assumed that the positions and momenta of the particles in a classical ideal gas were independent random variables, so that we could examine their contributions to the entropy independently. In Chapter 4 we calculated the contributions from the positions of the particles, and in Chapter 6 we calculated the contributions from the momenta of the particles. Now we are in a position to put them together to find the total entropy of a composite system.

7.1 Entropy of a Composite System of Classical Ideal Gases

We have seen that the contributions to the total entropy of the composite system from the positions of the particles in eq. (4.15) and their momenta in eq. (6.32) simply add to give the total entropy. Since both the configurational and energy-dependent contributions to the entropy of the composite system are separable into contributions from each subsystem, the total entropy is also separable. It is easier (and requires less space) to look at the expression for the entropy of a single system at this point, rather than the composite system needed for the derivations in Chapters 4 and 6,

$$S_j(E_j, V_j, N_j) = kN_j \left[\ln \left(\frac{E_j^{3N_j/2-1}}{(3N_j/2)!} \right) + \ln \left(\frac{V_j^{N_j}}{N_j!} \right) + X' \right]. \tag{7.1}$$

This equation contains two constants that are still arbitrary at this point: k, which will be discussed in Chapter 8, and X', which we will discuss further in this chapter.

For any macroscopic system—that is, any system with more than about 10^{10} particles—Stirling's approximation is excellent. We can also replace $3N_j/2 - 1$ with $3N_j/2$, which has a relative error of only $3/2N_j$. The result is the entropy of a classical ideal gas,

$$S_j(E_j, V_j, N_j) = kN_j \left[\frac{3}{2} \ln \left(\frac{E_j}{N_j} \right) + \ln \left(\frac{V_j}{N_j} \right) + X \right], \tag{7.2}$$

An Introduction to Statistical Mechanics and Thermodynamics. Robert H. Swendsen, Oxford University Press (2020).
© Robert H. Swendsen. DOI: 10.1093/oso/9780198853237.001.0001

under the approximation that the widths of the energy- and particle-number-distributions are neglected. The constant $X = X' + 1$ comes from the replacement of $N_j!$ by Stirling's approximation.

The value of X in eq. (7.2) is completely arbitrary if we restrict ourselves to classical mechanics. However, there is a traditional value chosen for X,

$$X = \frac{3}{2} \ln \left(\frac{4\pi m}{3h^2} \right) + \frac{5}{2}. \tag{7.3}$$

The constant h in this expression is Planck's constant, taken from quantum mechanics. The presence of Planck's constant in eq. (7.3) makes it obvious that this value of X has nothing to do with classical mechanics. It is determined by solving for the entropy of a quantum mechanical gas—for which the additive constant does have a meaning—and taking the classical limit. We will carry out this procedure in Chapters 27, 28, and 29.

7.2   Equilibrium Conditions for the Ideal Gas

By definition, the entropy of any composite system should be a maximum at equilibrium. Eq. (7.2) satisfies that condition with respect to both energy and particle number. We will show this explicitly below—first for equilibrium with respect to energy, and then with respect to particle number.

7.2.1   Equilibrium with Respect to Energy

We want to confirm that the expression we have derived for the entropy predicts the correct equilibrium values for the energies of two subsystems. To do that, consider an experiment on a composite system with fixed subvolumes, V_j and V_k, containing N_j and N_k particles respectively. The total energy of the particles in the composite system is $E_{T,jk}$. There is a diathermal wall separating the two subvolumes, so that they can exchange energy.

We want to find the maximum of the entropy to confirm that it occurs at the equilibrium values of E_j and $E_k = E_{T,jk} - E_j$. We find the maximum in the usual way, by setting the partial derivative with respect to E_j equal to zero,

$$\frac{\partial}{\partial E_j} S(E_j, V_j, N_j; E_{T,jk} - E_j, V_k, N_k) = \frac{\partial}{\partial E_j} \left[S_j(E_j, V_j, N_j) + S_k(E_{T,jk} - E_j, V_k, N_k) \right]$$

$$= \frac{\partial}{\partial E_j} S_j(E_j, V_j, N_j) + \frac{\partial E_k}{\partial E_j} \frac{\partial}{\partial E_k} S_k(E_k, V_k, N_k)$$

$$= \frac{\partial}{\partial E_j} S_j(E_j, V_j, N_j) - \frac{\partial}{\partial E_k} S_k(E_k, V_k, N_k)$$

$$= 0. \tag{7.4}$$

This gives us an equilibrium condition that will play a very important role in the development of thermodynamics,

$$\frac{\partial S_j}{\partial E_j} = \frac{\partial S_k}{\partial E_k}. \tag{7.5}$$

Since the partial derivative of the entropy with respect to energy is

$$\frac{\partial S_j}{\partial E_j} = kN_j \frac{\partial}{\partial E_j} \left[\frac{3}{2} \ln\left(\frac{E_j}{N_j}\right) + \ln\left(\frac{V_j}{N_j}\right) + X \right]$$

$$= \frac{3}{2} kN_j \frac{\partial}{\partial E_j} \left[\ln E_j - \ln N_j \right]$$

$$= \frac{3kN_j}{2E_j} \tag{7.6}$$

the equilibrium condition is predicted to be

$$\frac{E_j}{N_j} = \frac{E_k}{N_k} \tag{7.7}$$

which is the same equilibrium condition found in eq. (6.26).

7.2.2   Equilibrium with Respect to the Number of Particles

We now demonstrate that our expression for the entropy predicts the correct equilibrium values of the numbers of particles in each subsystem. Consider an experiment on a composite system with fixed subvolumes, V_j and V_k. The total number of particles is $N_{T,jk}$, but there is a hole in the wall between the two subvolumes, so that they can exchange both energy and particles. The total energy of the particles in the composite system is $E_{T,jk}$.

Since the two subsystems can exchange particles, they can certainly exchange energy. The derivation in the previous subsection is still valid, so that the entropy is correctly maximized at the equilibrium value of the energy; that is, the same average energy per particle in the two subsystems.

Now we want to find the maximum of the entropy with respect to N_j to confirm that it occurs at the equilibrium values of N_j and $N_k = N_{T,jk} - N_j$. We find the maximum in the usual way, by setting the partial derivative with respect to N_j equal to zero,

$$\frac{\partial}{\partial N_j} S(E_j, V_j, N_j; E_k, V_j, N_{T,jk} - N_j) = \frac{\partial}{\partial N_j} \left[S_j(E_j, V_j, N_j) + S_k(E_k, V_k, N_{T,jk} - N_j) \right]$$

$$= \frac{\partial}{\partial N_j} S_j(E_j, V_j, N_j) + \frac{\partial N_k}{\partial N_j} \frac{\partial}{\partial N_k} S_k(E_k, V_k, N_k)$$

$$= \frac{\partial}{\partial N_j} S_j(E_j, V_j, N_j) - \frac{\partial}{\partial N_k} S_k(E_k, V_k, N_k)$$

$$= 0. \tag{7.8}$$

This gives us another equilibrium condition, in analogy to eq. (7.5), which will also be important in thermodynamics,

$$\frac{\partial S_j}{\partial N_j} = \frac{\partial S_k}{\partial N_k}. \tag{7.9}$$

The partial derivative of the entropy with respect to N_j is rather complicated:

$$\frac{\partial S}{\partial N_j} = \frac{\partial}{\partial N_j} \left[kN_j \left[\frac{3}{2} \ln\left(\frac{E_j}{N_j}\right) + \ln\left(\frac{V_j}{N_j}\right) + X \right] \right]$$

$$= k \left[\frac{3}{2} \ln\left(\frac{E_j}{N_j}\right) + \ln\left(\frac{V_j}{N_j}\right) + X \right] + kN_j \frac{\partial}{\partial N_j} \left[\frac{3}{2} \ln\left(\frac{E_j}{N_j}\right) + \ln\left(\frac{V_j}{N_j}\right) + X \right]$$

$$= k \left[\frac{3}{2} \ln\left(\frac{E_j}{N_j}\right) + \ln\left(\frac{V_j}{N_j}\right) + X \right] + kN_j \left[-\frac{3}{2N_j} - \frac{1}{N_j} \right]$$

$$= k \left[\frac{3}{2} \ln\left(\frac{E_j}{N_j}\right) + \ln\left(\frac{V_j}{N_j}\right) + X \right] - \frac{5}{2} k. \tag{7.10}$$

The condition for equilibrium is to set the partial derivatives with respect to the number of particles equal in the two subvolumes,

$$k \left[\frac{3}{2} \ln\left(\frac{E_j}{N_j}\right) + \ln\left(\frac{V_j}{N_j}\right) + X \right] - \frac{5}{2} k = k \left[\frac{3}{2} \ln\left(\frac{E_k}{N_k}\right) + \ln\left(\frac{V_k}{N_k}\right) + X \right] - \frac{5}{2} k. \tag{7.11}$$

Because of the equilibrium with respect to energy, we also have eq. (7.7),

$$\frac{E_j}{N_j} = \frac{E_k}{N_k}. \tag{7.12}$$

Combining eqs. (7.11) and eqs. (7.12), we find the equation for equilibrium with respect to N_j,

$$\frac{N_j}{V_j} = \frac{N_k}{V_k}. \tag{7.13}$$

As expected from eq. (4.8), the number of particles per unit volume is the same in both subsystems.

7.3 The Volume-Dependence of the Entropy

An interesting and important feature of the expression we have derived for the entropy of the classical ideal gas in eq. (7.2) is that it also correctly predicts the results for a kind of experiment that we have not yet discussed.

Consider a cylinder, closed at both ends, containing an ideal gas. There is a freely moving partition, called a piston, that separates the gas into two subsystems. The piston is made of a diathermal material, so it can transfer energy between the two systems, but it is impervious to the particles. There are N_j particles on one side of the piston and N_k particles on the other. Since the piston can move freely, the volumes of the subsystems can change, although the total volume, $V_{T,jk} = V_j + V_k$, is fixed.

Because the piston can conduct heat, the average energy per particle will be the same on both sides of the piston, as derived in Chapter 6:

$$\frac{\langle E_j \rangle}{N_j} = \frac{\langle E_k \rangle}{N_k}.$$ (7.14)

The maximum of the entropy with respect to the position of the piston is then found by setting the derivative of the total entropy with respect to V_j equal to zero:

$$\frac{\partial}{\partial V_j} S(E_j, V_j, N_j; E_k, V_{T,jk} - V_j, N_k) = \frac{\partial}{\partial V_j} \left[S_j(E_j, V_j, N_j) + S_k(E_k, V_{T,jk} - V_j, N_k) \right]$$

$$= \frac{\partial}{\partial V_j} S_j(E_j, V_j, N_j) + \frac{\partial V_k}{\partial V_j} \frac{\partial}{\partial V_k} S_k(E_k, V_k, N_k)$$

$$= \frac{\partial}{\partial V_j} S_j(E_j, V_j, N_j) - \frac{\partial}{\partial V_k} S_k(E_k, V_k, N_k)$$

$$= 0.$$ (7.15)

This gives us the equilibrium condition in a familiar form,

$$\frac{\partial S_j}{\partial V_j} = \frac{\partial S_k}{\partial V_k}.$$ (7.16)

Since the partial derivative of the entropy with respect to volume is

$$\frac{\partial S_j}{\partial V_j} = kN \frac{\partial}{\partial V_j} \left[\frac{3}{2} \ln \left(\frac{E_j}{N_j} \right) + \ln \left(\frac{V_j}{N_j} \right) + X \right]$$

$$= kN_j \frac{\partial}{\partial V_j} \ln\left(\frac{V_j}{N_j}\right)$$

$$= \frac{kN_j}{V_j} \tag{7.17}$$

and the equilibrium condition that the particle density is the same in both subsystems is correctly predicted,

$$\frac{N_j}{V_j} = \frac{N_k}{V_k}. \tag{7.18}$$

How did this happen? It is nice to know that it is true, but why should the maximum of the entropy with respect to the position of the piston produce the correct equilibrium volumes? The answer can be found by calculating the probability distribution for the position of the piston.

Under the usual assumption that all configurations are equally likely (subject to the constraints), the joint probability density for the positions of the particles *and* the position of the piston should be a constant, which we denote as Y,

$$P\left(V_j, \{\vec{r}_{1,i}|i = 1,\ldots,N_j\}, \{\vec{r}_{2,j}|j = 1,\ldots,N_k\}\right) = Y. \tag{7.19}$$

To find the marginal probability distribution $P(V_j)$, we simply integrate over the positions of every particle, noting that N_j particles are restricted to volume V_j, and N_k particles are restricted to volume V_k:

$$P(V_j) = \int_{p_j,p_k} P\left(V_j, \{\vec{r}_{1,i}|i = 1,\ldots,N_j\}, \{\vec{r}_{2,j}|j = 1,\ldots,N_k\}\right) d^{3N_j} r_j d^{3N_j} r_k$$

$$= YV_j^{N_j} V_k^{N_k}. \tag{7.20}$$

To find the maximum of $P(V_j)$ under the condition that $V_k = V_{T,jk} - V_j$, we differentiate with respect to V_j, and set it equal to zero,

$$\frac{\partial}{\partial V_j} P(V_j) = YN_j V_j^{N_j-1} V_k^{N_k} - YN_k V_j^{N_j} (V_{T,jk} - V_j)^{N_k-1} = 0. \tag{7.21}$$

Solving eq. (7.21), we find the hoped-for result:

$$\frac{N_j}{V_j} = \frac{N_k}{V_k}. \tag{7.22}$$

The key point is that the logarithm of the volume dependence of $P(V_j)$ is exactly the same as that of the entropy. In every case, defining the entropy as the logarithm of the probability (to within additive and multiplicative constants) gives the correct answer.

7.4   Asymmetric Pistons

The derivation of the entropy assumes that the total volume is conserved when two systems are connected with a moveable piston. This will be the case if the pistons have the same cross section in each system that they connect, although that cross section may be different for the pistons linking two other systems.

However, this is not the most general experimental situation. A piston connecting systems j and k might have a different cross-sectional area of A_j on one side and a cross-sectional area of A_k on the other side. This would lead to a change in the total volume if the piston moves.

We can show that the same form of the entropy will correctly predict equilibrium for the case of pistons with differing cross sections for each system, although the form of the equilibrium equations will be modified. The demonstration that this is the case is much simpler if we have the interpretation of the partial derivatives with respect to energy (giving the temperature) and volume (giving the pressure). Therefore, we will postpone the demonstration until Section 8.7.

7.5   Indistinguishable Particles

In Chapter 4 we calculated the configurational entropy of a classical ideal gas of distinguishable particles. On the other hand, from quantum mechanics we know that atoms and molecules of the same type are indistinguishable. It is therefore interesting to ask in what way the probability distribution, and consequently the entropy and other properties of the composite system, would change if we were to assume that the particles were indistinguishable. It turns out that nothing changes.

For the following derivation, recall that particles are regarded as distinguishable if and only if the exchange of two particles produces a different microscopic state. If the exchange of two particles does not produce a distinct microscopic state, the particles are indistinguishable.

The question is how to modify the derivation of eq. (4.4) in Chapter 4 to account for indistinguishability.

The first consequence of indistinguishability is that the binomial coefficient introduced to account for the permutations of the particles between the two subsystems must be eliminated. Since the particles are indistinguishable, the number of microscopic states generated by exchanging particles is equal to 1.

The second consequence is that the factors $(V_j/V)^{N_j}(1 - V_j/V)^{N_{T,jk}-N_j}$ in eq. (4.4) must also be modified.

Consider our basic assumption that the probability density for the positions of the particles must be a constant, with the value of the constant determined by the condition that the probability density be normalized to 1.

For N_j distinguishable particles in a three-dimensional volume V_j, the probability density is clearly $V_j^{-N_j}$. The probability of finding a specific set of N_j particles in subvolume V_j and the rest in subvolume V_k is therefore $V_j^{N_j} V_k^{N_{T,jk}-N_j} V_{T,jk}^{-N_{T,jk}}$.

However, the calculation for indistinguishable particles is somewhat different. We cannot label individual particles in a way that distinguishes between them. We can, however, label the particles on the basis of their positions. If two particles are exchanged, their labels would also be exchanged, and the state would be unchanged. A simple way of achieving this is to introduce a three-dimensional Cartesian coordinate system and label particles in order of their x-coordinates; that is, for any microscopic state, particles are labeled so that $x_{j,i} < x_{j,i+1}$ in subsystem j and $x_{k,\ell} < x_{k,\ell+1}$ in subsystem k. For simplicity, we will assume that the subsystems are rectangular, with edges parallel to the coordinate axes. The lengths of the subsystems in the x-direction are L_j and L_k, and the corresponding areas are A_j and A_k, so that the volumes are given by $V_j = A_j L_j$ and $V_k = A_k L_k$.

We will follow the same procedure established in Chapters 4 and 6 and assume the probability distribution in coordinate space to be a constant, which we will denote by $Y(N_{T,jk}, V_{T,jk})$,

$$P(\{\vec{r}_{j,i}\}, \{\vec{r}_{k\ell}\}) = Y(N_{T,jk}, V_{T,jk}). \tag{7.23}$$

The probability of finding N_j particles in subvolume V_j and $N_k = N_{T,jk} - N_j$ in subvolume V_k is found by integrating over coordinate space. The integrals in the y- and z-directions just give factors of the cross-sectional areas of the two subsystems. The integrals over the x-coordinates must be carried out consistently with the labeling condition that $x_{j,i} < x_{j,i+1}$ and $x_{k,\ell} < x_{k,\ell+1}$:

$$P(N_j, N_k) = Y(N_{T,jk}, V_{T,jk}) A_j^{N_j} A_k^{N_k} \int_0^{L_j} dx_{j,N_j} \int_0^{x_{j,N_j}} dx_{j,N_j-1} \cdots \int_0^{x_k} dx_{j,1}$$

$$\times \int_0^{L_k} dx'_{k,N_k} \int_0^{x'_{k,N_k}} dx'_{k,N_k-1} \cdots \int_0^{x'_k} dx'_{k,1}. \tag{7.24}$$

I have used primes to indicate the integrals over the x-coordinates in subsystem k.

The integrals in eq. (7.24) are easily carried out by iteration,

$$P(N_j, N_k) = Y(N_{T,jk}, V_{T,jk}) \left(\frac{V_j^{N_j}}{N_j!} \right) \left(\frac{V_k^{N_k}}{N_k!} \right). \tag{7.25}$$

This gives the probability distribution of identical particles, except for the determination of the normalization constant, $Y(N_{T,jk}, V_{T,jk})$.

It is clear that the dependence of $P(N_j, N_k)$ on its arguments for indistinguishable particles in eq. (7.25) is exactly the same as that for distinguishable particles in eq. (4.4). This means that we can determine the value of the normalization constant by comparison with the normalized binomial distribution in eq. (3.47):

$$Y(N_{T,jk}, V_{T,jk}) = \frac{N_{T,jk}!}{V_{T,jk}^{N_{T,jk}}}. \tag{7.26}$$

The full probability distribution for identical particles is found by substituting eq. (7.26) in eq. (7.25),

$$P(N_j, N_k) = \frac{N_{T,jk}!}{V_{T,jk}^{N_{T,jk}}} \left(\frac{V_j^{N_j}}{N_j!} \right) \left(\frac{V_k^{N_k}}{N_k!} \right). \tag{7.27}$$

This expression is identical to the probability distribution for distinguishable particles in eq. (4.4).

Since the entropy is given by the logarithm of the probability distribution, the entropy of a classical gas is exactly the same for distinguishable and indistinguishable particles.

The result that the entropy is exactly the same for classical systems with distinguishable and indistinguishable particles might be surprising to some readers, since most of the literature for the past century has claimed that they were different. My claim that the two models have the same entropy depends crucially on relating the entropy to the probability distribution. However, I think we might take as a general principle that two models with identical properties should have the same entropy. Since the probability distributions are the same, their properties are the same. It is hard to see how any definition that results in different entropies for two models with identical properties can be defended.

7.6   Entropy of a Composite System of Interacting Particles

Up to this point we have considered only an ideal gas, in which there are no interactions between particles. In real gases, particles do interact: particles can attract each other to form liquids and they can repel each other to make the liquid difficult to compress.

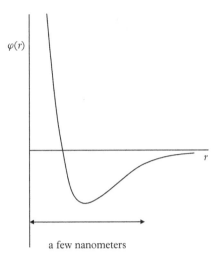

Fig. 7.1 *Schematic plot of typical interatomic potential. The potential, $\varphi(r)$, is significantly different from zero only over a few nanometers.*

Fig. 7.1 shows a typical interatomic potential, which is significantly different from zero only over very short distances.

Even though interatomic potentials are very short-ranged, they can have important effects, including the formation of liquid and solid phases, with phase transitions between them. We will return to this topic in Chapters 19 and 31.

To generalize our analysis of the entropy of the classical ideal gas to include interacting particles, it is useful to go back and re-derive the entropy of the ideal gas, without taking advantage of the independence of the positions and the momenta. Our basic assumption is still that of a uniform probability distribution in phase space, consistent with the constraints on the composite system.

7.6.1   The Ideal Gas Revisited

We begin with our assumption from Section 1.4 that the probability distribution in phase space is uniform, subject to the constraints. The physical constraints are:

1. N_j particles are in subsystem j and N_k in subsystem k.
2. Subsystem j has an energy E_j and subsystem k has an energy E_k.

Instead of doing the integrals separately for the configurations and the momenta, we can write the final result for $P(E_j, V_j, N_j; E_k, V_k, N_k)$ as an integral over all of phase space, with Dirac delta functions imposing the energy constraints as before. This does not simplify matters for the ideal gas, but it will make the generalization to real gases easier.

Calculating the probability distribution for the energies, volumes, and numbers of particles of each subsystem is straightforward, but the expression becomes a little unwieldy. While it is great fun for a teacher to write it across two or three blackboards so that students can see it as a whole, the constraints of a book make it somewhat more difficult to write out completely. To make it manageable, we will use the same compact notation that we used earlier for the integrals over phase space. The $3N_\alpha$-dimensional integral over the positions of the particles in subsystem α will be indicated as $\int dq_\alpha$, and similarly for the other variables.

Write the Hamiltonian of the subsystem α as

$$H_\alpha(q_\alpha, p_\alpha) = \sum_{i=1}^{3N_\alpha} \frac{|\vec{p}_{\alpha,i}|^2}{2m} \qquad (7.28)$$

where $\alpha = j$ or k to indicate the subsystems. To make the notation more compact we will write H_α to indicate $H_\alpha(q_\alpha, p_\alpha)$.

The marginal probability distribution for the energies, volumes, and number of particles of each subsystem is then found by integrating over all space:

$$P(E_j, V_j, N_j; E_k, V_k, N_k) =$$
$$\frac{N_{T,jk}!}{N_j! N_k!} \frac{\int dq_j \int dp_j \int dq_k \int dp_k\, \delta(E_j - H_j)\delta(E_k - H_k)}{\int dq \int dp\, \delta(E_{T,jk} - H_{T,jk})}. \qquad (7.29)$$

Because the particles in the two subsystems do not interact with each other, eq. (7.29) can be written in a more compact form by introducing a function

$$\Omega_\alpha(E_\alpha, V_\alpha, N_\alpha) = \frac{1}{h^{3N_\alpha} N_\alpha!} \int dq_\alpha \int dp_\alpha\, \delta(E_\alpha - H_\alpha(q_\alpha, p_\alpha)) \qquad (7.30)$$

where we have written $H_\alpha(q_\alpha, p_\alpha)$ for clarity. Note that we have included a factor of h^{-3N_α} in the expression for Ω_α. This factor is not required classically, but it is allowed since we can always multiply the right side of eq. (7.29) by $1 = h^{N_{T,jk}}/h^{N_j} h^{N_k}$. We will see later that this simple modification produces consistency between the classical result for the ideal gas and the classical limit of a quantum ideal gas.

With this notation, eq. (7.29) becomes

$$P(E_j, V_j, N_j; E_k, V_k, N_k) = \frac{\Omega_j(E_j, V_j, N_j)\Omega_k(E_k, V_k, N_k)}{\Omega(E_{T,jk}, V_{T,jk}, N_{T,jk})}. \qquad (7.31)$$

Taking the logarithm of the probability distribution in eq. (7.31), in analogy to eq. (4.17), we find the expression for the entropy of the composite system,

$$\begin{aligned} S_{tot}(E_j, V_j, N_j; E_k, V_k, N_k) &= k\ln\left[P(E_j, V_j, N_j; E_k, V_k, N_k)\right] + S(E_{T,jk}, V_{T,jk}, N_{T,jk}) \\ &= S_j(E_j, V_j, N_j) + S_k(E_k, V_k, N_k). \end{aligned} \qquad (7.32)$$

(We have neglected to include an arbitrary additive constant because it has no physical consequences.)

The entropy of an isolated system is given by the logarithm of Ω,

$$S(E,V,N) = k\ln\Omega(E,V,N), \qquad (7.33)$$

where E, V, and N are again generic variables, which can represent properties of subsystem j, subsystem k, or the entire composite system.

Eq. (7.32) shows that the total entropy of an ideal gas is an additive. It will be left as an exercise to show that this expression for the entropy of a classical ideal gas is identical to that in eq. (7.2)—including the value of the constant X given in eq. (7.3).

Note that we have introduced a new constant, h, into eq. (7.30). It is easy to confirm that this expression for the probability distribution is still equivalent to eq. (7.29) for any value of h, so that this new constant is completely arbitrary, and therefore meaningless within classical statistical mechanics. It has been introduced solely for the purpose of ensuring consistency with quantum statistical mechanics. It is a remarkable fact that the choice of h as Planck's constant produces agreement with quantum statistical mechanics in the classical limit. This probably seems rather mysterious at this point, but it will become clear when we return to this question in Chapters 27, 28, and 29.

7.6.2 Generalizing the Entropy to Interacting Particles

We do not need to make any changes in our fundamental assumptions about the probability distribution in phase space to find the entropy of a system with interacting particles. We still assume a uniform probability distribution in phase space, subject to the physical constraints of our composite system. The only difference is that the energy now includes interaction terms. However, because it is possible for a molecule in one subsystem to interact with a molecule in the other subsystem, the separability of the entropy of the composite system into the sum of the entropies for each subsystem is no longer trivial.

First write the Hamiltonian of the composite system composed of subsystems j and k in terms of sums over the particles,

$$H_{T,jk}(q,p) = \sum_{i=1}^{N_{T,jk}} \frac{|\vec{p}_i|^2}{2m} + \sum_{i=1}^{N_{T,jk}} \sum_{i'>i}^{N_{T,jk}} \phi(\vec{r}_i, \vec{r}_{i'}) \qquad (7.34)$$

or

$$H_{T,jk}(q,p) = \sum_{i=1}^{N_j} \frac{|\vec{p}_{j,i}|^2}{2m} + \sum_{\ell=1}^{N_k} \frac{|\vec{p}_{k,\ell}|^2}{2m}$$

$$+ \sum_{i=1}^{N_j} \sum_{i'>i}^{N_j} \phi(\vec{r}_{j,i}, \vec{r}_{j,i'}) + \sum_{\ell=1}^{N_k} \sum_{\ell'>\ell}^{N_k} \phi(\vec{r}_{k,\ell}, \vec{r}_{k,\ell'})$$

$$+ \sum_{i=1}^{N_j} \sum_{\ell=1}^{N_k} \phi(\vec{r}_{j,i}, \vec{r}_{k,\ell}). \tag{7.35}$$

This equation is written two ways. First, as a single, composite system, and then broken into pieces, representing the subsystems. The last term on the right in eq. (7.35) represents the interactions between particles in different subsystems. This is the term that causes difficulties in separating the entropy of the composite system into a sum of entropies of the subsystems.

We can define the Hamiltonian of subsystem j:

$$H_j(q_j, p_j) = \sum_{i=1}^{N_j} \frac{|\vec{p}_{j,i}|^2}{2m} + \sum_{i=1}^{N_j} \sum_{i'>i}^{N_j} \phi(\vec{r}_{j,i}, \vec{r}_{j,i'}). \tag{7.36}$$

The limit of $i' > i$ in the second sum over positions is there to prevent double counting. Similarly, we can define the Hamiltonian of subsystem k:

$$H_k(q_k, p_k) = \sum_{\ell=1}^{N_k} \frac{|\vec{p}_{k,\ell}|^2}{2m} + \sum_{\ell=1}^{N_k} \sum_{\ell'>\ell}^{N_k} \phi(\vec{r}_{k,\ell}, \vec{r}_{k,\ell'}). \tag{7.37}$$

However, because of the interaction term linking the two subsystems,

$$H_{jk}(q_j, q_k) = \sum_{j_1=1}^{N_j} \sum_{\ell=1}^{N_k} \phi(\vec{r}_{j,i}, \vec{r}_{k,\ell}) \tag{7.38}$$

the total Hamiltonian is not just the sum of H_j and H_k:

$$H_{T,jk}(q, p) = H_j(q_j, p_j) + H_k(q_k, p_k) + H_{jk}(q_j, q_k). \tag{7.39}$$

Fortunately, the range of the interactions between molecules in most materials (or between particles in colloids) is short in comparison with the size of the system. Therefore, the contributions to H_{jk} are only significant for molecules in different subsystems that are close to each other. Since the range of molecular interactions is only a few nanometers, this means that only pairs of molecules very close to the interface between the two subsystems contribute at all. The energy corresponding to these direct interactions between molecules in different systems is therefore proportional to the size of the interface, which scales as the surface area in each system, or $V_\alpha^{2/3}$.

In terms of the number of particles, the contribution of the direct interactions to molecules subsystem α with a different subsystem is proportional to $N_\alpha^{2/3}$. As a fraction of all molecules in subsystem α, this is $N_\alpha^{2/3}/N_\alpha = N_\alpha^{-1/3}$. However, the fraction of molecules close to the interface in another subsystem, α', is proportional to $N_{\alpha'}^{-1/3}$. This makes the total effect of the interface of the order of $N_\alpha^{-1/3}N_{\alpha'}^{-1/3}$. If N_α and $N_{\alpha'}$ are both of the order of 10^{21}, then the interaction term is of the order of $N_\alpha^{-1/3}N_{\alpha'}^{-1/3} \approx 10^{-7}10^{-7} = 10^{-14}$. Therefore, neglecting the interactions between molecules in different subsystems is a very good approximation.

Neglecting the term H_{jk} in eq. (7.39), we can again write the joint probability in terms of the product of two terms as in eq. (7.29). The difference is that now the Hamiltonians only include the interactions between particles within the same system,

$$P(E_j, V_j, N_j; E_k, V_k, N_k) =$$
$$\frac{N_{T,jk}!}{N_j!N_k!} \frac{\int dq_j \int dp_j \int dq_k \int dp_k \, \delta(E_j - H_j)\delta(E_k - H_k)}{\int dq \int dp \, \delta(E_{T,jk} - H_{T,jk})} \tag{7.40}$$

or

$$P(E_j, V_j, N_j; E_k, V_k, N_k) =$$
$$\frac{N_{T,jk}!}{N_j!N_k!} \frac{\int dq_j \int dp_j \, \delta(E_j - H_j) \int dq_k \int dp_k \, \delta(E_k - H_k)}{\int dq \int dp \, \delta(E_{T,jk} - H_{T,jk})}. \tag{7.41}$$

The factors in eq. (7.41) can be made explicit by introducing a generalized Ω function,

$$\Omega_\alpha(E_\alpha, V_\alpha, N_\alpha) = \frac{1}{h^{3N_\alpha}N_\alpha!} \int dq_\alpha \int dp_\alpha \, \delta(E_\alpha - H_\alpha) \tag{7.42}$$

and writing

$$P(E_j, V_j, N_j; E_k, V_k, N_k) = \frac{\Omega_j(E_j, V_j, N_j)\Omega_k(E_k, V_k, N_k)}{\Omega(E_{T,jk}, V_{T,jk}, N_{T,jk})}. \tag{7.43}$$

As before, the subscript α can take on the values of j or k to represent a subsystem, or be replaced by T, jk to represent the full system. The parameter h is still completely arbitrary within classical statistical mechanics. As mentioned above, h will be identified as Planck's constant in Chapters 27, 28, and 29 to ensure consistency with the classical limit of the corresponding quantum system.

Now that we have expressed the probability distribution for interacting classical particles as a product of terms, we can define the entropy by taking the logarithm and multiplying it by a constant, k. The result is formally the same as eq. (7.32) for ideal gases,

$$S_{tot}(E_j, V_j, N_j; E_k, V_k, N_k) = k \ln\left[P(E_j, V_j, N_j; E_k, V_k, N_k)\right] + S(E_{T,jk}, V_{T,jk}, N_{T,jk})$$
$$= S_j(E_j, V_j, N_j) + S_k(E_k, V_k, N_k). \tag{7.44}$$

The entropy terms of individual systems are again given by logarithm of the Ω functions,

$$S_\alpha(E_\alpha, V_\alpha, N_\alpha) = k \ln \Omega_\alpha(E_\alpha, V_\alpha, N_\alpha) \tag{7.45}$$

$$= k \ln\left[\frac{1}{h^{3N_\alpha} N_\alpha!} \int dq_\alpha \int dp_\alpha \delta(E_\alpha - H_\alpha)\right].$$

Now the Hamiltonians H_j and H_k include the intra-system interactions.

7.7 The Second Law of Thermodynamics

As was the case for ideal gases, the probability distributions for energy, volume, and number of particles in composite systems of interacting particles are very narrow—the width is generally proportional to the square root of the number of particles. The relative width of the probability distributions will therefore again be much smaller than experimental errors. For systems of interacting particles, the observed values of the energy, volume, and number of particles will agree with the location of the maxima of the corresponding probability distributions to within the accuracy of the experiment.

A comparison of eq. (7.41) for the probabilities in a compound system of interacting particles and eq. (7.43) for the entropy of the composite system shows that the location of the maximum of the entropy gives the equilibrium values of the quantity of interest when the corresponding internal constraints are released. This is an important result for two reasons.

First, it allows us to find the equilibrium values of the energy, volume, or number of particles for any experimental situation, once the entropy is known.

Second, if the entropy is maximized for unconstrained equilibrium, it necessarily follows that any constrained macroscopic state of the composite system must have an equal or lower total entropy. This is the Second Law of Thermodynamics.

If the composite system is in any constrained macroscopic state—that is, if there are any constraints on the energy, volume, or number of particles in any subsystem—the total entropy of the composite system cannot decrease if those constraints are removed.

7.8 Equilibrium between Subsystems

For a composite system composed of two general, classical subsystems, the total entropy is the sum of the entropies of the subsystems, as indicated in eq. (7.44),

$$S_{tot}(E_j, V_j, N_j; E_k, V_k, N_k) = S_j(E_j, V_j, N_j) + S_k(E_k, V_k, N_k). \tag{7.46}$$

Since the total entropy is maximized upon release of a constraint, we can determine the new equilibrium parameters by setting the appropriate partial derivative of the total entropy equal to zero.

For example, consider a situation in which the volumes and numbers of particles are held fixed in the subsystems, but the subsystems are connected by a diathermal wall. Equilibrium is found from the partial derivative with respect to E_j, with $E_k = E_{T,jk} - E_j$:

$$\frac{\partial}{\partial E_j} S_{tot}(E_j, V_j, N_j; E_{T,jk} - E_j, V_k, N_k)$$

$$= \frac{\partial}{\partial E_j} S_j(E_j, V_j, N_j) + \frac{\partial}{\partial E_j} S_k(E_{T,jk} - E_j, V_k, N_k) = 0. \tag{7.47}$$

Since $E_k = E_{T,jk} - E_j$, it must be that $\partial E_k / \partial E_j = -1$, and we can rewrite eq. (7.47) in a symmetric form:

$$\frac{\partial}{\partial E_j} S_j(E_j, V_j, N_j) = -\frac{\partial E_k}{\partial E_j} \frac{\partial}{\partial E_k} S_k(E_k, V_k, N_k) \tag{7.48}$$

or

$$\frac{\partial S_j}{\partial E_j} = \frac{\partial S_k}{\partial E_k}. \tag{7.49}$$

Eq. (7.49) is important. Beyond giving us an algorithm for finding the equilibrium value of E_j, it tells us that the condition of equilibrium is that a derivative of the entropy with respect to the energy has the same value for two subsystems in thermal contact.

Similar arguments show that the condition for equilibrium with respect to volume when two subsystems are separated by a moveable piston is given by

$$\frac{\partial S_j}{\partial V_j} = \frac{\partial S_k}{\partial V_k}, \tag{7.50}$$

and equilibrium with respect to particle number when a hole is made in the wall between two subsystems is given by

$$\frac{\partial S_j}{\partial N_j} = \frac{\partial S_k}{\partial N_k}. \tag{7.51}$$

These last three equations provide both a means of solving problems and the basis for establishing the connection between statistical mechanics and thermodynamics. The explicit relationships between these derivatives and familiar quantities like temperature and pressure will be the subject of the following chapter. However, the first and arguably the most important consequence of these equations is presented in the next section.

7.9 The Zeroth Law of Thermodynamics

An immediate consequence of eqs. (7.49), (7.50), and (7.51) is that if two systems are each in equilibrium with a third system, they must also be in equilibrium with each other.

Assume systems j and k are in equilibrium with respect to energy, then eq. (7.49) holds. If systems k and ℓ are also in equilibrium with respect to energy, then we have a similar equation for the partial derivatives of S_k and S_ℓ with respect to energy,

$$\frac{\partial S_j}{\partial E_j} = \frac{\partial S_k}{\partial E_k} = \frac{\partial S_\ell}{\partial E_\ell}. \tag{7.52}$$

Since the partial derivatives with respect to energy must be the same for systems j and ℓ, they must be in equilibrium with respect to energy.

Clearly, this argument is equally valid for the volume or the number of particles. We therefore have a general principle that if two systems are each in equilibrium with a third system, then they are in equilibrium with each other.

This is the Zeroth Law of Thermodynamics. It is valid both for equilibrium when the subsystems in thermal contact with each other, and for potential equilibrium if the subsystems are isolated from each other, but eq. (7.52) holds.

The numbering of the laws of thermodynamics might seem rather strange—especially the 'zeroth'. The need to state it explicitly as a law of thermodynamics was not recognized during the nineteenth century when thermodynamics was being developed, perhaps because it is so fundamental that it seemed obvious. When it was declared a law of thermodynamics, the numbers for the other laws had long been well established. Ralph H. Fowler (British physicist and astronomer, 1889–1944) is credited with calling it the Zeroth Law to place it in the leading position. At this point in the book we have encountered only the Second and Zeroth Laws of Thermodynamics. The First and Third Laws are yet to come.

7.10 Problems

....

PROBLEM 7.1
General forms for the entropy

We have derived the entropy of a classical ideal gas, which satisfies all the postulates of thermodynamics (except the Nernst postulate, which applies to quantum systems). However, when interactions are included, the entropy could be a very different function of U, V, and N. Here are some functions that are candidates for being the entropy of some system. [In each case, A is a positive constant.]

1. Which of the following equations for the entropy satisfy the postulates of thermodynamics (except the Nernst postulate)?

   1. $S = A(UVN)^{1/3}$

   2. $S = A\left(\dfrac{NU}{V}\right)^{2/3}$

   3. $S = A\left(\dfrac{UV}{N}\right).$

   4. $S = A\left(\dfrac{V^3}{NU}\right)$

2. For each of the valid forms of the entropy in the previous part of the problem, find the three equations of state.

PROBLEM 7.2

The 'traditional' expression for the entropy of an ideal gas of distinguishable particles

It is claimed in many textbooks on statistical mechanics that the correct entropy for a classical ideal gas of distinguishable particles differs from what we derived in class. The traditional expression is

$$S(E,V,N) = k_k N\left[\frac{3}{2}\ln\left(\frac{E}{N}\right) + \ln(V) + X\right].$$

Assume that you have two systems, j and k, with N_j and N_k particles and V_j and V_k volumes respectively, which are both described by an entropy of this form.

Consider an experiment in which the volumes are fixed, but a hole is made in the wall separating the two subsystems.

Calculate the location of the maximum of the total traditional entropy to determine its prediction for the equilibrium distribution of the energy and the particles between the two systems.

Naturally, your answer should depend on the relative sizes of the fixed volumes, V_j and V_k.

···

8

Temperature, Pressure, Chemical Potential, and All That

The most simple test of the equality of temperature of two bodies is that they remain in equilibrium when brought into thermal contact.

J. Willard Gibbs

In this chapter we will discuss the temperature, T, the pressure, P, and the chemical potential, μ. We will find explicit expressions for these three quantities in terms of partial derivatives of the entropy with respect to energy, volume, and number of particles. In doing so, we will complete the foundations of classical statistical mechanics.

In Chapter 9, which begins Part II, we will see how the structure of statistical mechanics leads to a set of postulates that provides a foundation for thermodynamics. The rest of the chapters in Part II develop the theory of thermodynamics from these postulates.

In Part III we shall return to classical statistical mechanics. While the equations developed in Part I are (in my opinion) the clearest way to understand the assumptions of statistical mechanics, the new methods developed in Part III are much more powerful for practical calculations.

8.1 Thermal Equilibrium

We have seen in Section 7.9 that if two objects are in thermal equilibrium with each other (equilibrium with respect to energy exchange), then the partial derivative of the entropy with respect to energy must have the same value in both systems. We also know that if two systems are in thermal equilibrium, they must be at the same temperature. Therefore, the partial derivative of the entropy with respect to energy must be a unique function of temperature. One purpose of this chapter is to determine the nature of that function.

After we have determined the relationship between the temperature and the partial derivative of the entropy with respect to energy, we will find similar relationships between other partial derivatives of the entropy and the pressure and the chemical potential.

An Introduction to Statistical Mechanics and Thermodynamics. Robert H. Swendsen, Oxford University Press (2020).
© Robert H. Swendsen. DOI: 10.1093/oso/9780198853237.001.0001

In all cases, the relationships linking T, P, and μ with partial derivatives of the entropy will be valid for all systems. This is a very powerful statement. It means that if we can calculate the entropy as a function of E, V, and N, we can calculate all thermal properties of the system. For this reason, the entropy as a function of energy, volume, and number of particles,

$$S = S(E, V, N), \tag{8.1}$$

is known as a 'fundamental relation'. We will see later that there are a number of other functions that also contain the same, complete thermodynamic information about a system, which makes them equivalent to eq. (8.1). Such functions are also representations of the fundamental relation.

8.2   What do we Mean by 'Temperature'?

For most of us, temperature is the reading we obtain from a thermometer. The basic property of a thermometer is that it undergoes some sort of physical change that can be measured when it is heated or cooled. The most readily available thermometer is our own body. It shivers when it is cold, and sweats when it is hot. These are subjective measures, but they do form the basis for our intuitive understanding of temperature.

To make the definition of temperature objective, we need to choose something that will provide a theoretical and experimental standard, as well as a numerical scale. We will use the ideal gas for this purpose. Since we can calculate all of the properties of an ideal gas, we can determine how its volume or pressure will change when it is heated or cooled. Since we can do experiments on real dilute gases, we can relate our definition of temperature to the real world.

The basic equation we will use to uniquely define the thermodynamic temperature is the ideal gas law:

$$PV = Nk_BT. \tag{8.2}$$

The pressure P in this equation is defined as the average force per unit area. The constant k_B is called Boltzmann's constant. It has units of Joules per degree, and relates the temperature scale to the energy scale. Eq. (8.2) is often written as

$$PV = nRT \tag{8.3}$$

where $n = N/N_A$ is the number of moles ($N_A = 6.0221415 \times 10^{23}$ being Avogadro's number, named after Amedeo Carlo Avogadro, Count of Quaregna and Cerreto, Italian scientist, 1776–1856), and $R = N_A k_B$ is known as the ideal gas constant. The experimental identification of the temperature as being proportional to the pressure for fixed volume, or the volume for fixed pressure, coincides with our notion that gas expands, or its pressure increases, when heated.

Although the ideal gas law is well known experimentally, to make contact with the entropy we must derive it from the properties of the ideal gas found in previous sections. What we will actually do in the next section is prove that the ratio PV/N is equal to a certain property of a thermal reservoir, and then use eq. (8.2) to associate that property with the temperature. This derivation will give us the universal relationship between the temperature and $\partial S/\partial E$.

8.3 Derivation of the Ideal Gas Law

The first step in the derivation of the ideal gas law is to derive the Maxwell–Boltzmann equation for the probability density of the momentum of one ideal gas particle. We will then use this probability density to determine the pressure due to collisions on the walls containing an ideal gas. This will lead us to the ideal gas law and the definition of the temperature in terms of the partial derivative of the entropy with respect to the energy.

8.3.1 The Maxwell–Boltzmann Equation

We are interested in the properties of a single particle in an ideal gas, and it does not matter which particle we choose, since they all have the same probability density. We can find that probability density as a marginal density of the full probability density in phase space. If the total energy of the system is E, the full probability density is given by

$$P(p,q) = \frac{1}{\Omega(E,V,N)} \frac{1}{h^{3N}N!} \delta(E - H(p,q)) \tag{8.4}$$

where

$$\Omega(E,V,N) = \frac{1}{h^{3N}N!} \int dp \int dq\, \delta(E - H(p,q)) \tag{8.5}$$

and we have used the compact notation $p = \{\vec{p}_i | i - 1, 2, \ldots, N\}$ and $q = \{\vec{r}_i | i = 1, 2, \ldots, N\}$.

As a first step, we will find the marginal probability density for both the momentum and position of particle 1 by integrating $P(p,q)$ in eq. (8.4) over all variables except $\vec{p}_1$ and $\vec{r}_1$:

$$
\begin{aligned}
P(\vec{p}_1, \vec{r}_1) &= \int d^3p_2 \ldots d^3p_N \int d^3r_2 \ldots d^3r_N P(p,q) \\
&= \frac{1}{\Omega(E,V,N)} \frac{1}{h^{3N}N!} \int d^3p_2 \ldots d^3p_N \int d^3r_2 \ldots d^3r_N \delta(E - H_N(p,q)).
\end{aligned}
\tag{8.6}
$$

In eq. (8.6), H_N is the Hamiltonian of the N-particle system.

We can write the integral in eq. (8.6) in terms of the function Ω for a system with $N - 1$ particles,

$$\Omega\left(E - |\vec{p}_1|^2/2m, V, N - 1\right) = \frac{1}{h^{3(N-1)}(N-1)!} \int dp_2 \ldots dp_{3N}$$

$$\times \int d^3 r_2 \ldots d^3 r_N \delta\left(E - |\vec{p}_1|^2/2m - \sum_{i=2}^{3N} p_i^2/2m\right). \quad (8.7)$$

Note that the energy term, $|\vec{p}_1|^2/2m$, has been separated from the rest of the Hamiltonian because it has not been integrated out. With this notation, eq. (8.6) can be written compactly,

$$P(\vec{p}_1, \vec{r}_1) = \frac{\Omega(E - |\vec{p}_1|^2/2m, V, N - 1)}{Nh^3 \Omega(E, V, N)}. \quad (8.8)$$

Since $P(\vec{p}_1, \vec{r}_1)$ does not depend explicitly on $\vec{r}_1$, we can easily integrate $\vec{r}_1$ out to obtain

$$P(\vec{p}_1) = \frac{V}{Nh^3} \frac{\Omega(E - |\vec{p}_1|^2/2m, V, N - 1)}{\Omega(E, V, N)}. \quad (8.9)$$

We would like to take advantage of the fact that $|\vec{p}_1|^2/2m$ is very small in comparison to E. We can exploit this by taking the logarithm of eq. (8.9), treating $|\vec{p}_1|^2/2m$ as a small perturbation, and keeping only the leading terms:

$$\ln P(\vec{p}_1) = \ln \Omega(E - |\vec{p}_1|^2/2m, V, N - 1)$$

$$- \ln \Omega(E, V, N) + \ln\left(\frac{V}{Nh^3}\right)$$

$$\approx \ln \Omega(E, V, N - 1) - \frac{|\vec{p}_1|^2}{2m} \frac{\partial}{\partial E} \ln \Omega(E, V, N - 1)$$

$$- \ln \Omega(E, V, N) + \ln\left(\frac{V}{Nh^3}\right). \quad (8.10)$$

The higher-order terms in the expansion are extremely small because the average value of $|\vec{p}_1|^2/2m$ is on the order of a factor of N smaller than E.

The first thing to note about eq. (8.10) is that only the second term depends on $\vec{p}_1$.

The second thing to note about eq. (8.10) is that the approximation

$$\frac{\partial}{\partial E} \ln \Omega(E, V, N - 1) \approx \frac{\partial}{\partial E} \ln \Omega(E, V, N) \quad (8.11)$$

is extremely good. The error is only of order $1/N$.

The derivative of $\ln \Omega$ with respect to energy turns out to be so important that it has been universally assigned the Greek letter β,

$$\beta \equiv \frac{\partial}{\partial E} \ln \Omega(E, V, N). \tag{8.12}$$

We can now rewrite eq. (8.10) in a much simpler form,

$$\ln P(\vec{p}_1) = -\beta |\vec{p}_1|^2 / 2m + \text{constants}, \tag{8.13}$$

where the 'constants' may depend on E, V, and N, but are independent of $\vec{p}_1$. Since the constants in this equation are determined by the normalization condition, we can complete the derivation of $P(\vec{p}_1)$ from our knowledge of the properties of Gaussian integrals:

$$P(\vec{p}_1) = \left(\frac{\beta}{2\pi m}\right)^{3/2} \exp\left(-\beta \frac{|\vec{p}_1|^2}{2m}\right). \tag{8.14}$$

This is the Maxwell–Boltzmann probability density for the momentum of a single particle.

> The expansion used in deriving $P(\vec{p}_1)$ is extremely valuable in statistical mechanics. This is the first time it appears in this book, but far from the last.

Note that the three components of the momentum are independent, so that the probability density of a single component can be found easily by integrating out the other two,

$$P(p_{1,x}) = \sqrt{\frac{\beta}{2\pi m}} \exp\left(-\beta \frac{p_{1,x}^2}{2m}\right). \tag{8.15}$$

8.3.2   The Pressure in an Ideal Gas

From the Maxwell–Boltzmann probability density for the momenta, we can calculate the pressure by integrating over the collisions that occur during a time Δt.

Consider a flat portion of the wall of the container of an ideal gas with area A. (The assumption of flatness is not necessary, but it makes the derivation much easier.) Define a coordinate system such that the x-axis is perpendicular to the wall. To simplify the problem, note that the y- and z-components of a particle's momentum do not change during a collision with the wall, so that they do not transfer any momentum to the wall and do not affect the pressure. This reduces the calculation to a one-dimensional problem.

By Newton's Second Law, the average force on the wall during a time-period Δt is given by the total momentum transferred to the wall during that period,

$$F\Delta t = \int_{\text{collisions}} \Delta p_x P(p_x, \Delta t) dp_x. \tag{8.16}$$

$P(p_x, \Delta t)$ is the probability of a particle hitting the wall during the time Δt, and the integral goes over all particles that hit the wall during that period of time. Since we intend to make Δt small, we do not have to include particles that might hit the wall a second time after bouncing off a different part of the container.

Let the wall be located at $x = 0$, with the particles confined to $x < 0$. Then particles will hit the wall only if $p_x > 0$ and the particles are within a distance $\Delta t v_x = \Delta t p_x / m$ of the wall, where m is the mass of a particle. Letting the area of the wall be denoted by A, the volume of particles that will hit the wall is $A\Delta t p_x / m$. The average number of particles in that volume is $NA\Delta t p_x / Vm$, where N is the total number of particles, and V is the total volume. If we multiply the number of particles in the volume by $P(p_x)$ (the probability density of the momentum from eq. (8.14)), then multiply that by the momentum transfer, $\Delta p_x = 2p_x$, and finally integrate over the momentum, we have the total momentum transfer to the wall during Δt, which is equal to the average force times Δt:

$$F\Delta t = \int_0^\infty \sqrt{\frac{\beta}{2\pi m}} \exp\left(-\beta \frac{p_x^2}{2m}\right) 2p_x \frac{NA\Delta t p_x}{Vm} dp_x. \tag{8.17}$$

Note that the integral extends only over positive momenta; particles with negative momenta are moving away from the wall.

Using the definition of the pressure as the force per unit area, $P = F/A$, eq. (8.17) becomes an equation for the pressure

$$P = \frac{2N}{Vm} \sqrt{\frac{\beta}{2\pi m}} \int_0^\infty \exp\left(-\beta \frac{p_x^2}{2m}\right) p_x^2 dp_x. \tag{8.18}$$

The integral in eq. (8.18) can be carried out by the methods discussed in Section 3.10. You could also look up the integral in a table, but it is more fun to work it out by yourself. The result is surprisingly simple,

$$PV = N\beta^{-1}. \tag{8.19}$$

Eq. (8.19) should be compared with the usual formulation of the ideal gas law,

$$PV = Nk_B T \tag{8.20}$$

where k_B is known as Boltzmann's constant. This comparison allows us to identify β with the inverse temperature,

$$\beta = \frac{1}{k_B T}.$$ (8.21)

Because of the Zeroth Law, we can use an ideal-gas thermometer to measure the temperature of anything, so that this expression for β must be universally true.

8.4 Temperature Scales

Although we are all familiar with the thermometers that we use every day, the temperature scale in the previous section is rather different than those commonly found in our homes. First of all, eq. (8.2) makes it clear that T cannot be negative (*for an ideal gas*), but both temperature scales in common use—Celsius and Fahrenheit—do include negative temperatures.

The Celsius temperature scale is defined by setting the freezing point of water equal to 0° C, and the boiling point of water equal to 100° C. The Celsius temperature is then taken to be a linear function of the temperature T given in eq. (8.2). Operationally, this could be roughly carried out by trapping a blob of mercury in a glass tube at high temperature, and measuring the position of the mercury in freezing water and in boiling water. I actually carried out this experiment in high school, before people were quite aware of the toxicity of mercury. I do not recommend trying it.

Using the Celsius temperature scale we can extrapolate to zero volume of the gas. Done carefully, it extrapolates to a temperature of −273.15° C, which is known as 'absolute zero' because it is the lowest possible temperature.

For the rest of this book we will use the Kelvin temperature scale, which is shifted from the Celsius scale to agree with eq. (8.2) and make zero temperature fall at absolute zero,

$$T(\text{K}) = T(^\circ\text{C}) + 273.15.$$ (8.22)

Although eq. (8.22) seems quite simple, it is easy to forget when doing problems. A very common error in examinations is to use the Celsius temperature scale when the Kelvin scale is required.

Since we are following tradition in defining the Celsius and Kelvin temperature scales, the value of Boltzmann's constant is fixed at $k_B = 1.380658 \times 10^{-23} J K^{-1}$. If we were to have thermometers that measure temperature in Joules, Boltzmann's constant would simply be 1.

8.5   The Pressure and the Entropy

From the explicit expression for the entropy of the classical ideal gas in eq. (7.2), we can find the derivative of the entropy with respect to the volume,

$$\left(\frac{\partial S}{\partial V}\right)_{E,N} = kN\frac{\partial}{\partial V}\left[\frac{3}{2}\ln\left(\frac{E}{N}\right) + \ln\left(\frac{V}{N}\right) + X\right] \qquad (8.23)$$

$$= kN\frac{1}{V}.$$

This relationship is only true for the ideal gas. By comparing eq. (8.23) with the ideal gas law, eq. (8.2), we find a simple expression for the partial derivative of entropy with respect to volume that is true for all macroscopic systems

$$\left(\frac{\partial S}{\partial V}\right)_{E,N} = \frac{kP}{k_B T}. \qquad (8.24)$$

At this point we will make the standard choice to set the constant k that we first introduced in eq. (4.15) equal to Boltzmann's constant, $k = k_B$. Eq. (8.24) now becomes

$$\left(\frac{\partial S}{\partial V}\right)_{E,N} = \frac{P}{T}. \qquad (8.25)$$

This is the formal thermodynamic relationship between entropy and pressure.

8.6   The Temperature and the Entropy

By comparing eqs. (7.30), (8.12), and (8.21), we can see that

$$\left(\frac{\partial S}{\partial E}\right)_{V,N} = k_B\beta = \frac{1}{T}. \qquad (8.26)$$

   Because of the Zeroth Law, this relationship must be valid for all thermodynamic systems. It is sometimes used as the definition of temperature in books on thermodynamics, although I have always felt that eq. (8.26) is too abstract to be a reasonable starting point for the study of thermal physics. One of the purposes of the first part of this book is to explain the meaning of eq. (8.26). If that meaning is clear to you at this point, you will be in a good position to begin the study of thermodynamics in the next part of the book.

Since we know the entropy of the classical ideal gas from eq. (7.2), we can evaluate the derivative in eq. (8.26) explicitly:

$$\left(\frac{\partial S}{\partial E}\right)_{V,N} = k_B N \frac{\partial}{\partial E}\left[\frac{3}{2}\ln\left(\frac{E}{N}\right) + \ln\left(\frac{V}{N}\right) + X\right]$$

$$= k_B N \frac{3}{2E} = \frac{1}{T}. \tag{8.27}$$

The last equality gives us the energy of the classical ideal gas as a function of temperature

$$E = \frac{3}{2}N k_B T. \tag{8.28}$$

An interesting consequence of eq. (8.28) is that the average energy per particle is just

$$\frac{E}{N} = \frac{3}{2}k_B T \tag{8.29}$$

which is independent of the particle mass. This is a simple example of the equipartition theorem, which we will return to in Chapter 19.

8.7 Equilibrium with Asymmetric Pistons, Revisited

There is an interesting and useful generalization of the equations for equilibrium that can be derived from the entropy. We had assumed that the total volume, $V_T = \sum_j V_j$, was conserved when systems exchanged volume through a piston. We do not need this restriction.

Suppose systems j and k are connected by pistons of unequal cross sections. To be specific, assume that the area of the piston on the 'j' side is A_j, while the area of the piston on the 'k' side is $A_k \neq A_j$. Now the sum of the changes in volume in the two systems does not vanish, $\Delta V_j + \Delta V_k = A_j \Delta x_j + A_k \wedge x_k \neq 0$, where Δx_α is the distance the piston moves outward. Naturally, $\Delta x_j = -\Delta x_k$. The force on the piston on the 'j' side is $A_j P_j$, where the pressure in system j is given by

$$\frac{P_j}{T_j} = \left(\frac{\partial S_j}{\partial V_j}\right)_{E_j, N_j} \tag{8.30}$$

and

$$\frac{1}{T_j} = \left(\frac{\partial S_j}{\partial E_j}\right)_{V_j, N_j}. \tag{8.31}$$

The corresponding force on the piston on the 'k' side is $A_k P_k$.

Assume that thermal equilibrium has been achieved, so that $T_j = T_k$.

In equilibrium, the forces must be equal, $A_j P_j = A_k P_k$, so that the pressures are not equal. This implies that

$$A_j \left(\frac{\partial S_j}{\partial V_j} \right)_{E_j, N_j} = A_j \frac{P_j}{T_j} = A_k \frac{P_k}{T_k} = A_k \left(\frac{\partial S_k}{\partial V_k} \right)_{E_k, N_k}. \tag{8.32}$$

The result is that instead of the usual equilibrium criterion, we have a generalized equilibrium criterion,

$$A_j \left(\frac{\partial S_j}{\partial V_j} \right)_{E_j, N_j} = A_k \left(\frac{\partial S_k}{\partial V_k} \right)_{E_k, N_k}. \tag{8.33}$$

When $A_j = A_k$ we recover the usual expression.

8.8   The Entropy and the Chemical Potential

The chemical potential, μ, is related to the number of particles in much the same way that the temperature is related to the energy and the pressure is related to the volume. However, while we do have thermometers and pressure gauges, we do not have a convenient way of directly measuring the chemical potential. This makes it difficult to develop our intuition for the meaning of chemical potential in the way we can for temperature and pressure.

On the other hand, it is straightforward to define the chemical potential in analogy with eqs. (8.25) and (8.26),

$$\left(\frac{\partial S}{\partial N} \right)_{E,V} = \frac{-\mu}{T}. \tag{8.34}$$

For the classical ideal gas, we can find an explicit equation for the chemical potential from the equation for the entropy,

$$\begin{aligned}
\mu &= -T \frac{\partial}{\partial N} \left[k_B N \left[\frac{3}{2} \ln \left(\frac{E}{N} \right) + \ln \left(\frac{V}{N} \right) + X \right] \right] \\
&= -k_B T \left[\frac{3}{2} \ln \left(\frac{E}{N} \right) + \ln \left(\frac{V}{N} \right) + X \right] - k_B T N \left[-\frac{3}{2N} - \frac{1}{N} \right] \\
&= -k_B T \left[\frac{3}{2} \ln \left(\frac{3}{2} k_B T \right) + \ln \left(\frac{V}{N} \right) + X - \frac{5}{2} \right].
\end{aligned} \tag{8.35}$$

The complexity of this expression probably adds to the difficulty of acquiring an intuitive feeling for the chemical potential. Unfortunately, that is as much as we can do at this point. We will revisit the question of what the chemical potential means several times in this book, especially when we discuss Fermi–Dirac and Bose–Einstein statistics in Chapters 27, 28, and 29.

8.9   The Fundamental Relation and Equations of State

We have seen that derivatives of the entropy with respect to energy, volume, and number of particles give us three equations for the three variables T, P, and μ, so that we have complete information about the behavior of the thermodynamic system. For this reason,

$$S = S(E, V, N) \tag{8.36}$$

is called a fundamental relation. Specifically, it is called the fundamental relation in the entropy representation, because the information is given in the functional form of the entropy. There are also other representations of the same complete thermodynamic information that we will discuss in Chapter 12.

The derivatives of the entropy give what are called 'equations of state', because they give information concerning the thermodynamic state of the system. Since there are three derivatives, there are three equations of state for a system containing only one kind of particle. The equations of state for the classical ideal gas are given in eqs. (8.2), (8.28), and (8.35).

Note that although it is quite common to hear $PV = Nk_BT$ referred to as 'the' equation of state, it is only valid for the classical ideal gas, and even then it is only one of three equations of state.

For a general thermodynamic system, the three equations of state are independent. If all three are known, the fundamental relation can be recovered (up to an integration constant).

On the other hand, as we will see in Chapter 13, only two of the equations of state are independent for extensive systems—that is, systems for which the entropy, energy, volume, and number of particles are all proportional to each other. In that case, only two equations of state are needed to recover the fundamental relation.

While the fundamental relation contains complete information about a thermodynamic system, a single equation of state does not. However, equations of state are still very important because, under the right circumstances, one equation of state might have exactly the information required to solve a problem.

8.10   The Differential Form of the Fundamental Relation

Much of thermodynamics is concerned with understanding how the properties of a system change in response to small perturbations. For this reason, as well as others that we will see in later chapters, the differential form of the fundamental relation is very important.

A small change in the entropy due to small changes in the energy, volume, and number of particles can be written formally in terms of partial derivatives

$$dS = \left(\frac{\partial S}{\partial E}\right)_{V,N} dE + \left(\frac{\partial S}{\partial V}\right)_{E,N} dV + \left(\frac{\partial S}{\partial N}\right)_{E,V} dN. \tag{8.37}$$

From the equations of state we can rewrite eq. (8.37) in a very useful form:

$$dS = \left(\frac{1}{T}\right) dE + \left(\frac{P}{T}\right) dV - \left(\frac{\mu}{T}\right) dN. \tag{8.38}$$

This is known as the differential form of the fundamental relation in the entropy representation. It is valid for all thermodynamic systems.

8.11 Thermometers and Pressure Gauges

As mentioned previously, the most important characteristic of a thermometer is that it shows a measurable and reproducible change when its temperature changes.

The next most important characteristic of a thermometer is that it should be small in comparison to the system in which you are interested. Energy will have to flow between the system and the thermometer, but the amount of energy should be small enough so that the temperature of the system of interest does not change.

A 'pressure gauge' must satisfy the same condition of 'smallness'. It must have some observable property that changes with pressure in a known way, and it must be small enough not to affect the pressure in the system being measured.

8.12 Reservoirs

In studying thermodynamics it will often be useful to invoke a thermal reservoir at a fixed temperature, so that objects brought into contact with the reservoir will also come to equilibrium at the same temperature. The most important characteristic of a 'reservoir' is that it must be very large in comparison to other systems of interest, so that its temperature does not change significantly when brought into contact with the system of interest.

We will also find it useful to extend the concept of a reservoir to include a source of particles at a fixed chemical potential, or a large container of gas that can maintain a fixed pressure when connected to the system of interest through a piston.

8.13 Problems

PROBLEM 8.1
Equilibrium between two systems

Consider a mysterious system X. The energy-dependence of entropy of system X has been determined to be

$$S_X(U_X, V_X, N_X) = k_B N_X \left[A \left(\frac{U_X}{V_X} \right)^{1/2} + B \right]$$

where A and B are positive constants.

Suppose that system X has been put into thermal contact with a monatomic ideal gas of N particles and the composite system has come to equilibrium.

1. Find an equation for U_X in terms of the energy of the ideal gas.
   [You do not have to solve the equation.]
2. What is the temperature of the ideal gas as a function of U_X, V_X, and N_X?

PROBLEM 8.2
Equilibrium between two ideal gases connected by an asymmetric piston

Consider two containers of ideal gas that are connected by a diathermal, asymmetrical piston. In system 1, the piston has a cross-sectional area of A_1, and in system 2 the other end of the piston has a cross-sectional area of A_2. The piston can move freely, changing the volumes of each of the systems. The energies, volumes, and particle numbers are E_1, V_1, N_1 and E_2, V_2, N_2.

1. Find two equations expressing the equilibrium conditions.
2. From the equations derived in the first part of the problem, find the relationships between the temperatures of the systems.
3. From the equations derived in the first part of the problem, find the relationships between the pressures of the systems.

Part II

Thermodynamics

9

The Postulates and Laws
of Thermodynamics

In this house, we OBEY the laws of thermodynamics!

Homer Simpson (Dan Castellaneta, American actor, and writer)

9.1 Thermal Physics

With this chapter we begin the formal study of thermodynamics, based on the properties of the entropy developed in Part 1 from the study of the foundations of classical statistical mechanics. We will return to statistical mechanics in Parts III and IV.

The theories of statistical mechanics and thermodynamics deal with the same physical phenomena. In many cases it is possible to derive exactly the same equations from either theory, although not necessarily with equal ease. In no case do the two theories contradict each other. Nevertheless, they do have different origins, and are based on very different assumptions. Because of the different points of view that accompanied the development of thermodynamics and statistical mechanics, it is easy to become confused about what is an assumption and what is a derived result.

The task of keeping assumptions and results straight is made more difficult by the traditional description of thermodynamic ideas by the terms 'postulates' and 'laws', since a postulate or law in thermodynamics might also be regarded as being derived from statistical mechanics.

9.1.1 Viewpoint: Statistical Mechanics

The foundations of statistical mechanics lie in the assumption that atoms and molecules exist and obey the laws of either classical or quantum mechanics. Since macroscopic measurements on thermal systems do not provide detailed information on the positions and momenta (or the many-particle wave function in quantum mechanics) of the 10^{20} or more atoms in a typical object, we must also use probability theory and make assumptions about a probability distribution in a many-dimensional space.

On the basis of the assumptions of statistical mechanics, we can write down a formal expression for the entropy as the logarithm of the probability of a composite system,

An Introduction to Statistical Mechanics and Thermodynamics. Robert H. Swendsen, Oxford University Press (2020).
© Robert H. Swendsen. DOI: 10.1093/oso/9780198853237.001.0001

as we did for classical statistical mechanics in Part I. In some simple cases, such as the classical ideal gas, we can even obtain a closed form expression for the entropy. In most cases we must make approximations.

In Part I, we found an explicit expression for the entropy of the classical ideal gas, which allows us to understand the properties of the entropy for this simple system. We also presented arguments that certain properties of the entropy should not change when interactions between particles are included in the calculation.

The point of view taken in this book is that statistical mechanics is the more fundamental theory. Thermodynamics is based on assumptions that can be understood in the context of statistical mechanics. What we will call the 'Postulates of Thermodynamics' and the 'Laws of Thermodynamics' are, if not theorems, at least plausible consequences of the theory of statistical mechanics.

Molecular theory and statistical mechanics provide a fundamental explanation and justification of thermodynamics. We might as well take advantage of them.

9.1.2   Viewpoint: Thermodynamics

Thermodynamics was defined in Chapter 1 as the study of everything connected with heat. The field was developed in the nineteenth century before the existence of atoms and molecules was accepted by most scientists. In fact, the thermodynamics of the nineteenth century was expressed in terms of the mass of a system, rather than the number of particles it contains, which is standard today.

Because thermodynamics was developed without benefit of the insight provided by molecular theory, its history is characterized by ingenious but rather convoluted arguments. These arguments eventually led Herbert Callen (American physicist, 1919–1993) and his advisor, László Tisza (Hungarian and American physicist, 1907–2009), to create a formal system of postulates based on the concept of entropy, from which all of thermodynamics could be derived.

From the point of view of thermodynamics, everything starts with the postulates presented later in this chapter. Even though these postulates can be traced back to statistical mechanics, they provide a reliable starting place for a self-consistent theory that produces complete agreement with experiments in the real world.

Thermodynamics can also be regarded as complementary to statistical mechanics. It gives different insights into physical properties. In many cases, thermodynamics is actually more efficient than statistical mechanics for deriving important equations.

Beginning the study of thermodynamics with the formal postulates, as we do in Part II of this book, avoids much of the complexity that nineteenth-century scientists had to deal with. The only drawback of the formal postulates is that they are rather abstract. A major purpose of Part I of this book was to make them less abstract by providing an explicit example of what entropy means and how it might be calculated.

In this chapter, we will make the connection between the statistical mechanics of Part I and the thermodynamics of Part II more explicit by discussing the concepts of 'state' and 'state function' in the two theories. We will then present the postulates and laws of thermodynamics, including references to the corresponding results from classical statistical mechanics that were obtained in Part I.

9.2 Microscopic and Macroscopic States

A microscopic state is a property of a system of particles. In classical mechanics, a microscopic state is characterized by specific values of the positions and momenta of every particle; that is, a point in phase space. In quantum mechanics, a microscopic state is characterized by a unique, many-particle wave function.

In thermodynamic experiments we restrict ourselves to 'macroscopic' measurements, which are characterized by their limited resolution; by definition, macroscopic measurements cannot resolve individual particles or microscopic length scales. It might be appropriate to remember that thermodynamics was developed before scientists even believed that molecules existed—and they certainly did not measure the behavior of individual molecules.

For both classical and quantum systems, the microscopic state is not experimentally accessible to macroscopic measurements. In the experiments we wish to describe, we can obtain some information about the microscopic state, but we cannot determine it completely.

A macroscopic state is not a property of a thermodynamic system; it is a description of a thermodynamic system based on macroscopic measurements. Due to the experimental uncertainty of macroscopic measurements, a macroscopic state is consistent with an infinity of microscopic states. A system can be in any microscopic state that would be consistent with the experimental observations.

Formally, this might seem to be a rather loose definition of what we mean by 'macroscopic state'. In practice, however, there is rarely a problem. Because microscopic fluctuations are so much smaller than experimental uncertainties, it is relatively easy to specify which microscopic states are consistent with macroscopic measurements.

9.3 Macroscopic Equilibrium States

An equilibrium state is a special kind of macroscopic state. Within the limits of experimental measurements, the properties of a system in an equilibrium state do not change with time, and there is no net transfer of energy or particles into or out of the system.

Equilibrium states can usually be described by a small number of measurements in the sense that all other properties are completely determined within experimental error. In the case of an ideal gas, discussed in Part I, we need to know only the number of particles, the volume, and the energy. Given those values, we can calculate the temperature, pressure, and chemical potential, along with a number of other quantities that we will discuss in the following chapters.

For more complicated systems we might need to know how many molecules of each type are present, and perhaps something about the shape and surface properties of the container. In all cases, the number of quantities needed to characterize an equilibrium system will be tiny in comparison with the number of particles in the system.

9.4   State Functions

The term 'state function' denotes any quantity that is a function of only the small number of variables needed to specify an equilibrium state.

The main substance of the thermodynamic postulates is that for every macroscopic system in thermal equilibrium there exists a state function called the entropy, which has certain properties listed in Section 9.5. Given these postulates, the entire mathematical structure of thermodynamics can be derived—which is the subject of the rest of Part II.

9.5   Properties and Descriptions

In order to understand thermodynamics it will be important to keep in mind the distinction between a *property* of a system and a *description* of a system. A 'property' is an exact characteristic of a system. If a system contains exactly 63749067496584764830091 particles, that is a property of the system. A 'description' is, at most, an estimate of a property. If measurements reveal that a system has $6.374907(2) \times 10^{23}$ particles, that is a description.

In Chapter 4 we calculated the average number of particles in a subsystem. That number is an excellent description of the system for human purposes because the relative statistical fluctuations are so small. However, it is not the true number of particles at any instant of time. In a system of 10^{20} particles. The true number of particles is a property of the system, but its exact value will fluctuate over time by roughly 10^{10} particles.

Because thermodynamics deals with descriptions of macroscopic systems, rather than their true properties, we will not belabor the distinction in Part II. All quantities will be descriptions of a system.

9.6   The Essential Postulates of Thermodynamics

I will present the postulates of thermodynamics in a somewhat different form than originally given by Callen. They will be separated into essential postulates, which must always be true, and optional postulates, which will not apply to every thermodynamic system, but which are very convenient when they do apply.

These four postulates must always be true for a complete theory of thermodynamics for general systems.

9.6.1 Postulate 1: Equilibrium States

Postulate 1: There exist equilibrium states of a macroscopic system that are characterized uniquely by a small number of extensive variables.

Recall that extensive variables are quantities that provide a measure of the size of the system. Examples include the total energy, the volume, and the number of particles. By contrast, intensive parameters are quantities that are independent of the size of the system. Examples include temperature, pressure, and chemical potential.

The remaining postulates all specify properties of a state function called the 'entropy'. These properties should be compared to what you know of the entropy of an ideal gas from Part I.

9.6.2 Postulate 2: Entropy Maximization

Postulate 2: The values assumed by the extensive parameters of an isolated composite system in the absence of an internal constraint are those that maximize the entropy over the set of all constrained macroscopic states.

The property of the entropy in the second postulate is by far the most important. It has two consequences that we will use repeatedly.

First, whenever we release a constraint on a composite system, the entropy will not decrease,

$$\Delta S \geq 0. \tag{9.1}$$

This follows directly from Boltzmann's idea that macroscopic systems go from less probable states to more probable states. The final state must be the most probable state and therefore have the highest entropy. This property is the Second Law of Thermodynamics.

Eq. (9.1) can be rather puzzling. It gives time a direction, sometimes called the 'arrow of time'. This can seem strange, since the underlying equations of motion for the molecules, either classical or quantum, are time-reversal invariant. There is not really a contradiction, and we will reconcile the apparent conflict in Chapter 22.

The second consequence of eq. (9.1) is that we can find the equilibrium values of the extensive parameters describing the amount of something in each (sub)system after a constraint has been released by maximizing the entropy. This will provide an effective approach to solving problems in thermodynamics.

9.6.3 Postulate 3: Additivity

Postulate 3: The entropy of a composite system is additive over the constituent subsystems.

Additivity means that

$$S_{j,k}(E_j, V_j, N_j; E_k, V_k, N_k) = S_j(E_j, V_j, N_j) + S_k(E_k, V_k, N_k), \qquad (9.2)$$

where S_j and S_k are the entropies of systems j and k.

Since we regard the composite system as fundamental in the definition of entropy, and the entropy of a composite system separates into the sum of the entropies of the subsystems, this postulate could be equally well called the postulate of separability. The term 'additivity' is generally preferred because of the concentration on the properties of simple systems in the history of thermodynamics.

Most interactions between molecules are short ranged. If we exclude gravitational interactions and electrical interactions involving unbalanced charges, the direct interaction between any two molecules is essentially negligible at distances of more than a few nanometers. As discussed at the end of Chapter 7, this leads to an approximate separation of the integrals defining the entropy for the subsystems, so that the entropy of the composite system is just the sum of the entropies of the subsystem.

Additivity for systems with interacting particles is an approximation in that it neglects direct interactions between particles in different subsystems (see the discussion in Section 7.6.2). However, deviations from additivity are usually very small due to the extremely short range of intermolecular interactions in comparison to the size of macroscopic systems.

9.6.4   Postulate 4: Continuity and Differentiability

Postulate 4: The entropy is a continuous and differentiable function of the extensive parameters.

The true number of particles is, of course, an integer. However, when we talk about the 'number of particles' or the value of N, we really mean a description of the number of particles that differs from the true number by less than the accuracy of our measurements. Since N is a description of the number of particles, it is not limited to being an integer.

Thermodynamics is much easier if we can work with functions that are continuous and differentiable. Conveniently, it is almost always true that we can.

This postulate is often given in terms of the assumed analyticity of the functions. However, the postulate of analyticity can break down. When it does, it usually signals some sort of instability or phase transition, which is particularly interesting. Stability and phase transitions are discussed in Chapters 16 and 17. We will consider specific examples of phase transitions in Chapters 28 and 31.

9.7   Optional Postulates of Thermodynamics

The following three postulates are not applicable to all thermodynamic systems. When they are applicable, they often simplify the analysis of thermodynamic behavior.

9.7.1   Optional Postulate 5: Extensivity

Postulate 5: The entropy is an extensive function of the extensive variables.

'Extensivity' is the property that the entropy is directly proportional to the size of the system; if the extensive parameters are all multiplied by some positive number λ, the entropy will be too,

$$S(\lambda U, \lambda V, \lambda N) = \lambda S(U, V, N). \tag{9.3}$$

Mathematically, eq. (9.3) means that the entropy is assumed to be an homogeneous, first-order function of the extensive variables.

The extensivity postulate is not true for all systems!

Even such a common system as a gas in a container violates this postulate, because the molecules of the gas can be adsorbed onto the inner walls of the container. Since the surface-to-volume ratio varies with the size of the container, the fraction of molecules adsorbed on the inner walls also varies, and such a system is not extensive.

For most of the book we will *not* assume extensivity, although in Chapter 13 we will see that extensivity can be extremely useful for investigating the properties of materials.

> The properties of additivity and extensivity are often confused. This is probably due to the fact that many textbooks restrict their discussion of thermodynamics to homogeneous systems, for which both properties are true. Additivity is the more fundamental property. It is true whenever the range of interactions between particles is small compared to the size of the system. Extensivity is only true to the extent that the surface of the system and the interface with its container can be neglected.

9.7.2   Optional Postulate 6: Monotonicity

Postulate 6: The entropy is a monotonically increasing function of the energy for equilibrium values of the energy.

In the example we discussed in Part I, the entropy of the classical ideal gas was shown to be a monotonically increasing function of the energy. This monotonicity is important, since it implies that the temperature is positive, as we have shown in Chapter 8 for the ideal gas. However, monotonicity is not true for some of the most important models in physics, as we will see in Chapter 31. These models exhibit negative temperatures in some energy regions.

The question of whether the temperature of certain thermodynamic systems can be negative has become a topic of some debate in recent years. I have gone on record as saying that negative temperatures are consistent with thermodynamics,[1] and I will

---

[1] R. H. Swendsen, 'Thermodynamics, Statistical Mechanics, and Entropy,' *Entropy*, 19:603 (2017).

develop the formalism of thermodynamics accordingly. What I say about positive temperatures is not controversial. What I say about negative temperatures is controversial.

> There are a number of models of great importance to statistical mechanics, for which $\partial S/\partial U = 1/T < 0$ in certain energy regions (see Chapter 31). For these models, Postulate 6 does not hold. On the other hand, this is not the usual case. Consequently, many of the applications of thermodynamics, like the theory of heat engines, assume that $T > 0$. These will be discussed in Chapter 11, which will include a brief section on the changes required for negative temperatures.

9.7.3   Optional Postulate 7: The Nernst Postulate

Postulate 7: The entropy of any system is non-negative.

The Nernst Postulate[2] is also known as the Third Law of Thermodynamics, adding to the confusion between laws and postulates. It only applies to quantum systems, but, since all real systems are quantum mechanical, it ultimately applies to everything. It has some important consequences, which are discussed in Chapter 18. The Nernst Postulate is listed as optional because classical systems are interesting in themselves and useful in providing a contrast with the special properties of quantum systems.

   The Third Law is the only law (or postulate) that cannot be understood on the basis of classical statistical mechanics. For classical models, the entropy always goes to negative infinity as the temperature goes to zero. The explanation of the Third Law is intrinsically quantum mechanical. Its consequences will be discussed in Chapter 18, but it will not be derived until Chapter 23 in Part IV.

   In quantum statistical mechanics there is a universally recognized convention for the absolute value of the entropy that will be discussed in Part IV. This convention implies that

$$\lim_{T \to 0} S(T) \geq 0. \tag{9.4}$$

> The Third Law (or Nernst Postulate) is often stated as $\lim_{T\to 0} S(T) = 0$, but that is incorrect. There are many examples of systems with disordered ground states that have a (positive) non-zero entropy at zero temperature.
>
>   The Nernst Postulate is also often credited with preventing the attainment of a temperature of absolute zero, but that is also incorrect; absolute zero would still be unattainable if the world obeyed classical statistical mechanics.

[2] Named after the German scientist Walther Hermann Nernst (1864–1941), who won the Nobel Prize in 1920 for its discovery.

9.8 The Laws of Thermodynamics

The 'laws' of thermodynamics were discovered considerably earlier than the formulation of thermodynamics in terms of 'postulates'. The peculiar numbering of the laws is due to fixing the First and Second laws before the importance of the 'zeroth' law was realized. The current numbering scheme was suggested by Ralph H. Fowler (British physicist and astronomer, 1889–1944) as a way of acknowledging the importance of the 'zeroth' law.

The laws of thermodynamics are:

Zeroth Law If two systems are each in equilibrium with a third system, they are also in equilibrium with each other.

First Law Heat is a form of energy, and energy is conserved.

Second Law After the release of a constraint in a closed system, the entropy of the system never decreases.

Third Law (Nernst Postulate) The entropy of any quantum mechanical system goes to a constant as the temperature goes to zero.

The Zeroth Law was derived from statistical mechanics in Section 7.9.

The First Law is simply conservation of energy. However, at the time it was formulated it had just been discovered that heat was a form of energy, rather than some sort of mysterious fluid as had been previously supposed.

The Second Law is essentially the same as Postulate 2 in Subsection 9.6.2.

The Third Law is listed as Postulate 7 in Section 9.7.3.

10

Perturbations of Thermodynamic State Functions

When you come to a fork in the road, take it.

Yogi Berra

As discussed in the Chapter 9, state functions specify quantities that depend only on the small number of variables needed to characterize an equilibrium state. It turns out that state functions are strongly restricted by the postulates of thermodynamics, so that we can derive a great many non-trivial equations linking observable quantities. The power of thermodynamics is that these relations are true for *all* systems, regardless of their composition!

> In Part I we used the symbol E to denote the energy of a system, which is customary in statistical mechanics. Unfortunately, it is customary in thermodynamics to use the symbol U for the energy. The reason for this deplorable state of affairs is that it is often convenient in thermodynamics to consider the energy per particle (or per mole), which is denoted by changing to lower case, $u \equiv U/N$. Since e might be confused with Euler's number, u and U have become standard. Technically, there is also a difference in meaning. The proper definition of U is $U = \langle E \rangle$. When we consider the microcanonical ensemble, for which the width of the energy distribution is zero, it does not make any difference. It will become important when we turn to quantum systems. We will follow thermodynamics convention in Part II and denote the energy by U. It is fortunate that students of thermal physics are resilient, and can cope with changing notation.

10.1 Small Changes in State Functions

In much of thermodynamics we are concerned with the consequences of small changes. Part of the reason is that large changes can often be calculated by summing or integrating the effects of many small changes, but in many cases the experimental questions of interest really do involve small changes.

An Introduction to Statistical Mechanics and Thermodynamics. Robert H. Swendsen, Oxford University Press (2020). © Robert H. Swendsen. DOI: 10.1093/oso/9780198853237.001.0001

Mathematically, we will approximate small changes by infinitesimal quantities. This, in turn, leads us to consider two distinct types of infinitesimals, depending on whether or not there exists a function that can be determined uniquely by integrating the infinitesimal.

10.2　Conservation of Energy

The great discovery by James Prescott Joule (English physicist, mathematician, and brewer, 1818–1889) that heat is form of energy, combined with the principle that energy is conserved, gives us the First Law of Thermodynamics.

We will usually apply the First Law to small changes in thermodynamic quantities, so that it can be written as

$$dU = đQ + đW \qquad (10.1)$$

if the number of particles in the system does not change. Although dU, $đQ$, and $đW$ are all infinitesimal quantities, there is an important distinction between those written with d and those written with $đ$ that is discussed in the next section.

The change in the energy of a system during some thermal process is denoted by dU, and is positive when energy is added to the system. The energy added to the system in the form of heat is denoted by $đQ$, and the work done on the system is denoted by $đW$. The sign convention is that all three of these differentials are positive when energy is added to the system.

> Our choice of sign convention for $đW$ is by no means universal. Taking an informal survey of textbooks, I have found that about half use our convention, while the other half take $đW$ to be positive when work is done by the system. Be careful when comparing our equations with those in other books.

In many applications of thermodynamics we will have to consider transfers of energy between two or more systems. In that case, the choice of the positive directions for $đQ$ and $đW$ will be indicated by arrows in a diagram.

10.3　Mathematical Digression on Exact and Inexact Differentials

Consider an infinitesimal quantity defined by the equation

$$dF = f(x)dx, \qquad (10.2)$$

where $f(x)$ is some known function. We can always—at least in principle—integrate this equation to obtain a function $F(x)$ that is unique to within an additive constant.

Now consider the infinitesimal defined by the equation

$$đF = f(x,y)dx + g(x,y)dy, \tag{10.3}$$

where $f(x,y)$ and $g(x,y)$ are known functions. Although the differential $đF$ can be integrated along any given path, the result might depend on the specific path, and not just on the initial and final points. In that case, no unique function $F(x,y)$ can be found by integrating eq. (10.3), and $đF$ is called an 'inexact' differential. The use of $đ$ instead of d indicates that an infinitesimal is an inexact differential.

Note that if two path integrals produce different values, the integral over the closed path going from the initial to the final point along one path and returning along the other will not vanish. Therefore, the failure of an integral around an arbitrary closed path to vanish is also characteristic of an inexact differential.

As an example, take the following inexact differential:

$$đF = x\,dx + x\,dy. \tag{10.4}$$

Consider the integral of this differential from $(0,0)$ to the point (x,y) by two distinct paths.

Path A: From $(0,0)$ to $(x,0)$ to (x,y).

Path B: From $(0,0)$ to $(0,y)$ to (x,y).

The path integral for Path A gives

$$\int_0^x x'\,dx' \bigg|_{y'=0} + \int_0^y x'\,dy' \bigg|_{x'=x} = \frac{1}{2}x^2 + xy \tag{10.5}$$

and the path integral for Path B gives

$$\int_0^y x'\,dy' \bigg|_{x'=0} + \int_0^x x'\,dx' \bigg|_{y'=y} = 0 + \frac{1}{2}x^2 = \frac{1}{2}x^2. \tag{10.6}$$

Since the two path integrals in eqs. (10.5) and (10.6) give different results for $y \neq 0$, there is no unique function $F(x,y)$ corresponding to the inexact differential in eq. (10.4).

10.3.1   Condition for a Differential to be Exact

We can easily determine whether a differential is exact or inexact without the necessity of performing integrals over all possible paths. If a function $F(x,y)$ exists, we can write its differential in terms of partial derivatives,

$$dF = \left(\frac{\partial F}{\partial x}\right) dx + \left(\frac{\partial F}{\partial y}\right) dy. \tag{10.7}$$

If the differential

$$dF = f(x,y)dx + g(x,y)dy \tag{10.8}$$

is exact, we can identify $f(x,y)$ and $g(x,y)$ with the corresponding partial derivatives of $F(x,y)$,

$$f(x,y) = \frac{\partial F}{\partial x} \tag{10.9}$$

$$g(x,y) = \frac{\partial F}{\partial y}. \tag{10.10}$$

By looking at partial derivatives of $f(x,y)$ and $g(x,y)$, we find that they must satisfy the following condition:

$$\frac{\partial f}{\partial y} = \frac{\partial^2 F}{\partial y \partial x} = \frac{\partial^2 F}{\partial x \partial y} = \frac{\partial g}{\partial x}. \tag{10.11}$$

The central equality follows from the fact that the order of partial derivatives can be switched without changing the result for well-behaved functions, which are the only kind we allow in physics textbooks.

10.3.2   Integrating Factors

There is an interesting and important relationship between exact and inexact differentials. Given an inexact differential,

$$đF = f(x,y)dx + g(x,y)dy \tag{10.12}$$

we can find a corresponding exact differential of the form

$$dG = r(x,y)đF \tag{10.13}$$

for a function $r(x,y)$, which is called an 'integrating factor'. The function $r(x,y)$ is not unique; different choices for $r(x,y)$ lead to different exact differentials that are all related to the same inexact differential.

The example given in eq. (10.4) is too easy. The function $r(x,y) = 1/x$ is obviously an integrating factor corresponding to $dG = dx + dy$.

A less trivial example is given by the inexact differential

$$đF = y\,dx + dy. \tag{10.14}$$

To find an integrating factor for eq. (10.14), first write the formal expression for the exact differential,

$$dG = r(x,y)\,dF = r(x,y)y\,dx + r(x,y)dy. \tag{10.15}$$

The condition that dG is an exact differential in eq. (10.11) gives

$$\frac{\partial}{\partial y}[r(x,y)y] = \frac{\partial}{\partial x}r(x,y) \tag{10.16}$$

or

$$r + y\frac{\partial r}{\partial y} = \frac{\partial r}{\partial x}. \tag{10.17}$$

Eq. (10.17) has many solutions, so let us restrict our search for an integrating factor to functions that only depend on x,

$$r(x,y) = r(x). \tag{10.18}$$

Eq. (10.17) then simplifies because the partial derivative with respect to y vanishes,

$$r = \frac{dr}{dx}. \tag{10.19}$$

Eq. (10.19) can be easily solved

$$r(x,y) = r(x) = \exp(x). \tag{10.20}$$

The final result for dG is

$$dG = \exp(x)\,dF = y\exp(x)\,dx + \exp(x)dy. \tag{10.21}$$

Naturally, any constant multiple of $\exp(x)$ is also an integrating factor.

10.4   Conservation of Energy Revisited

Since the energy U is a state function, the differential of the energy, dU, must be an exact differential. On the other hand, neither heat nor work is a state function, so both dQ and dW are inexact.

10.4.1 Work

For simplicity, first consider an infinitesimal amount of work done by using a piston to compress the gas in some container. Let the cross-sectional area of the piston be A, and let it move a distance dx, where positive dx corresponds to an increase in the volume, $dV = Adx$. The force on the piston due to the pressure P of the gas is $F = PA$, so that the work done on the gas is $-Fdx$ or

$$đW = -Fdx = -PdV. \tag{10.22}$$

The volume of the system is obviously a state function, so the differential dV is exact. Eq. (10.22) shows a relationship between an inexact and an exact differential, with the function $1/P$ playing the role of an integrating factor. To see that $đW$ is indeed an inexact differential, integrate it around any closed path in the P, V-plane. The result will not vanish.

10.4.2 Heat

The heat transferred to a system turns out to be closely related to the change in entropy of the system. Consider the change in entropy due to adding a small amount of heat $đQ$ to the system,

$$dS = S(E + đQ, V, N) - S(E, V, N) = \left(\frac{\partial S}{\partial E}\right)_{V,N} đQ = \frac{đQ}{T}. \tag{10.23}$$

Again we see an equation relating exact and inexact differentials. In this case, the function $1/T$ is the integrating factor.

Eq. (10.23) is extremely important for a number of reasons. It can be used to numerically integrate experimental data to calculate changes in entropy. It is also important for our purposes in the next section when written slightly differently,

$$đQ = T dS. \tag{10.24}$$

10.5 An Equation to Remember

Eqs. (10.22) and (10.24) can be used to rewrite eq. (10.1) in a very useful form,

$$dU = TdS - PdV. \tag{10.25}$$

This equation is, of course, only valid when there are no exchanges of particles with another system, so that the total number of particles in the system of interest is held constant.

Curiously enough, in Section 8.10 on the fundamental relation in the entropy representation, we have already derived the generalization of eq. (10.25) to include a change dN in the number of particles. Rewriting eq. (8.38) with dU instead of dE, we have

$$dS = \left(\frac{1}{T}\right) dU + \left(\frac{P}{T}\right) dV - \left(\frac{\mu}{T}\right) dN. \qquad (10.26)$$

Solving this equation for dU, we find one of the most useful equations in thermo-dynamics:

$$dU = TdS - PdV + \mu dN. \qquad (10.27)$$

The reason that the differential form of the fundamental relation is equivalent to the First Law of Thermodynamics (energy conservation) is that energy conservation was used in its derivation in Part I.

> If there is one equation in thermodynamics that you must memorize, it is eq. (10.27). In the following chapters you will see why this is so.

> Eq. (10.27) provides another interpretation of the chemical potential μ. It is the change in the energy of a system when a particle is added without changing either the entropy or the volume. The difficulty with this explanation of the meaning of μ is that it is far from obvious how to 'add a particle without changing the entropy'. Apologies.

10.6 Problems

....

PROBLEM 10.1
Integrating factors for inexact differentials

We showed that the inexact differential

$$\bar{d}F = y\, dx + dy$$

is related to an exact differential by an integrating function

$$r(x,y) = r_x(x).$$

1. Show that $đF$ is also related to another (different) exact differential by an integrating function of the form

$$r(x,y) = r_y(y).$$

2. Derive an explicit expression for $r_y(y)$ and show that the new differential is exact.
3. For the new differential and $x, y > 1$, calculate and compare the path integral from the point $(1,1)$ to $(1,y)$, from there to (x,y), with the path integral from the point $(1,1)$ to $(x,1)$, from there to (x,y).

11

Thermodynamic Processes

Nothing in life is certain except death, taxes and the second law of thermodynamics. All three are processes in which useful or accessible forms of some quantity, such as energy or money, are transformed into useless, inaccessible forms of the same quantity. That is not to say that these three processes don't have fringe benefits: taxes pay for roads and schools; the second law of thermodynamics drives cars, computers and metabolism; and death, at the very least, opens up tenured faculty positions.

Seth Lloyd, American professor of mechanical engineering and physics (1960–)

In this chapter we will discuss thermodynamic processes, which concern the consequences of thermodynamics for things that happen in the real world.

The original impetus for thermodynamics, aside from intellectual curiosity, was the desire to understand how steam engines worked so that they could be made more efficient. Later, refrigerators, air conditioners, and heat pumps were developed and were found to be governed by exactly the same principles as steam engines.

The main results in this chapter rely on the temperature being positive, which would be consistent with Optional Postulate 6: Monotonicity (Subsection 9.7.2). This is usually, but not always, true. We will call attention to the results that depend on $T > 0$ as they are derived, and later in the chapter, we will explain the changes necessary to take negative temperatures into account.

11.1 Irreversible, Reversible, and Quasi-Static Processes

If we release a constraint in a composite system, the new equilibrium state maximizes the total entropy so that the change in entropy is non-negative for all processes in an isolated system,

$$dS \geq 0. \tag{11.1}$$

The change in the total entropy is zero only when the system had already been in equilibrium.

When the total entropy increases during some isolated thermodynamic process, the process is described as 'irreversible'. Running the process backwards is impossible,

An Introduction to Statistical Mechanics and Thermodynamics. Robert H. Swendsen, Oxford University Press (2020).
© Robert H. Swendsen. DOI: 10.1093/oso/9780198853237.001.0001

since the total entropy in an isolated system can never decrease. All real processes are accompanied by an increase in the total entropy, and are irreversible.

On the other hand, if the change ΔX in some variable X is small, the change in entropy can also be small. Since the dependence of the entropy on a small change in conditions is quadratic near its maximum, the magnitude of the increase in entropy goes to zero quadratically, as $(\Delta X)^2$, while the number of changes grows linearly, as $1/\Delta X$. The sum of the small changes will be proportional to $(\Delta X)^2/\Delta X = \Delta X$, which goes to zero as ΔX goes to zero. Consequently, a series of very small changes will result in a small change of entropy for the total process. In the limit of infinitesimal steps, the increase of entropy can vanish. Such a series of infinitesimal steps is called a quasi-static process.

A quasi-static process is reversible. Since there is no increase in entropy, it could be run backwards and return to the initial state.

The concept of a quasi-static process is an idealization, but a very useful one. The first applications of thermodynamics in the nineteenth century concerned the design of steam engines. The gases used in driving steam engines relax very quickly because of the high speeds of the molecules, which can be of the order of $1000\,m/s$. Even though the approximation is not perfect, calculations for quasi-static processes can give us considerable insight into the way real engines work.

> Quasi-static processes are not merely slow. They must also take a system between two equilibrium states that differ infinitesimally. The classic example of a slow process that is not quasi-static occurs in a composite system with two subsystems at different temperatures, separated by a wall that provides good but not perfect insulation. Equilibration of the system can be made arbitrarily slow, but the total entropy will still increase. Such a process is not regarded as quasi-static.

Although the word 'dynamics' appears in the word 'thermodynamics', the theory is primarily concerned with transitions between equilibrium states. It must be confessed that part of the reason for this emphasis is that it is easier. Non-equilibrium properties are much more difficult to analyze, and we will only do so explicitly in Chapter 22, when we discuss irreversibility.

11.2   Heat Engines

As mentioned above, the most important impetus to the development of thermodynamics in the nineteenth century was the desire to make efficient steam engines. Beginning with the work of Sadi Carnot (French scientist, 1792–1832), scientists worked on the analysis of machines to turn thermal energy into mechanical energy for industrial purposes. Such machines are generally known as heat engines.

An important step in the analysis of heat engines is the conceptual separation of the engine itself from the source of heat energy. For this reason, a heat engine is defined to be 'cyclic'; whatever it does, it will return to its exact original state after going through a cycle. This definition ensures that there is no fuel hidden inside the heat engine—a condition that has been known to be violated by hopeful inventors of perpetual-motion machines. The only thing a heat engine does is to change energy from one form (heat) to another (mechanical work).

11.2.1   Consequences of the First Law

The simplest kind of heat engine that we might imagine would be one that takes a certain amount of heat $đQ$ and turns it directly into work $đW$. Since we are interested in efficiency, we might ask how much work can be obtained from a given amount of heat. The First Law of Thermodynamics (conservation of energy) immediately gives us a strict limit:

$$đW \leq đQ. \tag{11.2}$$

A heat engine that violated eq. (11.2) could be made into a perpetual-motion machine; the excess energy $(đW - đQ)$ could be used to run the factory, while the rest of the work would be turned into heat and fed back into the heat engine. Because such a heat engine would violate the First Law of Thermodynamics, it would be called a Perpetual Motion Machine of the First Kind.

Perpetual Motion Machines of the First Kind do not exist.

A pessimist might express this result as: 'You can't win.' Sometimes life seems like that.

11.2.2   Consequences of the Second Law

The limitations due to the Second Law are considerably more severe than those due to the First Law *if the temperature is positive.*

If a positive amount of energy in the form of heat $đQ$ is transferred to a heat engine, the entropy of the heat engine increases by $dS = đQ/T > 0$. If the heat engine now does an amount of work $đW = đQ$, the First Law (conservation of energy) is satisfied. However, the entropy of the heat engine is still higher than it was at the beginning of the cycle by an amount $đQ/T > 0$. The only way to remove this excess entropy so that the heat engine could return to its initial state would be to transfer heat out of the system (at a lower temperature), but this would cost energy, lowering the amount of energy available for work.

A machine that could transform heat directly into work, $đW = đQ$, could run forever, taking energy from the heat in its surroundings. It would be called a Perpetual Motion Machine of the Second Kind because it would violate the Second Law of Thermodynamics.

Perpetual Motion Machines of the Second Kind do not exist.

A pessimist might express this result as: 'You can't break even.' Sometimes life seems like that, too.

It is unfortunate that perpetual-motion machines of the second kind do not exist, since they would be very nearly as valuable as perpetual-motion machines of the first kind. Using one to power a ship would allow you to cross the ocean without the need of fuel, leaving ice cubes in your wake.

11.3 Maximum Efficiency

The limits on efficiency imposed by the First and Second Laws of Thermodynamics lead naturally to the question of what the maximum efficiency of a heat engine might be. The main thing to note is that after transferring heat into a heat engine, heat must also be transferred out again before the end of the cycle to bring the net entropy change back to zero. The trick is to bring heat in at a high temperature and take it out at a low temperature.

We do not need to be specific about the process, but we will assume that it is quasi-static (reversible). Simply require that heat is transferred into the heat engine from a reservoir at a high temperature T_H, and heat is removed at a low temperature $T_L < T_H$. If the net work done by the heat engine during an infinitesimal cycle is dW, conservation of energy gives us a relationship between the work done and the heat exchanged,

$$dW = dQ_H + dQ_L. \tag{11.3}$$

Sign conventions can be a tricky when there is more than one system involved. Here we have at least three systems: the heat engine and the two reservoirs. In eq. (11.3), work done by the heat engine is positive, and both dQ_H and dQ_L are positive when heat is transferred to the heat engine. In practice, the high-temperature reservoir is the source of energy, so $dQ_H > 0$, while wasted heat energy is transferred to the low-temperature reservoir, so $dQ_L < 0$.

Because a heat engine is defined to be cyclic, the total entropy change of the heat engine for a completed cycle must be zero:

$$dS = \frac{dQ_H}{T_H} + \frac{dQ_L}{T_L} = 0. \tag{11.4}$$

The assumption of a quasi-static process (reversibility) is essential to eq. (11.4). While $dS = 0$ in any case because of the cyclic nature of the heat engine, a violation of reversibility would mean that a third (positive) term must be added to $(dQ_H/T_H + dQ_L/T_L)$ to account for the additional generation of entropy.

If we use eq. (11.4) to eliminate dQ_L from eq. (11.3), we obtain a relationship between the heat in and the work done,

$$dW = \left(1 - \frac{T_L}{T_H}\right) dQ_H. \tag{11.5}$$

We can now define the efficiency of a heat engine, η,

$$\eta = \frac{dW}{dQ_H} = 1 - \frac{T_L}{T_H} = \frac{T_H - T_L}{T_H}. \tag{11.6}$$

The efficiency η is clearly less than 1, which is consistent with the limitation due to the Second Law of Thermodynamics. Actually, the Second Law demands that the efficiency given in eq. (11.6) is the maximum possible thermal efficiency of any heat engine, whether it is reversible (as assumed) or irreversible. The proof of this last statement will be left as an exercise.

11.4   Refrigerators and Air Conditioners

Since ideal heat engines are reversible, you can run them backwards to either cool something (refrigerator or air conditioner) or heat something (heat pump). The equations derived in the previous section do not change.

First consider a refrigerator. The inside of the refrigerator can be represented by the low-temperature reservoir, and the goal is to remove as much heat as possible for a given amount of work. The heat removed from the low-temperature reservoir is positive for a refrigerator, $dQ_L > 0$, and the work done *by* the heat engine is negative, $dW < 0$, because you have to use power to run the refrigerator.

We can define a coefficient of performance ϵ_R and calculate it from eqs. (11.3) and (11.4):

$$\epsilon_R = \frac{dQ_L}{-dW} = \frac{T_L}{T_H - T_L}. \tag{11.7}$$

This quantity can be much larger than 1, which means that you can remove a great deal of heat from the inside of your refrigerator with relatively little work; that is, you will need relatively little electricity to run the motor.

An air conditioner works the same way as a refrigerator, but dQ_L is the heat removed from inside your house or apartment. The coefficient of performance is again given by eq. (11.7).

The coefficient of performance for a refrigerator or an air conditioner is clearly useful when you are thinking of buying one. Indeed, it is now mandatory for new refrigerators and air conditioners to carry labels stating their efficiency. However, the labels do not carry the dimensionless quantity ϵ_R; they carry an 'Energy Efficiency Ratio' (EER) which is the same ratio, but using British Thermal Units (BTU) for dQ_L and Joules for dW. The result is that the EER is equal to ϵ_R times a factor of about $3.42\,BTU/J$. A cynic might think that these peculiar units are used to produce larger numbers and improve sales—especially since the units are sometimes not included on the label; I couldn't possibly comment.

11.5 Heat Pumps

Heat pumps also work the same way as refrigerators and air conditioners, but with a different purpose. They take heat from outside your house and use it to heat the inside of your house. The low-temperature reservoir outside your house usually takes the form of pipes buried outside in the yard. The goal is to heat the inside of your house as much as possible with a given amount of work.

We can define a coefficient of performance for a heat pump,

$$\epsilon_{HP} = \frac{-đQ_H}{-đW} = \frac{T_H}{T_H - T_L}. \tag{11.8}$$

This quantity can also be significantly larger than 1, which makes it preferable to use a heat pump rather than running the electricity through a resistive heater to heat your house.

11.6 The Carnot Cycle

The Carnot cycle is a specific model of how a heat engine might work quasistatically between high- and low-temperature thermal reservoirs. The Carnot heat engine is conceived as a closed piston containing gas—usually an ideal gas. To clarify the description of the cycle, we will include two figures: Fig. 11.1, which contains a plot of pressure vs. volume, and Fig. 11.2, which contains a plot of temperature vs. entropy for the same process.

We'll describe the cycle as beginning at pressure P_A and volume V_A in Fig. 11.1 and temperature T_A and entropy S_A in Fig. 11.2, both labelled as point A. The heat engine is assumed to be in contact with a (high-temperature) thermal reservoir at temperature T_A. It expands to a volume of V_B, while remaining in contact with the thermal reservoir and doing work. If Q_{AB} is the amount of heat transferred to the heat engine along the path AB, then the entropy of the heat engine grows by an amount Q_{AB}/T_A. The path from A to B in Fig. 11.2 is consequently a straight, horizontal line. Since the heat engine is expanding, the pressure is steadily reduced, according to the ideal gas law, $P = Nk_B T_A/V$. The heat engine is now at point B in both Fig. 11.1 and Fig. 11.2.

Next, the heat engine is removed from contact with the thermal reservoir. It then continues to expand and do work until it is reaches a volume of V_C. The heat engine has cooled further, according to the adiabatic equation, $P = (\text{constant}) V^{-5/2}$ reaching a temperature of T_C. Because there is no transfer of heat along the path BC, and the process is reversible, the entropy remains unchanged. The heat engine is now at point C.

At this point the heat engine is put in contact with a second (low-temperature) thermal reservoir, which is at temperature T_C. The heat engine is then compressed isothermally until it reaches a volume of V_D. Because it requires work to compress the heat engine, the heat engine does negative work on the path between point C and point D. The pressure grows during this part of the cycle according to the ideal gas law, $P = Nk_B T_C/V$, until it

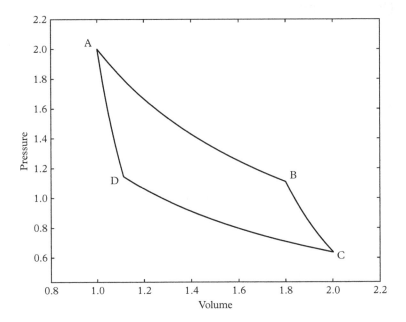

Fig. 11.1 *Pressure vs. volume plot for a Carnot cycle.*

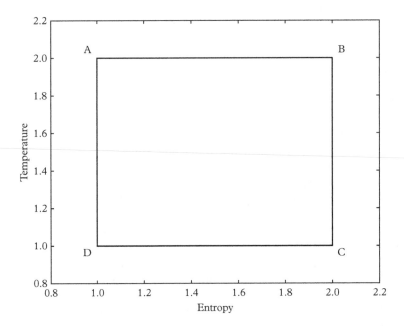

Fig. 11.2 *Temperature vs. entropy plot for a Carnot cycle.*

Table 11.1 *Carnot cycle.*

Path	Process	Temperature change	Volume change	Work done
A to B	isothermal	constant T	expand	positive work
B to C	adiabatic	cooling	expand	positive work
C to D	isothermal	constant T	contract	negative work
D to A	adiabatic	heating	contract	negative work

reaches a pressure of P_D. The entropy decreases because heat is removed from the heat engine. The heat engine is now at point D.

After reaching the point D, the heat engine is removed from the low-temperature reservoir. It continues to be compressed until it reaches a volume of V_A and a pressure of P_A, with temperature T_A and entropy S_A—which is where it started. The Carnot cycle is summarized in Table 11.1.

The Carnot cycle is historically of great importance. It continues to be of great importance to students taking standard national examinations. Its main drawback is that the analysis of the equations is somewhat tedious, and the physical insight is (in my opinion) less than that obtained from the differential forms presented earlier in this chapter.

11.7 Alternative Formulations of the Second Law

There have been nearly as many formulations of the second law as there have been discussions of it.

P. W. Bridgman, American physicist (1882–1961) 1946 Nobel Prize in Physics

Various alternate formulations of the Second Law of Thermodynamics have been suggested in the history of thermodynamics as scientists have wrestled with the meaning of entropy. They have provided simple examples of important consequences of thermodynamics in our lives.

1. The Clausius version of the Second Law
   The original version of the Second Law was phrased by Rudolf Clausius without reference to the entropy.

   Heat can never pass from a colder to a warmer body without some other change, connected therewith, occurring at the same time.

   This can be derived easily by considering a hypothetical heat engine (run in reverse) that extracts an amount of heat $đQ_L < 0$ from the lower temperature body, and adds an amount of heat $đQ_H > 0$ to the higher temperature body ($T_L < T_H$). If $đQ_L + đQ_H = 0$, this is essentially a transfer of heat from a colder to a warmer body, with $đQ_L = -đQ_H < 0$. However, the entropy of the heat engine must change by

$$dS = \frac{dQ_L}{T_L} + \frac{dQ_H}{T_H} = \frac{dQ_L}{T_L} - \frac{dQ_L}{T_H} = dQ_L\left(\frac{T_H - T_L}{T_L T_H}\right) < 0. \qquad (11.9)$$

This leaves us with an decrease in the entropy of the heat engine which would violate the Second Law.

2. The Kelvin version of the Second Law
   Lord Kelvin (Irish mathematical physicist and engineer, 1824–1907, 1st Baron Kelvin) phrased the second Law as follows.

   > *It is impossible, by means of inanimate material agency, to derive mechanical effect from any portion of matter by cooling it below the temperature of the coldest of the surrounding objects.*

   This is clear from the fact that entropy increases from adding heat at a lower temperature which cannot be compensated for by subtracting heat at a higher temperature.

3. The Planck version of the Second Law
   Max Planck's version of the second Law was again expressed in terms of a restriction on the construction of a refrigerator.

   > *It is impossible to construct an engine which will work in a complete cycle, and produce no effect except the raising of a weight and cooling of a heat reservoir.*

   Cooling a heat reservoir would add energy to the heat engine, thereby increasing its entropy. The work done by the heat engine would not decrease its entropy, so the process could not be cyclic.

4. The ter Haar and Wergeland version of the Second Law,
   Dirk ter Haar (Anglo-Dutch physicist, 1919–2002) and Harald Wergeland (Norwegian physicist, 1912–1987) put the Kelvin and Planck versions of the Second Law together in their textbook on thermodynamics.

   > *It is impossible to devise a cyclically operating device, the sole effect of which is to absorb energy in the form of heat from a single thermal reservoir and to deliver an equivalent amount of work.*

   This is again a process which would add heat to the heat engine, raising its entropy with no way to reduce it.

11.8 Positive and Negative Temperatures

If Optional Postulate 6 (Monotonicity, Subsection 9.7.2) does not apply to a particular system, things get considerably more interesting. If the entropy is decreasing in some range of energies, the expression for the temperature, Eq. (8.26), becomes negative.

Negative temperatures have been the subject of considerable debate in recent years. Some workers claim that the temperature is never negative, while others claim that negative temperature states exist, but are unstable. I claim that they exist and are quite interesting.

For many models in statistical mechanics, Optional Postulate 6 (Monotonicity, Subsection 9.7.2) is not valid. These models have an increasing density of states at low energies, but a decreasing density of states for high energies (see Chapter 31). Since the temperature can be expressed as the derivative of the logarithm of the density of states with respect to energy, the temperature is positive at low energies, but negative at high energies.

Although it might seem counterintuitive, negative temperatures are not colder than $T=0$, but instead hotter than $T=\infty$. This is easier to see if we consider the inverse temperature, $\beta = 1/k_B T$, which has units of inverse energy. Then $T=0$ corresponds to $\beta = \infty$. Making the system hotter *decreases* β until $T=\infty$ at $\beta = 0$. Making the system even hotter lowers β to negative values.

To describe systems that exhibit both positive and negative temperatures, it is convenient to use a dimensionless entropy, $\tilde{S} = S/k_B$. The differential form of the fundamental equation, expressed in the $\tilde{S}$-representation can be obtained from eq. (10.26),

$$d\tilde{S} = \beta dU + (\beta P)dV - (\beta \mu)dN. \tag{11.10}$$

The parentheses are put around (βP) and $(\beta \mu)$ as a reminder that these products are to be regarded a single variable. We will use this representation again in Chapter 12, Thermodynamic Potentials.

We consider two bodies, one at high temperature (H) and one at low temperature (L). The convention being used is that work done by the heat engine is positive, and heat coming out of any body is positive.

The entropy balance equation for the heat engine reflects the cyclic requirement and the quasi-static limit. In terms of β and $\tilde{S}$, this becomes

$$d\tilde{S} = \beta_H đQ_H + \beta_L đQ_L = 0, \tag{11.11}$$

or

$$\beta_H đQ_H = -\beta_L đQ_L. \tag{11.12}$$

This equation implies that if β_H and β_L have the same sign, $đQ_H$ and $đQ_L$ must have opposite signs. If β_H and β_L have opposite signs, $đQ_H$ and $đQ_L$ must have the same sign.

The energy balance equation is

$$đW = đQ_H + đQ_L. \tag{11.13}$$

In eqs. (11.12) and (11.13), there are three variables and two equations. Eliminating one variable leaves a relationship between the other two, which is what is needed to calculate various measures of efficiency. There are two cases that we will need.

1. Eliminate $đQ_L$.
   Using eq. (11.12) to eliminate $đQ_L$ from eq. (11.13), we find

$$đW = \left(1 - \frac{\beta_H}{\beta_L}\right)đQ_H = \left(\frac{\beta_L - \beta_H}{\beta_L}\right)đQ_H. \tag{11.14}$$

2. Eliminate $đQ_H$.
   Using eq. (11.12) to eliminate $đQ_H$ from eq. (11.13), we find

$$đW = \left(1 - \frac{\beta_L}{\beta_H}\right)đQ_L = \left(\frac{\beta_H - \beta_L}{\beta_H}\right)đQ_L. \tag{11.15}$$

Three ranges of temperatures will be treated: (1) both temperatures are positive, (2) the higher temperature is negative, but the lower temperature is positive, and (3) both temperatures are negative. A summary of the directions of energy flow is given in Table 11.2.

11.8.1   Both temperatures positive, $0 < \beta_H < \beta_L$

11.8.1.1   *Heat engines*

A heat engine does work, so that $đW > 0$. Consequently, $đQ_H > 0$ and eq. (11.12) tells us that $đQ_L < 0$, which represents the heat exhaust. The signs of these energy flows are indicated in the second line of Table 11.2.

The work performed, $đW$, and the heat extracted from the higher-temperature body are both positive. Their ratio gives a measure of the efficiency of the heat engine:

Table 11.2 *Signs of work and heat flow for various inverse temperature ranges. If $đW > 0$, work is being performed. If $đQ_H > 0$, heat is being extracted from the higher-temperature body. The forward direction of operation is take as that which generates positive work.*

Range of β's	$đW$	$đQ_H$	$đQ_L$	Direction
$0 < \beta_H < \beta_L$	+	+	−	forward
	−	−	+	reverse
$\beta_H < 0 < \beta_L$	+	+	+	forward
	−	−	−	reverse
$\beta_H < \beta_L < 0$	+	−	+	forward
	−	+	−	reverse

$$\eta = \frac{dW}{dQ_H} = 1 - \frac{\beta_H}{\beta_L} = \frac{\beta_L - \beta_H}{\beta_L} < 1. \tag{11.16}$$

For positive temperatures, as the higher of the two temperatures, T_H, increases, β_H decreases, and the efficiency η approaches one.

11.8.1.2 Refrigerators and air conditioners

The essential property of a refrigerator is that heat is removed from the lower-temperature system, $dQ_L > 0$. For positive temperatures, this requires the heat engine to run backwards, as indicated in the third line of Table 11.2. From eq. (11.15), we find the ratio of the heat extracted from the lower-temperature system, dQ_L, to the work done on the refrigerator, $-dW$, which is the coefficient of performance for a refrigerator,

$$\epsilon_R = \frac{dQ_L}{-dW} = \frac{\beta_H}{\beta_L - \beta_H} > 0. \tag{11.17}$$

Coefficients of performance are not limited to a maximum value of one. In fact, ϵ_R can be very large if β_H is close in value to β_L.

11.8.1.3 Heat pumps

For positive temperatures, heat pumps are also heat engines running backwards, but with the purpose of adding heat to the higher-temperature body, $dQ_H < 0$. The coefficient of performance for a heat pump is given by the amount of heat added to the higher-temperature body, $-dQ_H$, divided by the work performed on the engine, $-dW$,

$$\epsilon_{HP} = \frac{-dQ_H}{-dW} = \frac{\beta_L}{\beta_L - \beta_H} > 0. \tag{11.18}$$

This quantity can also be significantly larger than 1 if β_H is close to β_L, which makes it preferable to use a heat pump rather than running electricity through a resistive heater. It remains larger than one for all $\beta_L > \beta_H > 0$.

11.8.1.4 Summary for $0 < \beta_H < \beta_L$ (positive temperatures)

Table 11.3 *Thermal efficiencies for $0 < \beta_H < \beta_L$ (positive temperatures).*

Heat engine	$\eta \equiv \dfrac{dW}{dQ_H}$	$\eta = \dfrac{\beta_L - \beta_H}{\beta_L} = \dfrac{T_H - T_L}{T_H}$
Refrigerator	$\epsilon_R \equiv \dfrac{dQ_L}{-dW}$	$\epsilon_R = \dfrac{\beta_H}{\beta_L - \beta_H} = \dfrac{T_L}{T_H - T_L}$
Heat pump	$\epsilon_{HP} \equiv \dfrac{-dQ_H}{-dW}$	$\epsilon_{HP} = \dfrac{\beta_L}{\beta_L - \beta_H} = \dfrac{T_H}{T_H - T_L}$

11.8.2   The lower temperature is positive, but the higher temperature is negative, $\beta_H < 0 < \beta_L$

If $\beta_H < 0$ and $\beta_L > 0$, eq. (11.12) shows that $đQ_H$ and $đQ_L$ have the same sign. If heat is extracted from the higher-temperature body, it must also be extracted from the lower-temperature body. This is indicated in the fourth line in Table 11.2, which also shows that the work is positive, $đW > 0$.

11.8.2.1   Heat engines, $\beta_H < 0 < \beta_L$

The nominal efficiency of a heat engine is $(1 - \beta_H/\beta_L)$, which is greater than one if β_H and β_L have different signs. This gives the (false) impression that the efficiency is greater than one. Actually, energy from the lower-temperature body is also contributing to the work done,

$$đW_L = đQ_L = -\left(\frac{\beta_H}{\beta_L}\right)đQ_H > 0, \tag{11.19}$$

since $\beta_H/\beta_L < 0$.

If both sources of heat are taken into account, the true efficiency is

$$\eta_{true} = \frac{đW}{đQ_H + đQ_L} = 1. \tag{11.20}$$

11.8.2.2   Refrigerators and air conditioners, $\beta_H < 0 < \beta_L$

Since the process we just described in Subsection 11.8.2.1 also extracts heat from the lower-temperature body, identifying a refrigerator as a heat engine running backwards is incorrect in this case. The process generates work, rather than requiring it ($đW > 0$). Consequently, the definition of a coefficient of efficiency is not appropriate.

11.8.2.3   Heat pumps, $\beta_H < 0 < \beta_L$

A heat pump requires $đQ_H < 0$ (heat is added to the higher-temperature body), which means that the heat engine must run in reverse ($đW < 0$ and $đQ_H < 0$), as indicated in the fifth line in Table 11.2. The coefficient of performance for a heat engine is then

$$\epsilon_{HP} = \frac{-đQ_H}{-đW} = \left(\frac{\beta_L}{\beta_L - \beta_H}\right), \tag{11.21}$$

which is exactly the same as eq. (11.18) for positive temperatures. However, $0 < \epsilon_{HP} < 1$ because the temperatures have opposite signs.

11.8.2.4 Summary for $\beta_L > 0 > \beta_H$

Table 11.4 *Thermal efficiencies for $\beta_L > 0 > \beta_H$. The normal definition of the efficiency of a heat engine gives a value greater than one, as shown on the first line of the table. A definition that includes both sources of energy from the bodies is given on the second line.*

Heat engine	If 'η' $\equiv \dfrac{\dABar W}{\dABar Q_H}$	'η' $= \left(1 - \dfrac{\beta_H}{\beta_L}\right) > 1$
	If $\eta_{true} \equiv \dfrac{\dABar W}{\dABar Q_H + \dABar Q_L}$	$\eta_{true} = 1$
Refrigerator	$\dABar W > 0$ and $\dABar Q_L > 0$	No work required for refrigeration
Heat pump	$\epsilon_{HP} \equiv \dfrac{-\dABar Q_H}{-\dABar W}$	$\epsilon_{HP} = \left(\dfrac{\beta_L}{\beta_L - \beta_H}\right) < 1.$

11.8.3 Both temperatures negative, $\beta_H < \beta_L < 0$

If both temperatures are negative, whether the engine is used as a heat engine, a refrigerator, or a heat pump, it is always run in the positive direction, that is, it always generates work, ($\dABar W > 0$, $\dABar Q_H < 0$, and $\dABar Q_L > 0$). This is shown in the second to last line in Table 11.2. Running a heat engine in the reverse direction for this temperature range, as shown in the last line of Table 11.2, does not appear to be useful.

11.8.3.1 Heat engines, $\beta_H < \beta_L < 0$

In this temperature range, eq. (11.12) implies that $\dABar Q_H$ and $\dABar Q_L$ have opposite signs. Since $|\beta_H| > |\beta_L|$, eq. (11.14) requires $\dABar Q_H < 0$. If the heat engine performs work ($\dABar W > 0$), the higher-temperature body must absorb energy rather than supplying it. The energy to run the heat engine comes from the lower-temperature body!

A reasonable measure of the efficiency of this process is given by $\dABar W/\dABar Q_L$, which can be obtained from eq. (11.15),

$$\eta_L = \frac{\dABar W}{\dABar Q_L} = \frac{\beta_H - \beta_L}{\beta_H} = 1 - \frac{\beta_L}{\beta_H} < 1. \qquad (11.22)$$

11.8.3.2 Refrigerators, $\beta_H < \beta_L < 0$

The operation of a refrigerator in this temperature range is strange, as eq. (11.15) shows,

$$\dABar W = \left(\frac{\beta_H - \beta_L}{\beta_H}\right)\dABar Q_L = \left(\frac{|\beta_H| - |\beta_L|}{|\beta_H|}\right)\dABar Q_L > 0. \qquad (11.23)$$

This refrigerator extracts heat from the lower-temperature body, *and* does work. It would not make sense to try to define a refrigerator coefficient of efficiency.

11.8.3.3 *Heat pumps,* $\beta_H < \beta_L < 0$

If the engine is operated as a heat pump, $\partial Q_H < 0$, it must still do work, $\partial W > 0$. All of the energy to heat the higher-temperature body comes from the lower-temperature body. It would not make sense to try to define a heat pump coefficient of efficiency.

11.8.3.4 *Summary,* $\beta_H < \beta_L < 0$

Table 11.5 *Thermal efficiency for $\beta_H < \beta_L < 0$. The operation of the engine as a heat engine, a refrigerator, and a heat pump is exactly the same. The work to drive the heat engine comes from the low-temperature body, so a reasonable measure of efficiency uses of heat from the low-temperatures body, η_L, which is shown in the table. Neither a refrigerator nor a heat pump requires work as a source of energy, so it does not make sense to define a coefficient of efficiency.*

Heat engine	$\eta_L \equiv \dfrac{\partial W}{\partial Q_L}$	$\eta_L = \dfrac{\beta_H - \beta_L}{\beta_H}$
Refrigerator	$\partial W > 0$ and $\partial Q_L > 0$	No work required for refrigeration
Heat pump	$\partial W > 0$ and $\partial Q_H < 0$	No work required for heat pump

11.9 Problems

....

PROBLEM 11.1
Efficiency of real heat engines

We showed that the efficiency of an ideal heat engine is given by

$$\eta_{\text{ideal}} = \frac{T_H - T_L}{T_H} = 1 - \frac{T_L}{T_H}.$$

Real heat-engines must be run at non-zero speeds, so they cannot be exactly quasi-static. Prove that the efficiency of a real heat engine is less than that of an ideal heat engine,

$$\eta < \eta_{\text{ideal}}.$$

PROBLEM 11.2
Maximum work from temperature differences

Suppose we have two buckets of water with constant heat capacities C_A and C_B, so that the relationship between the change in temperature of bucket A and the change in energy is

$$dU_A = C_A dT$$

with a similar equation for bucket B.

The buckets are initially at temperatures $T_{A,0}$ and $T_{B,0}$.

The buckets are used in conjunction with an ideal heat engine, guaranteed not to increase the world's total entropy (FBN Industries, patent applied for).

1. What is the final temperature of the water in the two buckets?
2. What is the maximum amount of work that you can derive from the heat energy in the buckets of water?
3. If you just mixed the two buckets of water together instead of using the heat engine, what would be the final temperature of the water?
4. Is the final temperature in this case higher, lower, or the same as when the heat engine is used? Explain your answer.
5. What is the change in entropy when the water in the two buckets is simply mixed together?

PROBLEM 11.3
Work from finite heat reservoirs

1. Suppose we have N objects at various initial temperatures. The objects have constant but different heat capacities $\{C_j | j = 1, \ldots, N\}$. The objects are at initial temperatures $\{T_{j,0} | j = 1, \ldots, N\}$.

   If we again have access to an ideal heat-engine, what is the maximum work we can extract from the thermal energy in these objects?

   What is the final temperature of the objects?
2. Suppose that the heat capacities of the objects were not constant, but proportional to the cube of the absolute temperature

$$\left\{ C_j(T_j) = A_j T_j^3 | j = 1, \ldots, N \right\}$$

   where the A_j are constants.

   What is the maximum work we can extract from the thermal energy in these objects?

   What is the final temperature of the objects?

12

Thermodynamic Potentials

Nobody knows why, but the only theories which work are the mathematical ones.

Michael Holt, in *Mathematics in Art*

Although the fundamental relation in either the entropy, $S = S(U,V,N)$, or energy, $U = U(S,V,N)$ (for monotonic entropy functions), representations contains all thermodynamic information about the system of interest, it is not always easy to use that information for practical calculations.

For example, many experiments are done at constant temperature, with the system in contact with a thermal reservoir. It would be very convenient to have the fundamental relation expressed in terms of the intensive parameters T and P, instead of the extensive parameters S and V. It turns out that this can be done, but it requires us to introduce new functions that are generally known as thermodynamic potentials. The mathematics required involves Legendre transforms (named after the French mathematician Adrien-Marie Legendre (1752–1833)), which will be developed in the next section. The rest of the chapter is devoted to investigating the properties and advantages of various thermodynamic potentials.

Of particular interest are Massieu functions (Section 12.6), which are Legendre transforms of the entropy.

12.1 Mathematical Digression: The Legendre Transform

Before we go through the details of the Legendre transform, it might be useful to discuss why we need it and what problems it solves.

12.1.1 The Problem of Loss of Information

The basic reason for representing the fundamental relation in terms of T, P, or μ, instead of U, V, or N, is that we want to use the derivative of a function as the independent variable. For simplicity, consider some function

$$y = y(x) \tag{12.1}$$

An Introduction to Statistical Mechanics and Thermodynamics. Robert H. Swendsen, Oxford University Press (2020).
© Robert H. Swendsen. DOI: 10.1093/oso/9780198853237.001.0001

and its derivative

$$p = \frac{dy}{dx} = p(x).$$
(12.2)

We could, of course, invert eq. (12.2) to find $x = x(p)$, and then find $y = y(x(p)) = y(p)$. Unfortunately, this procedure results in a loss of information. The reason is illustrated in Fig. 12.1, which shows that all functions of the form $y(x - x_0)$ give exactly the same function $y(p)$. If we only have $y(p)$, we have lost all information about the value of x_0.

12.1.2 Point Representations and Line Representation

To solve the problem of lost information, we can turn to an alternative representation of functions.

When we write $y = y(x)$, we are using a 'point' representation of the function; for each value of x, $y(x)$ specifies a value of y, and the two together specify a point in the x,y-plane. The set of all such points specifies the function.

However, we could also specify a function by drawing tangent lines at points along the curve, as illustrated in Fig. 12.2. The envelope of the tangent lines also reveals the function. This is known as a 'line' representation of a function.

12.1.3 Direct Legendre Transforms

The idea of a Legendre transform is to calculate the equations of the tangent lines that carry the full information about the function. The information we need for each tangent line is clearly the slope, which we want to be our new variable, and the y-intercept.

Consider a function, $y = y(x)$, which is plotted as a curve in Fig. 12.3. We can construct a straight-line tangent to any point (x, y) along the curve, which is also shown in Fig. 12.3.

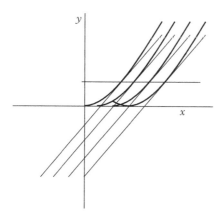

Fig. 12.1 *Illustration of distinct functions $y = y(x)$ that have the same slope at a given value of y.*

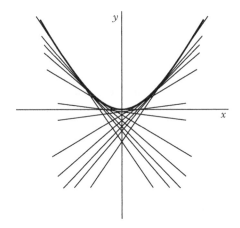

Fig. 12.2 *Illustration of how to represent a function by the envelope of tangent lines.*

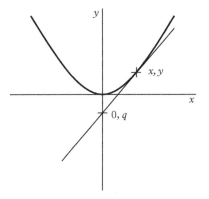

Fig. 12.3 *Graphical representation of a general Legendre transform. The straight line is tangent to the curve $y = y(x)$. The y-intercept of the tangent line is located at $(0, q)$.*

If the y-intercept is located at $(0, q)$, we can calculate the slope of the tangent line,

$$p = \frac{dy}{dx} = \frac{y - q}{x - 0} = p(x). \tag{12.3}$$

Assuming that $p = p(x)$ is monotonic, we can invert it to obtain $x = x(p)$. We can then eliminate x in favor of p in eq. (12.3) and solve for $q = q(p)$,

$$q = y - px = q(p) \equiv y[p]. \tag{12.4}$$

The square brackets in $y[p]$ indicate that $y[p] = y - px$ is the Legendre transform of $y(x)$, with p as the new independent variable.

12.1.4 Inverse Transformations

The inverse Legendre transformation proceeds in the same way. In this case,

$$x = -\frac{dq}{dp} = x(p). \tag{12.5}$$

Assuming that $x = x(p)$ is monotonic, we can invert it to obtain $p = p(x)$. Eq. (12.4) can be solved for y,

$$y = q - (-x)p = q + xp. \tag{12.6}$$

We can then eliminate p in favor of x in eq. (12.3) and solve for $y = y(x)$. There is no loss of information in either the direct or inverse Legendre transform.

The Legendre transform and its inverse are summarized in Table 12.1.

12.1.5 Legendre Transform of Infinitesimals

It will be very useful to write the information in eqs. (12.3) and (12.5) using infinitesimals,

$$dy = p\,dx \tag{12.7}$$
$$dq = -x\,dp. \tag{12.8}$$

Note that going between the original infinitesimal and the Legendre transform merely involves switching x and p and changing the sign. This simple observation will make life much easier when working with the thermodynamic potentials discussed in the rest of this chapter—assuming that you have memorized eq. (10.27).

Table 12.1 *General form of a Legendre transform.*

Direct transform	Inverse transform
$y = y(x)$	$q = q(p)$
$dy = p\,dx$	$dq = -x\,dp$
$p = \dfrac{dy}{dx}$	$-x = \dfrac{dq}{dp}$
$q = y - px$	$y = q + xp$
$q = q(p) = y[p]$	$y = y(x)$
$dq = -x\,dp$	$dy = p\,dx$

12.2 Helmholtz Free Energy

Many experiments are carried out at constant temperature, and we have thermometers to measure the temperature. However, we do not have an easy way of measuring entropy, so it would be very useful to express the fundamental relation, which contains all thermodynamic information about a system, in terms of the temperature.

We begin with the fundamental relation in the energy representation, $U = U(S, V, N)$. We wish to replace the entropy S with the temperature T as the independent variable

$$T = \left(\frac{\partial U}{\partial S} \right)_{V,N}. \tag{12.9}$$

We follow exactly the same procedure as in Section 12.1, except that we use partial derivatives and hold V and N constant. The Legendre transform of the energy with respect to temperature is called the Helmholtz free energy, and is denoted by the symbol F in this book, although you will also see it called A in some other texts,

$$F(T, V, N) \equiv U[T] = U - TS. \tag{12.10}$$

Since you have memorized the differential form of the fundamental relation in the energy representation, eq. (10.27),

$$dU = TdS - PdV + \mu dN \tag{12.11}$$

you can easily find the differential form of the fundamental relation in the Helmholtz free energy representation by exchanging S and T in the first term and reversing the sign:

$$dF = -SdT - PdV + \mu dN. \tag{12.12}$$

From eq. (12.12) we can read off the partial derivative of F with respect to the new independent variable T,

$$S = -\left(\frac{\partial F}{\partial T} \right)_{V,N}. \tag{12.13}$$

The other partial derivatives look much the same as before, but it is important to keep track of what is being held constant. For example, in the energy representation, the pressure is found from the partial derivative

$$P = -\left(\frac{\partial U}{\partial V} \right)_{S,N} \tag{12.14}$$

Table 12.2 *Helmholtz free energy.*

Direct transform	Inverse transform
$U = U(S,V,N)$	$F = F(T,V,N)$
$dU = TdS - PdV + \mu dN$	$dF = -SdT - PdV + \mu dN$
$T = \left(\dfrac{\partial U}{\partial S}\right)_{V,N}$	$-S = \left(\dfrac{\partial F}{\partial T}\right)_{V,N}$
$F = U - TS$	$U = F + TS$
$F = F(T,V,N) = U[T]$	$U = U(S,V,N)$
$dF = -SdT - PdV + \mu dN$	$dU = TdS - PdV + \mu dN$

where the entropy is held constant. In the Helmholtz free energy representation, the pressure is found from a similar partial derivative

$$P = -\left(\frac{\partial F}{\partial V}\right)_{T,N} \tag{12.15}$$

but now the temperature is held constant.

The Legendre transforms, direct and inverse, for the Helmholtz free energy are summarized in Table 12.2.

12.3 Enthalpy

The Legendre transform of the energy with respect to pressure is called the enthalpy, and is denoted by the symbol $H = U[p]$. It is often convenient because it takes the pressure as one of the independent variables. It is very widely used in chemistry for that reason, since most chemical experiments are carried out at atmospheric pressure.

We again begin with the fundamental relation in the energy representation, $U = U(S,V,N)$. This time we will replace the volume V with the pressure P as the independent variable,

$$P = -\left(\frac{\partial U}{\partial V}\right)_{S,N}. \tag{12.16}$$

To find the enthalpy we follow exactly the same procedure as in the previous section,

$$H(S,P,N) \equiv U[p] = U + PV. \tag{12.17}$$

The differential form for the enthalpy is found from the differential form of the fundamental relation in the energy representation, eq. (12.11), but we now exchange V and P and change the sign of that term,

$$dH = TdS + VdP + \mu dN. \qquad (12.18)$$

From eq. (12.18) we can read off the partial derivative of H with respect to the new independent variable P,

$$V = \left(\frac{\partial H}{\partial P}\right)_{S,N}. \qquad (12.19)$$

The Legendre transforms, direct and inverse, for the enthalpy are summarized in Table 12.3.

In some chemistry texts, the enthalpy is also referred to as the 'heat content', even though $đQ$ is not an exact differential, and we cannot properly speak of the amount of 'heat' in a system. The reason for this usage can be seen by referring to eq. (12.18). Most chemical experiments are carried out under conditions of constant pressure, with a fixed number of particles. If we recall the relationship between heat transfer and entropy change in eq. (10.24), we find that the change in enthalpy is equal to the heat transferred into the system under these conditions:

$$\begin{aligned} dH &= TdS + VdP + \mu dN \\ &= TdS + 0 + 0 \\ &= đQ. \end{aligned} \qquad (12.20)$$

This equation is a bit peculiar in that it sets an exact differential equal to an inexact differential. This came about by specifying the path of integration with the conditions

Table 12.3 *Enthalpy.*

Direct transform	Inverse transform
$U = U(S,V,N)$	$H = H(S,P,N)$
$dU = TdS - PdV + \mu dN$	$dH = TdS + VdP + \mu dN$
$P = -\left(\dfrac{\partial U}{\partial V}\right)_{S,N}$	$V = \left(\dfrac{\partial H}{\partial P}\right)_{S,N}$
$H = U + PV$	$U = H - PV$
$H = H(S,V,N) = U[p]$	$U = U(S,V,N)$
$dH = TdS + VdP + \mu dN$	$dU = TdS - PdV + \mu dN$

$dP = 0$ and $dN = 0$, which made the path one-dimensional. Since one-dimensional differentials are always exact, $dH = đQ$ along the specified path.

12.4   Gibbs Free Energy

The Gibbs free energy is the Legendre transform of the energy with respect to both the temperature and the pressure. It is denoted with the letter G, which seems remarkably logical after seeing the standard choices for the Helmholtz free energy (F or A) and the enthalpy (H). There are three ways to calculate it:

1. Find the Legendre transform of the Helmholtz free energy with respect to pressure.
2. Find the Legendre transform of the enthalpy with respect to temperature.
3. Start with the energy and carry out both Legendre transforms at the same time.

Since the first two possibilities should be self-evident, I will just give the third option in the form of Table 12.4.

Table 12.4 *Gibbs free energy.*

Direct transform	Inverse transform
$U = U(S, V, N)$	$G = G(T, P, N)$
$dU = TdS - PdV + \mu dN$	$dG = -SdT + VdP + \mu dN$
$T = \left(\dfrac{\partial U}{\partial S}\right)_{V,N}$	$-S = \left(\dfrac{\partial G}{\partial T}\right)_{P,N}$
$P = -\left(\dfrac{\partial U}{\partial V}\right)_{S,N}$	$V = \left(\dfrac{\partial G}{\partial P}\right)_{T,N}$
$G = U - TS + PV$	$U = G + TS - PV$
$G = G(T, P, N) = U[T, P]$	$U = U(S, V, N)$
$dG = -SdT + VdP + \mu dN$	$dU = TdS - PdV + \mu dN$

12.5   Other Thermodynamic Potentials

Since there are three thermodynamic variables for even the simplest system (U, V, and N), and we can construct a Legendre transform from any combination of the corresponding partial derivatives, there are $2^3 = 8$ different thermodynamic potentials. Each of these potentials provides a representation of the fundamental relation. They can all be derived by the same procedures that we used for F, H, and G.

12.6 Massieu Functions

In addition to those discussed previously, there is another class of thermodynamic potentials, known as Massieu functions, which are generated from the fundamental relation in the entropy representation, $S = S(U, V, N)$. The differential form for the fundamental relation in the entropy representation was found in eq. (10.26),

$$dS = \left(\frac{1}{T}\right) dU + \left(\frac{P}{T}\right) dV - \left(\frac{\mu}{T}\right) dN. \tag{12.21}$$

As can be seen from eq. (12.21), the natural variables for a Legendre transform are $1/T$, P/T, and μ/T. Although these variables appear rather strange, working with them is essentially the same as for the more usual thermodynamic potentials.

It is easier to use a different notation. If we use the inverse temperature $\beta = 1/k_B T$ and a dimensionless entropy $\tilde{S} = S/k_B$, as we did in Section 11.8, the differential form for the fundamental relation in the dimensionless entropy representation becomes

$$d\tilde{S} = \beta dU + (\beta P) dV - (\beta \mu) dN. \tag{12.22}$$

In this equation, we have put parentheses around (βP) and $(\beta \mu)$ to emphasize that each of these products is to be regarded as a single variable.

The Legendre transform of the dimensionless entropy with respect to β is then $\tilde{S}[\beta]$, with the differential form given as

$$d\tilde{S}[\beta] = -U d\beta + (\beta P) dV - (\beta \mu) dN. \tag{12.23}$$

Similarly, the Legendre transform with respect to both β and (βP) is given by

$$d\tilde{S}[\beta, (\beta P)] = -U d\beta - V d(\beta P) - (\beta \mu) dN. \tag{12.24}$$

We will find that both of these Massieu functions are quite useful when we discuss partition functions in Chapters 19, 20, 21, 23, and 24.

12.7 Summary of Legendre Transforms

The usefulness of Legendre transforms and thermodynamic potentials is not limited to providing alternative representations of the fundamental relation, although that is very important. We will see in Chapter 14 that they also play a crucial role in deriving thermodynamic identities; that is, equations relating measurable quantities that are true for *all* thermodynamic systems. In Chapter 15 we will see that thermodynamic potentials can be used to determine equilibrium conditions from extremum principles similar to the maximization principle for the entropy. Finally, in Chapter 16 we will find that they are essential to deriving stability criteria that are also valid for all systems.

12.8 Problems

PROBLEM 12.1
Legendre transform of the energy with respect to temperature for the classical ideal gas

Starting with the fundamental equation in the entropy representation (that is, the entropy of the classical ideal gas—which you have memorized):

$$S = S(U,V,N) = k_B N \left(\frac{3}{2} \ln\left(\frac{U}{N}\right) + \ln\left(\frac{V}{N}\right) + \ln X \right).$$

1. Derive the fundamental relation in the energy representation ($U = U(S,V,N)$).
2. Derive the temperature as a function of entropy, volume, and number of particles (one of the three equations of state).
3. Find the entropy as a function of temperature, volume, and number of particles. Is this an equation of state or a fundamental relation?
4. Derive the Helmholtz free energy of the classical ideal gas, $F(T,V,N) = U - TS$.

PROBLEM 12.2
More Legendre transforms

1. Starting with the fundamental equation in the Helmholtz free energy representation ($F = F(T,V,N)$) that you derived for the previous assignment, derive the fundamental relation in the Gibbs free energy representation ($G = G(T,P,N)$).
2. Find the volume of the ideal gas as a function of temperature, pressure, and number of particles by taking a derivative of the Gibbs free energy.

PROBLEM 12.3
General thermodynamic function

Suppose we know some terms in a series expansion of the Gibbs free energy as a function of T and P for some region near the point (T_0, P_0),

$$G = AT + BT^2 + CP + DP^2 + ETP$$

where $A, B, C, D,$ and E are constants.

1. Find the volume of the system as a function of T and P.
2. The isothermal compressibility is defined as

$$\kappa_T = -\frac{1}{V}\left(\frac{\partial V}{\partial P}\right)_{T,N}.$$

Find κ_T for this system.
3. Find the entropy as a function of T and P for this system.

PROBLEM 12.4
Enthalpy

The enthalpy of a material in a certain range of temperature and pressure is well approximated by the expression:

$$H = A + BT + CP + DT^2 + EP^2 + FTP.$$

1. Is this expression for H an approximation to the fundamental relation, or is it an equation of state? Explain your answer for credit.
2. Calculate the specific heat at constant pressure, c_P, within the approximation for H given previously.

13

The Consequences of Extensivity

Every mathematician knows it is impossible to understand an elementary course in thermodynamics.

V. I. Arnold, Soviet and Russian mathematician, 1937–2010

In this chapter we return to the thermodynamic postulates and consider the consequences of extensivity. As defined in Section 9.7.1, the entropy is extensive if

$$S(\lambda U, \lambda V, \lambda N) = \lambda S(U, V, N) \tag{13.1}$$

for any positive value of the parameter λ. The postulate of extensivity means that S, U, V, and N are all proportional to the size of the system.

The properties of additivity and extensivity are often confused. However, they are not the same. As long as the molecular interactions are short-ranged, the entropy of the macroscopic system will be additive, that is, the total entropy of systems A and B is given by the sum of the individual entropies,

$$S_{A,B}(U_A, V_A, N_A; U_B, V_B, N_B) = S_A(U_A, V_A, N_A) + S_B(U_B, V_B, N_B). \tag{13.2}$$

It is common for books on thermodynamics to assume from the beginning that all systems are homogeneous. If a system is homogeneous, then additivity automatically implies extensivity. This is the source of the confusion.

Most real systems are not extensive. A simple example is given by a container of gas that can adsorb molecules on its walls. As the size of the container is varied, the surface-to-volume ratio varies, and the fraction of the molecules adsorbed on the walls also varies. The properties of the system will depend on the surface-to-volume ratio, and eq. (13.1) will not be satisfied.

Even crystals or liquids with free boundary conditions will have contributions to the energy and free energies from surfaces and interfaces. It is important to be able to treat such systems, so that surface properties can be unambiguously defined and studied.

On the other hand, we are often interested in the bulk properties of a material and would like to be able to investigate its thermodynamic behavior without concerning ourselves with surfaces or interfaces. In these circumstances it is reasonable to consider

An Introduction to Statistical Mechanics and Thermodynamics. Robert H. Swendsen, Oxford University Press (2020).

a homogeneous and extensive system consisting entirely of the material of interest. Such a system would satisfy the postulate of extensivity.

13.1   The Euler Equation

If a system is extensive its energy is a homogeneous first-order function, and satisfies eq. (13.1), $\lambda U(S,V,N) = U(\lambda S, \lambda V, \lambda N)$, for any value of λ.

If we differentiate eq. (13.1) with respect to λ, we find

$$U(S,V,N) = \frac{\partial U(\lambda S, \lambda V, \lambda N)}{\partial(\lambda S)} \frac{\partial(\lambda S)}{\partial \lambda}$$
$$+ \frac{\partial U(\lambda S, \lambda V, \lambda N)}{\partial(\lambda V)} \frac{\partial(\lambda V)}{\partial \lambda}$$
$$+ \frac{\partial U(\lambda S, \lambda V, \lambda N)}{\partial(\lambda N)} \frac{\partial(\lambda N)}{\partial \lambda}. \qquad (13.3)$$

Setting $\lambda = 1$, this becomes

$$U(S,V,N) = \frac{\partial U(S,V,N)}{\partial S} S + \frac{\partial U(S,V,N)}{\partial V} V + \frac{\partial U(S,V,N)}{\partial N} N. \qquad (13.4)$$

If we now substitute for the partial derivatives, we find

$$U = TS - PV + \mu N \qquad (13.5)$$

which is known as the Euler equation.

Note that the Euler equation is trivial to remember if you have memorized eq. (10.27), which states: $dU = TdS - PdV + \mu dN$.

The Euler equation can also be expressed in terms of the entropy,

$$S = \left(\frac{1}{T}\right) U + \left(\frac{P}{T}\right) V - \left(\frac{\mu}{T}\right) N. \qquad (13.6)$$

or, using $\tilde{S}$ from Section 12.6,

$$\tilde{S} = \beta U + (\beta P) V - (\beta \mu) N. \qquad (13.7)$$

This equation can be found directly from the homogeneous first-order property of the entropy given in eq. (13.1), or more simply by rewriting eq. (13.5).

The most important consequence of extensivity is the Euler equation. It can be extremely useful, but it must be kept in mind that it is valid only for extensive systems.

13.2   The Gibbs–Duhem Relation

If a system is extensive, the three intensive parameters, T, P, and μ are not independent. The Gibbs–Duhem relation, which can be derived from the Euler equation, makes explicit the connection between changes in these parameters.

First, we can write the complete differential form of eq. (13.5),

$$dU = TdS - PdV + \mu dN + SdT - VdP + Nd\mu. \tag{13.8}$$

Comparing eq. (13.8) to the differential form of the First Law, eq. (10.27),

$$dU = TdS - PdV + \mu dN \tag{13.9}$$

and subtracting the one from the other, we find the Gibbs–Duhem relation:

$$0 = SdT - VdP + Nd\mu. \tag{13.10}$$

Another way of writing this is to exhibit the change in μ as a function of the changes in T and P

$$d\mu = -\left(\frac{S}{N}\right)dT + \left(\frac{V}{N}\right)dP. \tag{13.11}$$

This form of the Gibbs–Duhem relation involves the entropy per particle and the volume per particle, which might not be the most convenient quantities to work with. If we carry out the same derivation starting with the fundamental relation in the entropy representation, $S = S(U, V, N)$, we find an alternative formulation of the Gibbs–Duhem relation:

$$d\left(\frac{\mu}{T}\right) = \left(\frac{U}{N}\right)d\left(\frac{1}{T}\right) + \left(\frac{V}{N}\right)d\left(\frac{P}{T}\right). \tag{13.12}$$

Using $\tilde{S}$ and β, this becomes

$$d(\beta\mu) = \left(\frac{U}{N}\right)d\beta + \left(\frac{V}{N}\right)d(\beta P). \tag{13.13}$$

For our simple example of a one-component system, we can see that only two parameters are free. For any changes in the temperature or pressure, the change in the chemical potential is fixed by eq. (13.11). The number of free parameters needed to specify a thermodynamic system is called the number of *thermodynamic degrees of freedom* for the system. A simple, homogeneous, one-component system has two degrees of freedom.

The Euler equation and the Gibbs–Duhem relation can both be generalized to r components by including a chemical potential term $\mu_j dN_j$ for each component j in the differential form of the fundamental relation

$$dU = TdS - PdV + \sum_{j=1}^{r} \mu_j dN_j. \tag{13.14}$$

The Gibbs–Duhem relation becomes

$$0 = SdT - VdP + \sum_{j=1}^{r} N_j d\mu_j. \tag{13.15}$$

A direct consequence is that if we have an extensive system with r components, it will have $r+1$ thermodynamic degrees of freedom.

13.3   Reconstructing the Fundamental Relation

We have seen in Section 8 that although an equation of state contains thermodynamic information, the information is not complete. For a general system we can only reconstruct the fundamental relation and obtain complete information if we know all three equations of state for a one-component system, or all $r+2$ equations of state for a system with r components.

An important consequence of the Gibbs–Duhem relation is that for extensive systems we only need $r+1$ equations of state for an r-component system to recover the fundamental relation and with it access to all thermodynamic information.

As an example, consider the classical, one-component ideal gas. Suppose we were to know only two of the three equations of state,

$$PV = Nk_B T \tag{13.16}$$

and

$$U = \frac{3}{2} Nk_B T. \tag{13.17}$$

If we can calculate the third equation of state, we can substitute it into the Euler equation to find the fundamental relation.

To carry out this project, it is convenient to introduce the volume per particle $v = V/N$ and the energy per particle $u = U/N$. The ideal gas law in the form

$$\frac{P}{T} = k_B v^{-1}$$
(13.18)

gives us

$$d\left(\frac{P}{T}\right) = -k_B v^{-2} dv$$
(13.19)

and the energy equation

$$\frac{1}{T} = \frac{3}{2} k_B u^{-1}$$
(13.20)

gives us

$$d\left(\frac{1}{T}\right) = -\frac{3}{2} k_B u^{-2} du.$$
(13.21)

Inserting eqs. (13.19) and (13.21) into the Gibbs–Duhem relation, eq. (13.12),

$$d\left(\frac{\mu}{T}\right) = ud\left(\frac{1}{T}\right) + v d\left(\frac{P}{T}\right),$$
(13.22)

we find

$$d\left(\frac{\mu}{T}\right) = -u\frac{3}{2} k_B u^{-2} du - v k_B v^{-2} dv = -\frac{3}{2} k_B u^{-1} du - k_B v^{-1} dv.$$
(13.23)

Integrating eq. (13.23) we obtain the third equation of state,

$$\frac{\mu}{T} = -\frac{3}{2} k_B \ln(u) - k_B \ln(v) + X.$$
(13.24)

An arbitrary constant of integration, X, is included in this equation, which reflects the arbitrary constant in classical expressions for the entropy. Putting these results together to find a fundamental relation is left as an exercise.

13.4 Thermodynamic Potentials

The Euler equation (13.5) puts strong restrictions on thermodynamic potentials introduced in Chapter 12 for extensive systems in the form of alternative expressions. For the thermodynamic potentials F, H, and G, we have the following identities for extensive systems:

$$F = U - TS = -PV + \mu N \tag{13.25}$$
$$H = U + PV = TS + \mu N \tag{13.26}$$
$$G = U - TS + PV = \mu N. \tag{13.27}$$

Note that eq. (13.27) expresses the chemical potential as the Gibbs free energy per particle for extensive systems

$$\mu = \frac{G}{N}. \tag{13.28}$$

The thermodynamic potential $U[T,P,\mu]$ vanishes for extensive systems,

$$U[T,P,\mu] = U - TS + PV - \mu N = 0. \tag{13.29}$$

For this reason, $U[T,P,\mu]$ is often omitted in books on thermodynamics that restrict the discussion to extensive systems. However, $U[T,P,\mu]$ does not vanish for general thermodynamic systems, and it is sensitive to surface and interface properties.

14

Thermodynamic Identities

Thermodynamics is a funny subject. The first time you go through it, you don't understand it at all. The second time you go through it, you think you understand it, except for one or two small points. The third time you go through it, you know you don't understand it, but by that time you are so used to it, it doesn't bother you any more.

Arnold Sommerfeld

14.1 Small Changes and Partial Derivatives

Many of the questions that arise in thermodynamics concern the effects of small perturbations on the values of the parameters describing the system. Even when the interesting questions concern large changes, the best approach to calculations is usually through adding up a series of small changes. For this reason, much of thermodynamics is concerned with the response of a system to small perturbations. Naturally, this takes the mathematical form of calculating partial derivatives.

Ultimately, the values of some partial derivatives must be found either from experiment or from the more fundamental theory of statistical mechanics. However, the power of thermodynamics is that it is able to relate different partial derivatives through general identities that are valid for all thermodynamic systems. Then, we might be able to relate a partial derivative that is difficult to calculate or measure to one that is easy to calculate or measure. In this chapter we will develop the mathematical tools needed to derive such thermodynamic identities.

14.2 A Warning about Partial Derivatives

For much of physics, partial derivatives are fairly straightforward: if we want to calculate the partial derivative with respect to some variable, we treat all other variables as constants and take the derivative in the usual way.

However, in thermodynamics it is rarely obvious what the 'other' variables are. Consequently, it is *extremely* important to specify explicitly which variables are being held constant when taking a partial derivative.

An Introduction to Statistical Mechanics and Thermodynamics. Robert H. Swendsen, Oxford University Press (2020).
© Robert H. Swendsen. DOI: 10.1093/oso/9780198853237.001.0001

For example, consider the innocent-looking partial derivative, $\partial U / \partial V$, for a classical ideal gas. If we hold T constant,

$$\left(\frac{\partial U}{\partial V}\right)_{T,N} = \frac{\partial}{\partial V}\left(\frac{3}{2}Nk_BT\right) = 0, \tag{14.1}$$

but if we hold P constant,

$$\left(\frac{\partial U}{\partial V}\right)_{P,N} = \frac{\partial}{\partial V}\left(\frac{3}{2}PV\right) = \frac{3}{2}P. \tag{14.2}$$

When S is held constant we can use the equation $dU = TdS - PdV$ to see that

$$\left(\frac{\partial U}{\partial V}\right)_{S,N} = -P. \tag{14.3}$$

We find three different answers—one positive, one negative, and one zero—just from changing what we are holding constant!

Don't forget to write explicitly what is being held constant in a partial derivative!

14.3   First and Second Derivatives

Since you memorized eq. (10.27) long ago, you know the first derivatives of the fundamental relation in the energy representation,

$$\left(\frac{\partial U}{\partial S}\right)_{V,N} = T \tag{14.4}$$

$$\left(\frac{\partial U}{\partial V}\right)_{S,N} = -P \tag{14.5}$$

$$\left(\frac{\partial U}{\partial N}\right)_{S,V} = \mu. \tag{14.6}$$

Since you also know how to find the Legendre transform of the differential form of a fundamental relation, you know the first derivatives of the Gibbs free energy:

$$\left(\frac{\partial G}{\partial T}\right)_{P,N} = -S \tag{14.7}$$

$$\left(\frac{\partial G}{\partial P}\right)_{T,N} = V \tag{14.8}$$

$$\left(\frac{\partial G}{\partial N}\right)_{T,P} = \mu. \tag{14.9}$$

Essentially, we find various subsets of U, S, T, V, P, μ, and N when we take a first partial derivative of the fundamental relation. Second derivatives provide information about how these quantities change when other quantities are varied.

There are clearly many different second derivatives that can be constructed from all the variables we have introduced—especially when we remember that holding different quantities constant produces different partial derivatives. However, the number of *independent* second derivatives is limited. This is the fundamental reason for the existence of thermodynamic identities.

If we consider the fundamental relation in the energy representation for a simple system with a single component, first derivatives can be taken with respect to the three independent variables: S, V, and N. That means that there are a total of six independent second derivatives that can be formed. All other second derivatives must be functions of these six.

In many applications the composition of the system is fixed, so that there are only two independent variables. This means that even if we are dealing with a multi-component system, there are only three independent second derivatives. For example, if we consider $U(S, V, N)$ with N held constant, the independent second derivatives are:

$$\left(\frac{\partial^2 U}{\partial S^2}\right), \tag{14.10}$$

$$\left(\frac{\partial^2 U}{\partial S \partial V}\right) = \left(\frac{\partial^2 U}{\partial V \partial S}\right), \tag{14.11}$$

and

$$\left(\frac{\partial^2 U}{\partial V^2}\right). \tag{14.12}$$

This is a key observation, because it implies that only three measurements are needed to predict the results of *any* experiment involving small changes. This can be crucial if the quantity you want to determine is very difficult to measure directly. By relating it to things that are easy to measure, you can both increase accuracy and save yourself a lot of work.

On the other hand, knowing that everything is related is not the same as knowing what that relationship is. The subject of thermodynamic identities is the study of how to reduce any partial derivative to a function of a convenient set of standard partial derivatives.

14.4   Standard Set of Second Derivatives

The three standard second derivatives are:

The coefficient of thermal expansion

$$\alpha = \frac{1}{V}\left(\frac{\partial V}{\partial T}\right)_{P,N}.$$

(14.13)

The isothermal compressibility

$$\kappa_T = -\frac{1}{V}\left(\frac{\partial V}{\partial P}\right)_{T,N}.$$

(14.14)

The minus sign in the definition is there to make κ_T positive, since the volume decreases when pressure increases.

The specific heat per particle at constant pressure

$$c_P = \frac{T}{N}\left(\frac{\partial S}{\partial T}\right)_{P,N}.$$

(14.15)

The specific heat per particle at constant volume

$$c_V = \frac{T}{N}\left(\frac{\partial S}{\partial T}\right)_{V,N}.$$

(14.16)

Well, yes, I have listed four of them.

However, because there are only three independent second derivatives, we will be able to find a universal equation linking them. Usually, the first three are the easiest to measure, and they are taken as fundamental. Trying to measure c_V for iron is difficult, but c_P is easy. However, measuring c_V for a gas might be easier than measuring c_P, so it has been included.

If we are interested in the properties of a particular system we might also use the 'heat capacity', denoted by a capital C; that is, $C_P = N c_P$ and $C_V = N c_V$.

It should be noted that I have defined c_P and c_V as specific heats per particle. It is also quite common to speak of the specific heat per mole, for which division by the number of moles would appear in the definition instead of division by N. It is also sometimes useful to define a specific heat per unit mass.

Our goal for the rest of the chapter is to develop methods for reducing all possible thermodynamic second derivatives to combinations of the standard set.

The four second derivatives listed in eqs. (14.13) through (14.16) should be memorized. While they are not as important as eq. (10.27), it is useful not to have to consult them when deriving thermodynamic identities. The brief time spent memorizing them will be repaid handsomely.

14.5 Maxwell Relations

The first technique we need is a direct application of what you learned about exact differentials in Section 10.3.1. Since every differential representation of the fundamental relation is an exact differential, we can apply eq. (10.11) to all of them.

For example, start with the differential form of the fundamental relation in the energy representation, eq. (10.27), which you have long since memorized,

$$dU = TdS - PdV + \mu dN. \tag{14.17}$$

If N is held constant, this becomes

$$dU = TdS - PdV. \tag{14.18}$$

Applying the condition for dU being an exact differential produces a new identity,

$$\left(\frac{\partial T}{\partial V}\right)_{S,N} = -\left(\frac{\partial P}{\partial S}\right)_{V,N}. \tag{14.19}$$

Eq. (14.19) and all other identities derived in the same way are called Maxwell relations. They all depend on the condition in eq. (10.11) that the differential of a thermodynamic potential be an exact differential.

The thing that makes Maxwell relations so easy to use is that the differential forms of the fundamental relation in different representations are all simply related. For example, to find dF from dU, simply switch T and S in eq. (14.17) and change the sign of that term,

$$dF = -SdT - PdV + \mu dN. \tag{14.20}$$

From this equation, we find another Maxwell relation,

$$-\left(\frac{\partial S}{\partial V}\right)_{T,N} = -\left(\frac{\partial P}{\partial T}\right)_{V,N}. \tag{14.21}$$

It is even easy to find exactly the right Maxwell relation, given a particular partial derivative that you want to transform, by finding the right differential form of the fundamental relation.

For example, suppose you want a Maxwell relation to transform the following partial derivative:

$$\left(\frac{\partial T}{\partial P}\right)_{S,\mu} = \left(\frac{\partial ?}{\partial ?}\right)_{?,?}. \tag{14.22}$$

Begin with eq. (14.17) and choose the Legendre transform that

- leaves T in front of dS, so that T is being differentiated;
- changes $-PdV$ to VdP, so that the derivative is with respect to P; and
- changes μdN to $-Nd\mu$, so that μ is held constant.

The result is the differential form of the fundamental equation in the $U[P,\mu]$ representation

$$dU[P,\mu] = TdS + VdP - Nd\mu. \tag{14.23}$$

Now apply eq. (10.11) with $d\mu = 0$,

$$\left(\frac{\partial T}{\partial P}\right)_{S,\mu} = \left(\frac{\partial V}{\partial S}\right)_{P,\mu}. \tag{14.24}$$

Note that you do not even have to remember which representation you are transforming into to find Maxwell relations. The thermodynamic potential $U[P,\mu]$ is not used sufficiently to have been given a separate name, but you do not need to know its name to derive the Maxwell relation in eq. (14.24).

You do have to remember that the Legendre transform changes the sign of the differential term containing the transformed variables. But that is easy.

14.6 Manipulating Partial Derivatives

Unfortunately, Maxwell relations are not sufficient to derive all thermodynamic identities. We will still need to manipulate partial derivatives to put them into more convenient forms. A very elegant way of doing this is provided by the use of Jacobians.

We do have to introduce some new mathematics at this point, but the result will be that thermodynamic identities become quite easy to derive—perhaps even fun.

14.6.1 Definition of Jacobians

Jacobians are defined as the determinant of a matrix of derivatives:

$$\frac{\partial(u,v)}{\partial(x,y)} = \begin{vmatrix} \dfrac{\partial u}{\partial x} & \dfrac{\partial u}{\partial y} \\ \dfrac{\partial v}{\partial x} & \dfrac{\partial v}{\partial y} \end{vmatrix} = \frac{\partial u}{\partial x}\frac{\partial v}{\partial y} - \frac{\partial u}{\partial y}\frac{\partial v}{\partial x}. \tag{14.25}$$

Since we have only the two variables, x and y, in this definition, I have suppressed the explicit indication of which variables are being held constant. This is an exception that should not be imitated.

The definition in eq. (14.25) can be extended to any number of variables, but there must be the same number of variables in the numerator and denominator; the determinant must involve a square matrix.

14.6.2   Symmetry of Jacobians

The Jacobian changes sign when any two variables in the numerator or the denominator are exchanged, because the sign of a determinant changes when two rows or two columns are switched. (Variables *cannot* be exchanged between numerator and denominator.)

$$\frac{\partial (u, v)}{\partial (x, y)} = -\frac{\partial (v, u)}{\partial (x, y)} = \frac{\partial (v, u)}{\partial (y, x)} = -\frac{\partial (u, v)}{\partial (y, x)}. \tag{14.26}$$

This also works for larger Jacobians, although we will rarely need them,

$$\frac{\partial (u, v, w, s)}{\partial (x, y, z, t)} = -\frac{\partial (v, u, w, s)}{\partial (x, y, z, t)} = \frac{\partial (v, u, w, s)}{\partial (y, x, z, t)}. \tag{14.27}$$

14.6.3   Partial Derivatives and Jacobians

Consider the Jacobian

$$\frac{\partial (u, y)}{\partial (x, y)} = \begin{vmatrix} \dfrac{\partial u}{\partial x} & \dfrac{\partial u}{\partial y} \\ \dfrac{\partial y}{\partial y} & \dfrac{\partial y}{\partial y} \\ \dfrac{\partial y}{\partial x} & \dfrac{\partial y}{\partial y} \end{vmatrix}. \tag{14.28}$$

Since

$$\frac{\partial y}{\partial x} = 0 \tag{14.29}$$

and

$$\frac{\partial y}{\partial y} = 1 \tag{14.30}$$

we have

$$\frac{\partial (u, y)}{\partial (x, y)} = \left(\frac{\partial u}{\partial x} \right)_y. \tag{14.31}$$

This is the link between partial derivatives in thermodynamics and Jacobians.

We will also see it in an extended form with more variables held constant,

$$\frac{\partial(u, y, z)}{\partial(x, y, z)} = \left(\frac{\partial u}{\partial x}\right)_{y,z}. \tag{14.32}$$

For example:

$$\frac{\partial(V, T, N)}{\partial(P, T, N)} = \left(\frac{\partial V}{\partial P}\right)_{T,N}. \tag{14.33}$$

14.6.4   A Chain Rule for Jacobians

The usual chain rule for derivatives takes on a remarkably simple form for Jacobians:

$$\frac{\partial(u, v)}{\partial(x, y)} = \frac{\partial(u, v)}{\partial(r, s)} \frac{\partial(r, s)}{\partial(x, y)}. \tag{14.34}$$

To prove this, it is convenient to introduce a more compact notation,

$$\frac{\partial u}{\partial x} \equiv u_x. \tag{14.35}$$

Now we can write

$$
\begin{aligned}
\frac{\partial(u, v)}{\partial(r, s)} \frac{\partial(r, s)}{\partial(x, y)} &= \begin{vmatrix} u_r & u_s \\ v_r & v_s \end{vmatrix} \begin{vmatrix} r_x & r_y \\ s_x & s_y \end{vmatrix} \\
&= \begin{vmatrix} u_r r_x + u_s s_x & u_r r_y + u_s s_y \\ v_r r_x + v_s s_x & v_r r_y + v_s s_y \end{vmatrix} \\
&= \begin{vmatrix} u_x & u_y \\ v_x & v_y \end{vmatrix} \\
&= \frac{\partial(u, v)}{\partial(x, y)}
\end{aligned} \tag{14.36}
$$

which proves eq. (14.34).

14.6.5   Products of Jacobians

The following identities are very useful:

$$\frac{\partial(u, v)}{\partial(x, y)} \frac{\partial(a, b)}{\partial(c, d)} = \frac{\partial(u, v)}{\partial(c, d)} \frac{\partial(a, b)}{\partial(x, y)} = \frac{\partial(a, b)}{\partial(x, y)} \frac{\partial(u, v)}{\partial(c, d)}. \tag{14.37}$$

To prove them, we use eq. (14.34) which we showed to be an identity in the previous subsection,

$$
\begin{aligned}
\frac{\partial (u,v)}{\partial (x,y)} \frac{\partial (a,b)}{\partial (c,d)} &= \frac{\partial (u,v)}{\partial (r,s)} \frac{\partial (r,s)}{\partial (x,y)} \frac{\partial (a,b)}{\partial (r,s)} \frac{\partial (r,s)}{\partial (c,d)} \\
&= \frac{\partial (u,v)}{\partial (r,s)} \frac{\partial (r,s)}{\partial (c,d)} \frac{\partial (a,b)}{\partial (r,s)} \frac{\partial (r,s)}{\partial (x,y)} \\
&= \frac{\partial (u,v)}{\partial (c,d)} \frac{\partial (a,b)}{\partial (x,y)}.
\end{aligned}
\tag{14.38}
$$

For many purposes, eq. (14.37) allows expressions of the form $\partial (x,y)$ to be manipulated almost as if they were algebraic factors.

14.6.6 Reciprocals of Jacobians

Reciprocals of Jacobians are both simple and useful.

Write the identity in eq. (14.34) with (x,y) set equal to (u,v),

$$
\frac{\partial (u,v)}{\partial (u,v)} = \frac{\partial (u,v)}{\partial (r,s)} \frac{\partial (r,s)}{\partial (u,v)}
\tag{14.39}
$$

But

$$
\frac{\partial (u,v)}{\partial (u,v)} = \begin{vmatrix} u_u & u_v \\ v_u & v_v \end{vmatrix} = \begin{vmatrix} 1 & 0 \\ 0 & 1 \end{vmatrix} = 1,
\tag{14.40}
$$

so that

$$
\frac{\partial (u,v)}{\partial (r,s)} = 1 \Big/ \frac{\partial (r,s)}{\partial (u,v)}.
\tag{14.41}
$$

Using eq. (14.41) for reciprocals, we can express the chain rule from the previous section in a particularly useful alternative form

$$
\frac{\partial (u,y)}{\partial (x,y)} = \frac{\partial (u,y)}{\partial (r,s)} \Big/ \frac{\partial (x,y)}{\partial (r,s)}.
\tag{14.42}
$$

Note that eq. (14.41), combined with eq. (14.31), gives us a useful identity for partial derivatives

$$
\left(\frac{\partial u}{\partial x} \right)_y = 1 \Big/ \left(\frac{\partial x}{\partial u} \right)_y.
\tag{14.43}
$$

An important application of eq. (14.43) can occur when you are looking for a Maxwell relation for a derivative like the following:

$$\left(\frac{\partial V}{\partial S}\right)_{T,N}. \tag{14.44}$$

If you try to find a transform of the fundamental relation, $dU = TdS - Pdv + \mu dN$, you run into trouble because S is in the denominator of eq. (14.44) and T is being held constant. However, we use eq. (14.41) to take the reciprocal of eq. (14.44), for which a Maxwell relation can be found,

$$1 \Bigg/ \left(\frac{\partial V}{\partial S}\right)_{T,N} = \left(\frac{\partial S}{\partial V}\right)_{T,N} = \left(\frac{\partial P}{\partial T}\right)_{V,N}. \tag{14.45}$$

14.7   Working with Jacobians

As a first example of how to use Jacobians to derive thermodynamic identities, consider the partial derivative of pressure with respect to temperature, holding V and N constant,

$$\left(\frac{\partial P}{\partial T}\right)_{V,N}. \tag{14.46}$$

To simplify the notation, we will suppress the explicit subscript N.

The first step is always to take the partial derivative you wish to simplify, and express it as a Jacobian using eq. (14.31),

$$\left(\frac{\partial P}{\partial T}\right)_V = \frac{\partial (P, V)}{\partial (T, V)}. \tag{14.47}$$

The next step is usually to insert $\partial(P, T)$ into the Jacobian, using either eq. (14.34) or eq. (14.42). The reason is that the second derivatives given in Section 14.4—which constitute the standard set—are all derivatives with respect to P, holding T constant, or with respect to T, holding P constant. Inserting $\partial(P, T)$ is a step in the right direction in either case,

$$\frac{\partial (P, V)}{\partial (T, V)} = \frac{\partial (P, V)}{\partial (P, T)} \Bigg/ \frac{\partial (T, V)}{\partial (P, T)}. \tag{14.48}$$

Next, we have to line up the variables in the Jacobians to produce the correct signs, remembering to change the sign of the Jacobian every time we exchange variables,

$$\frac{\partial (P, V)}{\partial (T, V)} = -\frac{\partial (P, V)}{\partial (P, T)} \Bigg/ \frac{\partial (V, T)}{\partial (P, T)} = -\frac{\partial (V, P)}{\partial (T, P)} \Bigg/ \frac{\partial (V, T)}{\partial (P, T)}. \tag{14.49}$$

The minus sign came from switching T and V in the second factor. I also switched both top and bottom in the first factor to bring it into the same order as the form given above. This is not really necessary, but it might be helpful as a memory aid.

Next, we can transform back to partial derivatives,

$$\left(\frac{\partial P}{\partial T}\right)_V = -\left(\frac{\partial V}{\partial T}\right)_P \bigg/ \left(\frac{\partial V}{\partial P}\right)_T. \tag{14.50}$$

Using the standard expressions for the coefficient of thermal expansion

$$\alpha = \frac{1}{V}\left(\frac{\partial V}{\partial T}\right)_{P,N} \tag{14.51}$$

and the isothermal compressibility

$$\kappa_T = -\frac{1}{V}\left(\frac{\partial V}{\partial P}\right)_{T,N} \tag{14.52}$$

eq. (14.50) becomes

$$\left(\frac{\partial P}{\partial T}\right)_V = -V\alpha\big/(-V\kappa_T) = \frac{\alpha}{\kappa_T}. \tag{14.53}$$

To see why eq. (14.53) might be useful, consider the properties of lead, which has a fairly small coefficient of expansion

$$\alpha(Pb) = 8.4 \times 10^{-5}\,K^{-1}. \tag{14.54}$$

Lead also has a small isothermal compressibility

$$\kappa_T(Pb) = 2.44 \times 10^{-6}\,Atm^{-1}. \tag{14.55}$$

Inserting these values into eq. (14.53), we find the change in pressure with temperature for a constant volume for lead

$$\left(\frac{\partial P}{\partial T}\right)_{V,N} = 34.3\,Atm/K. \tag{14.56}$$

As might be expected, it is not easy to maintain lead at a constant volume as the temperature increases. A direct experiment to measure the change in pressure at constant volume would be extremely difficult. However, since α and κ_T are known, we can obtain the result in eq. (14.56) quite easily.

This is a relatively simple case. The most common additional complexity is that you might have to use a Maxwell relation after you have transformed the derivatives, but it must be confessed that some identities can be quite challenging to derive.

14.8   Examples of Identity Derivations

The following two examples illustrate further techniques for proving thermodynamic identities using Jacobians. They are of interest both for the methods used and for the identities themselves.

14.8.1   The Joule–Thomson Effect

In a 'throttling' procedure, gas is continuously forced through a porous plug with a high-pressure P_A on one side and a low-pressure P_B on the other. The initial and final states are in equilibrium. If we consider a volume V_A of gas on the left, it will take up a volume V_B after it passes to the right. The energy of this amount of gas on the right is determined by conservation of energy:

$$U_B = U_A + P_A V_A - P_B V_B. \tag{14.57}$$

Rearranging this equation, we find that the enthalpy is unchanged by the process, even though the process is clearly irreversible,

$$H_A = U_A + P_A V_A = U_B + P_B V_B = H_B. \tag{14.58}$$

If the pressure change is small, the temperature change is given by a partial derivative

$$dT = \left(\frac{\partial T}{\partial P}\right)_{H,N} dP. \tag{14.59}$$

The partial derivative in eq. (14.59) is called the Joule–Thomson coefficient, μ_{JT}.[1]

$$\mu_{JT} = \left(\frac{\partial T}{\partial P}\right)_{H,N}. \tag{14.60}$$

We can express the Joule–Thomson coefficient in terms of the standard set of derivatives using Jacobians

---

[1] James Prescott Joule (1818–1889) was a British physicist who had worked on the free expansion of gases, and William Thomson (1824–1907) was an Irish physicist who continued along the lines of Joule's investigations and discovered the Joule–Thomson effect in 1852. Thomson was later elevated to the peerage with the title Baron Kelvin, and the Joule–Thomson effect is therefore sometimes called the Joule–Kelvin effect.

$$\left(\frac{\partial T}{\partial P}\right)_{H,N} = \frac{\partial(T,H)}{\partial(P,H)}$$

$$= \frac{\partial(T,H)}{\partial(P,T)} \bigg/ \frac{\partial(P,H)}{\partial(P,T)}$$

$$= -\left(\frac{\partial H}{\partial P}\right)_{T,N} \bigg/ \left(\frac{\partial H}{\partial T}\right)_{P,N}. \tag{14.61}$$

The derivative with respect to P can be evaluated by using the differential form of the fundamental relation in the enthalpy representation, (The term μdN is omitted because the total number of particles is fixed.)

$$dH = TdS + VdP \tag{14.62}$$

$$\left(\frac{\partial H}{\partial P}\right)_{T,N} = T\left(\frac{\partial S}{\partial P}\right)_{T,N} + V\left(\frac{\partial P}{\partial P}\right)_{T,N}$$

$$= -T\left(\frac{\partial V}{\partial T}\right)_{P,N} + V$$

$$= -TV\alpha + V$$

$$= -V(T\alpha - 1). \tag{14.63}$$

The proof of the Maxwell relation used in this derivation will be left as an exercise.

We can also transform the other partial derivative in eq. (14.61) using eq. (14.62),

$$\left(\frac{\partial H}{\partial T}\right)_{P,N} = T\left(\frac{\partial S}{\partial T}\right)_{P,N} + V\left(\frac{\partial P}{\partial T}\right)_{P,N}. \tag{14.64}$$

$$= Nc_P.$$

Putting eqs. (14.61), (14.63), and (14.64) together, we find our final expression for the Joule–Thomson coefficient

$$\mu_{JT} = \frac{V}{Nc_P}(T\alpha - 1). \tag{14.65}$$

The Joule–Thomson effect is central to most refrigeration and air-conditioning systems. When the Joule–Thomson coefficient is positive, the drop in pressure across the porous plug produces a corresponding drop in temperature, for which we are often grateful in the summer.

14.8.2 c_P and c_V

The thermodynamic identity linking c_P and c_V has already been alluded to in Section 14.4. It is central to the reduction to the *three* standard partial derivatives by eliminating c_V. This derivation is one of the relatively few times that an expansion of the Jacobian turns out to be useful in proving thermodynamic identities.

Begin the derivation with c_V, suppressing the subscript N for simplicity,

$$
\begin{aligned}
c_V &= \frac{T}{N}\left(\frac{\partial S}{\partial T}\right)_V \\
&= \frac{T}{N}\frac{\partial(S,V)}{\partial(T,V)} \\
&= \frac{T}{N}\frac{\partial(S,V)}{\partial(T,P)}\bigg/\frac{\partial(T,V)}{\partial(T,P)}.
\end{aligned}
\tag{14.66}
$$

The denominator of the last expression can be recognized as the compressibility,

$$
\frac{\partial(T,V)}{\partial(T,P)} = \frac{\partial(V,T)}{\partial(P,T)} = \left(\frac{\partial V}{\partial P}\right)_T = -V\kappa_T.
\tag{14.67}
$$

The Jacobian in the numerator of the last expression in eq. (14.66) can be expressed by expanding the defining determinant

$$
\frac{\partial(S,V)}{\partial(T,P)} = \left(\frac{\partial S}{\partial T}\right)_P\left(\frac{\partial V}{\partial P}\right)_T - \left(\frac{\partial S}{\partial P}\right)_T\left(\frac{\partial V}{\partial T}\right)_P.
\tag{14.68}
$$

Now we can identify each of the four partial derivatives on the right of eq. (14.68):

$$
\left(\frac{\partial S}{\partial T}\right)_P = \frac{Nc_P}{T}
\tag{14.69}
$$

$$
\left(\frac{\partial V}{\partial P}\right)_T = -V\kappa_T
\tag{14.70}
$$

$$
\left(\frac{\partial S}{\partial P}\right)_T = -\left(\frac{\partial V}{\partial T}\right)_P = -\alpha V
\tag{14.71}
$$

$$
\left(\frac{\partial V}{\partial T}\right)_P = \alpha V.
\tag{14.72}
$$

In eq. (14.71), a Maxwell relation has been used to relate the partial derivative to α.

Putting the last six equations together, we obtain

$$
c_P = c_V + \frac{\alpha^2 TV}{N\kappa_T}
\tag{14.73}
$$

which is the required identity.

Although it should not be necessary to memorize eq. (14.73), you should be able to derive it.

14.9 General Strategy

This section summarizes a useful strategy to reduce a given partial derivative to an algebraic expression containing only the three partial derivatives given in Section 14.4. These steps might be needed for attacking the most difficult derivations, but not all steps are necessary in most cases.

1. Express the partial derivative as a Jacobian.
2. If there are any thermodynamic potentials in the partial derivative, bring them to the numerator by applying eq. (14.42). Unless you have a good reason to do something else, insert $\partial(T, P)$ in this step.
3. Eliminate thermodynamic potentials if you know the derivative. For example:

$$\left(\frac{\partial F}{\partial T}\right)_{V,N} = -S.$$

4. Eliminate thermodynamic potentials by using the differential form of the fundamental relation. For example: if you want to evaluate

$$\left(\frac{\partial F}{\partial P}\right)_{T,N}$$

use

$$dF = -SdT - PdV + \mu dN$$

to find

$$\left(\frac{\partial F}{\partial P}\right)_{T,N} = -S\left(\frac{\partial T}{\partial P}\right)_{T,N} - P\left(\frac{\partial V}{\partial P}\right)_{T,N} + \mu\left(\frac{\partial N}{\partial P}\right)_{T,N}.$$

Note that the first and last terms on the right in this example vanish.

5. If the system is extensive, bring μ to the numerator and eliminate it using the Gibbs–Duhem relation, eq. (13.11)

$$d\mu = -\left(\frac{S}{N}\right)dT + \left(\frac{V}{N}\right)dP.$$

For example:

$$\left(\frac{\partial \mu}{\partial V}\right)_{S,N} = -\left(\frac{S}{N}\right)\left(\frac{\partial T}{\partial V}\right)_{S,N} + \left(\frac{V}{N}\right)\left(\frac{\partial P}{\partial V}\right)_{S,N}.$$

6. Move the entropy to the numerator using Jacobians and eliminate it by either identifying the partial derivative of the entropy as a specific heat if the derivative is with respect to T, or using a Maxwell relation if the derivative is with respect to pressure.

7. Bring V into the numerator and eliminate the partial derivative in favor of α or κ_T.

8. Eliminate c_V in favor of c_P, using the identity in eq. (14.73), derived in the previous section. (This last step is not always needed, since c_V is sometimes easier to measure than c_P.)

14.10   Problems

PROBLEM 14.1
Maxwell relations

Note: This assignment is useless if you just look up the answers in a textbook. You will not have a book during examinations, so you should do this assignment with book and notes closed.

Transform the following partial derivatives using Maxwell relations:

1.
$$\left(\frac{\partial \mu}{\partial V}\right)_{S,N} = \left(\frac{\partial ?}{\partial ?}\right)_{?,?}$$

2.
$$\left(\frac{\partial \mu}{\partial V}\right)_{T,N} = \left(\frac{\partial ?}{\partial ?}\right)_{?,?}$$

3.
$$\left(\frac{\partial S}{\partial P}\right)_{T,N} = \left(\frac{\partial ?}{\partial ?}\right)_{?,?}$$

4.
$$\left(\frac{\partial V}{\partial S}\right)_{P,N} = \left(\frac{\partial ?}{\partial ?}\right)_{?,?}$$

5.
$$\left(\frac{\partial N}{\partial P}\right)_{S,\mu} = \left(\frac{\partial ?}{\partial ?}\right)_{?,?}$$

6.
$$\left(\frac{\partial P}{\partial T}\right)_{S,N} = \left(\frac{\partial ?}{\partial ?}\right)_{?,?}$$

7.
$$\left(\frac{\partial N}{\partial P}\right)_{S,V} = \left(\frac{\partial ?}{\partial ?}\right)_{?,?}.$$

PROBLEM 14.2
A thermodynamic identity

Express the following thermodynamic derivative in terms of α, κ_T, c_V, and c_P using Jacobians.

$$\left(\frac{\partial T}{\partial P}\right)_{S,N} = ?.$$

PROBLEM 14.3
Prove the following thermodynamic identity

$$\left(\frac{\partial c_V}{\partial V}\right)_{T,N} = \frac{T}{N}\left(\frac{\partial^2 P}{\partial T^2}\right)_{V,N}.$$

PROBLEM 14.4
Express the following partial derivative in terms of the usual standard quantities

$$\left(\frac{\partial F}{\partial S}\right)_{T,N}.$$

PROBLEM 14.5
Yet another thermodynamic identity

Prove the following thermodynamic derivative and determine what should replace the question marks. Note that you do not have to reduce this to the standard expressions, but can leave it in terms of partial derivatives

$$c_V = -\frac{T}{N}\left(\frac{\partial P}{\partial T}\right)_{V,N}\left(\frac{\partial ?}{\partial ?}\right)_{S,N}.$$

PROBLEM 14.6
Compressibility identity

In analogy to the isothermal compressibility,

$$\kappa_T = -\frac{1}{V}\left(\frac{\partial V}{\partial P}\right)_{T,N}$$

we can define the adiabatic compressibility,

$$\kappa_S = -\frac{1}{V}\left(\frac{\partial V}{\partial P}\right)_{S,N}.$$

The name is due to the fact that $TdS = đQ$ for quasi-static processes.

In analogy to the derivation for the difference in the specific heats at constant pressure and constant temperature, derive the following thermodynamic identity:

$$\kappa_S = \kappa_T - \frac{TV\alpha^2}{Nc_P}.$$

PROBLEM 14.7
Callen's horrible example of a partial derivative

In his classic book on thermodynamics, Herbert Callen used the following derivative as an example:

$$\left(\frac{\partial P}{\partial U}\right)_{G,N}.$$

However, he did not complete the derivation. Your task is to reduce this partial derivative to contain only the standard partial derivatives α, κ_T, c_P, and c_V, and, of course, any of the first derivatives, T, S, P, V, μ, or N.

This derivation will require all the techniques that you have learned for transforming partial derivatives. Have fun!

PROBLEM 14.8
A useful identity

Prove the identity:

$$\frac{\partial (V,S)}{\partial (T,P)} = \frac{NVc_Vc_V\kappa_T}{T}.$$

PROBLEM 14.9
The *TdS* equations

Prove the following three *TdS* equations. [N is held constant throughout.]

1. First *TdS* equation

$$TdS = Nc_V dT + (T\alpha/\kappa_T)dV.$$

2. Second *TdS* equation
   Prove:

$$TdS = Nc_P dT - TV\alpha dP.$$

3. Third *TdS* equation
   Three forms:

$$TdS = Nc_P \left(\frac{\partial T}{\partial V}\right)_P dV + Nc_V \left(\frac{\partial T}{\partial P}\right)_V dP$$

   or

$$TdS = \frac{Nc_P}{V\alpha}dV + Nc_V \left(\frac{\partial T}{\partial P}\right)_V dP$$

or

$$TdS = \frac{Nc_P}{V\alpha}dV + \frac{Nc_V\kappa T}{\alpha}dP.$$

PROBLEM 14.10
Another useful identity

Since we will be using it in statistical mechanics, prove the following identity:

$$\left(\frac{\partial(\beta F)}{\partial\beta}\right)_{V,N} = U$$

where $\beta = \dfrac{1}{k_B T}$.

15

Extremum Principles

All correct reasoning is a grand system of tautologies, but only God can make direct use of that fact.

> Herbert Simon, political scientist, economist, and psychologist (1916–2001),
> Nobel Prize in Economics, 1978

In Part I we defined the entropy, derived a formal expression for it in classical statistical mechanics, and established that it is a maximum in equilibrium for an isolated composite system with constant total energy, volume, and particle number. This last result is important for many reasons, including providing us with a systematic way of calculating the equilibrium values of quantities when a constraint is released.

There are several alternative principles, called extremum principles, that also allow us to calculate equilibrium properties under different thermodynamic conditions. The derivation of these principles is the main purpose of this chapter.

> In this chapter, we derive extremum principles with respect to internal variables in a composite system. For example, energy could be exchanged through a diathermal wall, and the extremum condition would tell us how much energy was in each subsystem in equilibrium.
>
> All extremum principles in this chapter are with respect to extensive variables; that is, variables that describe how much of something is in a subsystem: for a simple system, these are just the values of U_α, V_α, or N_α for subsystem α.
>
> This chapter does not contain any extremum principles with respect to the intensive variables, such as T, P, or μ.
>
> In Chapter 16 we will examine thermodynamic stability between subsystems, which will lead to conditions on second derivatives with respect to the total volume or number of particles of the subsystems.

To begin with, corresponding to the principle that the entropy is maximized when the energy is constant, there is an 'energy minimum' principle, which states that the energy is minimized when the entropy is held constant.

An Introduction to Statistical Mechanics and Thermodynamics. Robert H. Swendsen, Oxford University Press (2020).
© Robert H. Swendsen. DOI: 10.1093/oso/9780198853237.001.0001

15.1 Energy Minimum Principle

To derive the energy minimum principle, start with the entropy maximum principle.

Consider the entropy, S, of a composite system as a function of its energy, U, and some other parameter, X, which describes the distribution of the amount of something in the composite system. For example, X could be the number of particles in a subsystem, N_A. When a hole is made in the wall between subsystems A and B, the value of $X = N_A$ at equilibrium would maximize the entropy. In other examples we have looked at, X could be the energy of a subsystem or the volume of a subsystem that might vary with the position of a piston. It could not be the temperature, pressure, or chemical potential of a subsystem, because they are not extensive; that is, they do not represent the amount of something.

In equilibrium, the entropy is at a maximum with respect to variations in X, holding U constant. This means that the first partial derivative must vanish, and the second partial derivative must be negative

$$\left(\frac{\partial S}{\partial X} \right)_U = 0 \tag{15.1}$$

$$\left(\frac{\partial^2 S}{\partial X^2} \right)_U < 0. \tag{15.2}$$

Near equilibrium, a small change in the entropy can be represented to leading order by changes in U and X

$$dS \approx \frac{1}{2} \left(\frac{\partial^2 S}{\partial X^2} \right)_U (dX)^2 + \left(\frac{\partial S}{\partial U} \right)_X dU. \tag{15.3}$$

Let us first consider the case of $T > 0$. We know that

$$\left(\frac{\partial S}{\partial U} \right)_X = \frac{1}{T} > 0, \tag{15.4}$$

so eq. (15.3) can be written as

$$dS \approx \frac{1}{2} \left(\frac{\partial^2 S}{\partial X^2} \right)_U (dX)^2 + \left(\frac{1}{T} \right) dU. \tag{15.5}$$

Now comes the subtle part.

For the analysis of the entropy maximum principle, we isolated a composite system and released an internal constraint. Since the composite system was isolated, its total energy remained constant. The composite system went to the most probable macro-scopic state after release of the internal constraint, and the total entropy went to its maximum. Because of the increase in entropy, the process was irreversible.

Now we are considering a quasi-static process without heat exchange with the rest of the universe, so that the entropy of the composite system is constant. However, for the process to be quasi-static we cannot simply release the constraint, as this would initiate an irreversible process, and the total entropy would increase. Outside forces are required to change the constraints slowly to maintain equilibrium conditions. This means that the energy of the composite system for this process is *not* constant.

The distinction between the entropy maximum and energy minimum principles can be illustrated by considering two experiments with a thermally isolated cylinder containing a piston that separates two volumes of gas. The two experiments are illustrated in Fig. 15.1.

Entropy maximum experiment: Here the piston is simply released, as illustrated in the upper picture in 15.1. The total energy is conserved, so that entropy is maximized at constant energy.

Energy minimum experiment: Here the piston is connected by a rod to something outside the cylinder, as illustrated in the lower picture in 15.1. The piston is moved quasi-statically to a position at which the net force due to the pressure from the gas in the two

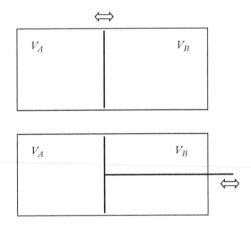

Fig. 15.1 *Piston experiments. The two diagrams each represent a cylinder that is sealed at both ends and is divided into two subvolumes by a moveable piston. Both systems are thermally isolated, and cannot exchange heat energy with the rest of the universe. When the piston in the upper cylinder is released it will move freely until it comes to an equilibrium position that maximizes the entropy. In the lower cylinder, the piston is attached to a rod that can be pushed or pulled from outside the cylinder. The piston is moved to a position at which there is no net force on the piston from the gas in the two subvolumes of the cylinder, minimizing the energy.*

subvolumes is zero. The total energy of the cylinder has been reduced. However, since the process is quasi-static and there has been no heat exchange between the cylinder and the rest of the universe, the total entropy of the system is unchanged. Energy is minimized at constant entropy.

To determine the change in energy for the process in which the piston is held and moved quasi-statically to the equilibrium position, we can turn eq. (15.5) around to find the leading behavior of dU as a function of dS and dX,

$$dU \approx -\frac{T}{2}\left(\frac{\partial^2 S}{\partial X^2}\right)_U (dX)^2 + T dS. \qquad (15.6)$$

Eq. (15.6) gives us a new way of expressing the equilibrium conditions:

$$\left(\frac{\partial U}{\partial X}\right)_S = 0, \qquad (15.7)$$

and

$$\left(\frac{\partial^2 U}{\partial X^2}\right)_S = -\frac{T}{2}\left(\frac{\partial^2 S}{\partial X^2}\right)_U > 0. \qquad (15.8)$$

Since the first partial derivative vanishes, and the second partial derivative is positive, the energy is a minimum at equilibrium for constant entropy (still assuming that the temperature is positive). This is another consequence of the Second Law of Thermodynamics. Indeed, it is equivalent to the maximization of the entropy in an isolated system as an expression of the Second Law.

A consequence of eq. (15.8) is that the maximum work that you can extract from a system in a process with fixed entropy is given by the change in energy of the system. To see this, note that maximum efficiency is always achieved by quasi-static processes. The differential form of the fundamental relation is valid for quasi-static processes

$$dU = T dS - P dV + \mu dN = T dS + \mathchar'26\mkern-12mu dW + \mu dN. \qquad (15.9)$$

We can see that if the entropy and the number of particles are held constant, the change in the energy of the system is equal to the work done on it,

$$dU = \mathchar'26\mkern-12mu dW. \qquad (15.10)$$

If the temperature is negative, the equations are unchanged, but the sign of $\partial^2 U/\partial X^2$ becomes negative. This is a reflection of the upper limit of the energy for systems that can support negative temperatures.

For the rest of the chapter, we will limit the discussion to positive temperatures.

15.2   Minimum Principle for the Helmholtz Free Energy

Both the entropy maximum principle and the energy minimum principle apply to a situation in which the composite system is thermally insulated from the rest of the universe. Now consider a different situation, in which the composite system (and its constituent subsystems) are in contact with a thermal reservoir, so that the temperature of the composite system is held constant. The entropy maximum and energy minimum principles no longer apply, because we can transfer energy and entropy in and out of the thermal reservoir. Nevertheless, we can find an extremum principle under these conditions.

Under constant temperature conditions, it is natural to use the Helmholtz free energy, which is the Legendre transform of U with respect to T:

$$F = U[T] = U - TS. \tag{15.11}$$

To analyze the new situation at constant temperature, consider the explicit case of the composite system of interest in contact with a thermal reservoir at temperature T_R, but thermally isolated from the rest of the universe. At equilibrium with respect to some thermodynamic variable X (subsystem volume, number of particles, and so on), the total energy is a minimum. This gives us the equations

$$\frac{\partial}{\partial X}(U + U_R) = 0 \tag{15.12}$$

and

$$\frac{\partial^2}{\partial X^2}(U + U_R) > 0 \tag{15.13}$$

subject to the condition that the entropy is constant (called the 'isentropic' condition)

$$\frac{\partial}{\partial X}(S + S_R) = 0. \tag{15.14}$$

The only energy exchange between the reservoir and the system is in the form of heat. The differential form of the fundamental relation for the reservoir simplifies to $dU_R = T_R dS_R$, which gives us an equation for the partial derivatives,

$$\frac{\partial U_R}{\partial X} = T_R \frac{\partial S_R}{\partial X}. \tag{15.15}$$

The partial derivative of the Helmholtz free energy of the system can now be transformed to show that it vanishes in equilibrium

$$\frac{\partial F}{\partial X} = \frac{\partial}{\partial X}(U - TS) \tag{15.16}$$

$$= \frac{\partial}{\partial X}(U - T_R S)$$

$$= \frac{\partial U}{\partial X} - T_R \frac{\partial S}{\partial X}$$

$$= \frac{\partial U}{\partial X} + T_R \frac{\partial S_R}{\partial X}$$

$$= \frac{\partial U}{\partial X} + \frac{\partial U_R}{\partial X}$$

$$= \frac{\partial}{\partial X}(U + U_R)$$

$$= 0.$$

A similar derivation shows that F is a minimum at equilibrium

$$\frac{\partial^2 F}{\partial X^2} = \frac{\partial^2}{\partial X^2}(U - TS) \tag{15.17}$$

$$= \frac{\partial^2}{\partial X^2}(U - T_R S)$$

$$= \frac{\partial^2 U}{\partial X^2} - T_R \frac{\partial^2 S}{\partial X^2}$$

$$= \frac{\partial^2 U}{\partial X^2} + T_R \frac{\partial^2 S_R}{\partial X^2}$$

$$= \frac{\partial^2 U}{\partial X^2} + \frac{\partial^2 U_R}{\partial X^2}$$

$$= \frac{\partial^2}{\partial X^2}(U + U_R)$$

$$> 0.$$

The maximum work that can be obtained from a system in contact with a thermal reservoir is not given by the change in U for the system, because energy can be extracted from the reservoir. However, the maximum work is given by the change in Helmholtz free energy.

As in the previous section, we note that maximum efficiency is always obtained with quasi-static processes. The differential form of the fundamental relation in the Helmholtz free energy representation is valid for quasi-static processes at constant temperature:

$$dF = -SdT - PdV + \mu dN = -SdT + đW + \mu dN. \tag{15.18}$$

We can see that if the temperature and the number of particles are held constant, the change in the Helmholtz free energy of the system is equal to the work done on it,

$$dF = đW. \tag{15.19}$$

As was the case for the corresponding eq. (15.10), the work referred to in eq. (15.19) is more general than $-PdV$ for a composite system.

As was the case for the energy minimum principle, the minimum principle for the Helmholtz free energy refers to processes in which a constraint on the amount of something in a subsystem is released quasi-statically. It is therefore not valid to substitute the temperature or pressure of a subsystem for X in eq. (15.17), even though it might be tempting. We will discuss second derivatives with respect to temperature, pressure, and chemical potential in Chapter 16.

15.3 Minimum Principle for the Enthalpy

When the pressure is held constant, minimizing the enthalpy gives us the condition of equilibrium. The proof is similar to those in the previous sections.

Let the system of interest be in contact with a constant pressure reservoir at pressure P_R. At equilibrium with respect to some thermodynamic variable X, the total energy is again a minimum, subject to the constant total volume condition

$$\frac{\partial}{\partial X}(V + V_R) = 0. \tag{15.20}$$

The only energy exchange between the reservoir and the system is in the form of work. The differential form of the fundamental relation for the reservoir simplifies to $dU_R = -P_R dV_R$, which gives us an equation for the partial derivatives:

$$\frac{\partial U_R}{\partial X} = -P_R\frac{\partial V_R}{\partial X}. \tag{15.21}$$

The partial derivative of the enthalpy of the system can now be transformed to show that it vanishes in equilibrium:

$$\frac{\partial H}{\partial X} = \frac{\partial}{\partial X}(U + PV) \tag{15.22}$$

$$= \frac{\partial}{\partial X}(U + P_R V)$$

$$= \frac{\partial U}{\partial X} + P_R\frac{\partial V}{\partial X}$$

$$= \frac{\partial U}{\partial X} - P_R\frac{\partial V_R}{\partial X}$$

$$= \frac{\partial U}{\partial X} + \frac{\partial U_R}{\partial X}$$

$$= \frac{\partial}{\partial X}(U + U_R)$$

$$= 0.$$

A similar derivation shows that H is a minimum at equilibrium:

$$\frac{\partial^2 H}{\partial^2 X} = \frac{\partial^2}{\partial^2 X}(U+PV) \tag{15.23}$$

$$= \frac{\partial^2}{\partial X^2}(U+P_R V)$$

$$= \frac{\partial^2 U}{\partial X^2} + P_R \frac{\partial^2 V}{\partial X^2}$$

$$= \frac{\partial^2 U}{\partial X^2} - P_R \frac{\partial^2 V_R}{\partial X^2}$$

$$= \frac{\partial^2 U}{\partial X^2} + \frac{\partial^2 U_R}{\partial X^2}$$

$$= \frac{\partial^2}{\partial X^2}(U+U_R)$$

$$> 0.$$

Changes in the enthalpy can be related to the heat added to a system. For reversible processes holding both P and N constant, $TdS = đQ$, and

$$dH = TdS + VdP + \mu dN \tag{15.24}$$

becomes

$$dH = đQ. \tag{15.25}$$

Because of eq. (15.25), the enthalpy is often referred to as the 'heat content' of a system. Although this terminology is common in chemistry, I do not recommend it. It can be quite confusing, since you can add arbitrarily large amounts of heat to a system without altering its state, as long as you also remove energy by having the system do work. Heat is not a state function and $đQ$ is not an exact differential. Eq. (15.25) equates an exact and an inexact differential only with the restrictions of $dN = 0$ and $dP = 0$, which makes variations in the system one-dimensional.

15.4 Minimum Principle for the Gibbs Free Energy

If a system is in contact with both a thermal reservoir and a pressure reservoir, the Gibbs free energy is minimized at equilibrium,

$$\frac{\partial G}{\partial X} = 0 \tag{15.26}$$

$$\frac{\partial^2 G}{\partial X^2} > 0. \tag{15.27}$$

The proof of these equations is left as an exercise.

For a composite system in contact with both a thermal reservoir and a pressure reservoir, the Gibbs free energy also allows us to calculate the maximum work that can be obtained by a reversible process. If a composite system can do work on the outside world through a rod, as in the bottom diagram in Fig. 15.1, the differential form of the fundamental relation will be modified by the addition of a term $đW_X$,

$$dG = -SdT + VdP + đW_X + \mu dN. \tag{15.28}$$

For a reversible process with no leaks ($dN = 0$), constant pressure ($dP = 0$), and constant temperature ($dT = 0$), the change in the Gibbs free energy is equal to the work done on the system,

$$dG = đW_X. \tag{15.29}$$

15.5 Exergy

In engineering applications it is common to introduce another thermodynamic potential called the 'exergy', E, which is of considerable practical value.[1] Exergy is used in the common engineering situation in which the environment provides a reversible reservoir at temperature, T_o, and pressure, P_o, which are usually taken to be the atmospheric temperature and pressure.

> In engineering textbooks, the exergy is often denoted by the letter E, which could lead to confusion with the energy. This is especially true for books in which the energy is denoted by E, making the distinction rely on a difference in font. Reader beware!

The exergy is defined as the amount of useful work that can be extracted from the system and its environment. If we use X to denote the internal variable (or variables) that change when work is done by the system, we can write the exergy as

$$E(T_o, P_o, X) = -W(X) \tag{15.30}$$

[1] The concept of exergy was first introduced in the nineteenth century by the German chemist Friedrich Wilhelm Ostwald (1853–1932), who in 1909 was awarded the Nobel Prize in Chemistry. Ostwald divided energy into 'Exergie' (exergy = useful energy) and 'Anergie' (anergy = wasted energy).

where W is the work done *on* the system. (Note that engineering textbooks often use the opposite sign convention for W.)

Clearly, exergy is used in essentially the same situation described in the previous section on the Gibbs free energy, so it is not surprising that the exergy and the Gibbs free energy are closely related. The main difference is that the exergy is defined so that it is zero in the 'dead state', in which G takes on its minimum value and X takes on the corresponding value X_o. The dead state is characterized by the impossibility of extracting any more useful work. If we subtract the value taken on by the Gibbs free energy in the dead state, we can make explicit the connection to the exergy,

$$\mathrm{E}(T_o, P_o, X) = G(T_o, P_o, X) - G(T_o, P_o, X_o). \tag{15.31}$$

The equation for the exergy in engineering texts is usually written without reference to the Gibbs free energy,

$$\mathrm{E}(T_o, P_o, X) = (U - U_o) + P_o(V - V_o) - T_o(S - S_o) + KE + PE. \tag{15.32}$$

Note that the kinetic energy, KE, and the potential energy, PE, are included explicitly in this equation, while they are implicit in eq. (15.31).

It is a common engineering convention that although the kinetic and potential energy are included in the exergy, they are not included in the internal energy of the system, as we have done in the rest of this book. Note that the zero of potential energy is also assumed to be determined by the environment.

Since the exergy is the maximum amount of work that can be obtained from a system exposed to the atmospheric temperature and pressure, it is an extremely useful concept in engineering. An engineer is concerned with how much work can be obtained from a machine in a real environment, and this is exactly what the exergy provides.

15.6 Maximum Principle for Massieu Functions

Just as minimization principles can be found for F, H, and G, which are Legendre transforms of U, maximization principles can be found for the Massieu functions, which are Legendre transforms of the entropy.

The maximization principles can be seen most quickly from the relations

$$S[1/T] = -F/T \tag{15.33}$$

and

$$S[1/T, P/T] = -G/T \tag{15.34}$$

and a comparison with the minimization principles for F and G.

15.7 Summary

The extremum principles derived in this chapter are central to the structure of thermo-dynamics for several reasons.

1. We will use them immediately in Chapter 16 to derive stability conditions for general thermodynamic systems.
2. When stability conditions are violated, the resulting instability leads to a phase transition, which is the topic of Chapter 17.
3. Finally, in Parts III and IV, we will often find that the most direct calculations in statistical mechanics lead to thermodynamic potentials other than the energy or the entropy. The extremum principles in the current chapter are essential for finding equilibrium conditions directly from such statistical mechanics results.

15.8 Problems

....

PROBLEM 15.1
Extremum conditions

1. Helmholtz free energy for the ideal gas
   From the expression for the Helmholtz free energy of the ideal gas, calculate ΔF for an isothermal expansion from volume V_A to V_B when the system is in equilibrium with a thermal reservoir at temperature T_R. Compare this to the work done on the system during the same expansion.
2. Gibbs free energy minimum principle
   Consider a composite system with a property X that can be varied externally. [For example, X could be the volume of a subsystem, as discussed in class.]
   Prove that at equilibrium,

$$\left(\frac{\partial G}{\partial X}\right)_{T,P,N} = 0$$

and

$$\left(\frac{\partial^2 G}{\partial X^2}\right)_{T,P,N} > 0.$$

3. Let $đW$ be the work done on the system by manipulating the property X from outside the system.
   Prove that $dG = đW$, when the system is in contact with a reservoir that fixes the temperature at T_R and the pressure at P_R.

16

Stability Conditions

No one welcomes chaos, but why crave stability and predictability?

Hugh Mackay, Australian psychologist

So far, we have implicitly assumed that the systems we have studied are stable. For example, we have been assuming that the density of a gas will remain uniform, rather than having most of the particles clump together in one part of the container, leaving the rest of the volume nearly empty. Gases are usually well behaved in this respect, but we all know from experience that molecules of H_2O can clump together, form drops, and rain on us.

The question of stability therefore leads naturally to a consideration of phase transitions, in which the properties of a material change drastically in response to small changes in temperature or pressure.

In this chapter we will discuss the stability of thermodynamic systems. We will find that certain inequalities must be satisfied for a system to be stable. The violation of these inequalities signals a phase transition and a number of interesting and important phenomena—including rain.

The methods used in this chapter are based on the extremum principles derived in Chapter 15. Those extremum principles were valid for composite systems, in which the energy, volume, or particle number of the subsystems was varied.

In this chapter we look inside a composite system to see what conditions a subsystem must satisfy as a consequence of the extremum principles in Chapter 15. These are called the 'intrinsic' stability conditions. Since any thermodynamic system could become a subsystem of some composite system, the stability conditions we derive will be valid for all thermodynamic systems.

16.1 Intrinsic Stability

Several approaches have been developed during the history of thermodynamics to derive the intrinsic stability conditions. The one used here has the advantage of being mathematically simple. Its only drawback is that it might seem to be a special case, but it really does produce the most general intrinsic stability conditions.

An Introduction to Statistical Mechanics and Thermodynamics. Robert H. Swendsen, Oxford University Press (2020).
© Robert H. Swendsen. DOI: 10.1093/oso/9780198853237.001.0001

Although all stability conditions can be obtained from any of the extremum principles derived in Chapter 15, there are advantages to using different thermodynamic potentials for different stability conditions to simplify the derivations, as we will see in the following sections.

16.2 Stability Criteria based on the Energy Minimum Principle

Consider two arbitrary thermodynamic systems. Denote the properties of the first as S, U, V, N, and so on, and distinguish the properties of the second with a hat, $\widehat{S}$, $\widehat{U}$, $\widehat{V}$, $\widehat{N}$, and so forth. Combine the two systems to form a composite system, which is thermally isolated from the rest of the universe.

Allow the subsystems to interact with each other through a partition, which can be fixed or moveable, diathermal or adiabatic, impervious to particles or not, as you choose. Denote the quantity being exchanged as X. As in Chapter 15, X can denote any variable that describes how much of something is contained in the system. For example, the wall might be diathermal and X the energy of a subsystem, or the wall might be moveable and X the volume of a subsystem. However, X must be an extensive variable; it cannot be the temperature or pressure.

16.2.1 Stability with Respect to Volume Changes

As we saw in Section 15.1, the total energy is a minimum at equilibrium for constant total volume. Suppose we have a moveable wall (piston) separating the subsystems, as in the lower diagram in Fig. 15.1, and we use it to increase V to $V + \Delta V$, while decreasing $\widehat{V}$ to $\widehat{V} - \Delta V$. The energy minimum principle would then demand that the change in total energy must be non-negative,

$$\Delta U_{total} = U(S, V+\Delta V, N) - U(S, V, N)$$
$$+ \widehat{U}(\widehat{S}, \widehat{V} - \Delta V, \widehat{N}) - \widehat{U}(\widehat{S}, \widehat{V}, \widehat{N})$$
$$\geq 0. \tag{16.1}$$

The equality will hold only when $\Delta V = 0$. The volume transferred need not be small; the inequality is completely general.

Now take the special case that the properties of the two systems are identical. Eq. (16.1) can then be simplified to give an equation that must be true for a single system,

$$U(S, V+\Delta V, N) + U(S, V-\Delta V, N) - 2U(S, V, N) \geq 0. \tag{16.2}$$

This equation must also hold for arbitrary values of ΔV; it is not limited to small changes.

Now divide both sides of eq. (16.2) by $(\Delta V)^2$ and take the limit of $\Delta V \to 0$,

$$\lim_{\Delta V \to 0} \left[\frac{U(S, V + \Delta V, N) + U(S, V - \Delta V, N) - 2U(S, V, N)}{(\Delta V)^2} \right] = \left(\frac{\partial^2 U}{\partial V^2} \right)_{S,N} \geq 0. \tag{16.3}$$

This gives us a stability condition on a second derivative.
Since we know that

$$\left(\frac{\partial U}{\partial V} \right)_{S,N} = -P, \tag{16.4}$$

eq. (16.3) can also be expressed as a stability condition on a first derivative,

$$-\left(\frac{\partial P}{\partial V} \right)_{S,N} \geq 0. \tag{16.5}$$

If we define an isentropic (constant entropy) compressibility in analogy to the isothermal compressibility in eq. (14.14),

$$\kappa_S = -\frac{1}{V} \left(\frac{\partial V}{\partial P} \right)_{S,N} \tag{16.6}$$

we find that κ_S must also be positive for stability,

$$-\left(\frac{\partial P}{\partial V} \right)_{S,N} = \frac{-1}{\left(\frac{\partial V}{\partial P} \right)_{S,}} = \frac{V}{\kappa_S} \geq 0 \tag{16.7}$$

or, since $V > 0$

$$\kappa_S \geq 0. \tag{16.8}$$

This means that if you increase the pressure on a system, its volume will decrease. Eq. (16.8) is true for all thermodynamic systems.

The stability conditions in this chapter are expressed as inequalities for second partial derivatives of the thermodynamic potentials. This makes them look very much like the extremum principles in Chapter 15. However, these inequalities do not represent extremum principles, because the corresponding first derivatives are not necessarily zero, as they must be at an extremum.

16.2.2   Stability with Respect to Heat Transfer

We can use the same method as in Subsection 16.2.1 to consider the consequences of transferring some amount of entropy between the subsystems. The energy minimum principle again demands that the change in total energy be non-negative,

$$
\begin{aligned}
\Delta U_{total} &= U(S + \Delta S, V, N) - U(S, V, N) \\
&\quad + \widehat{U}(\widehat{S} - \Delta S, \widehat{V}, \widehat{N}) - \widehat{U}(\widehat{S}, \widehat{V}, \widehat{N}) \\
&\geq 0.
\end{aligned}
\tag{16.9}
$$

The equality will hold only when $\Delta S = 0$. The entropy transferred need not be small; this inequality is also completely general.

Now again take the special case that the properties of the two systems are identical. Eq. (16.9) can then be simplified to give an equation that must be true for a single system,

$$
U(S + \Delta S, V, N) + U(S - \Delta S, V, N) - 2U(S, V, N) \geq 0.
\tag{16.10}
$$

This equation must also hold for arbitrary values of ΔS; it is not limited to small changes in entropy.

Divide both sides of eq. (16.10) by $(\Delta S)^2$ and take the limit of $\Delta S \to 0$,

$$
\lim_{\Delta S \to 0} \left[\frac{U(S + \Delta S, V, N) + U(S - \Delta S, V, N) - 2U(S, V, N)}{(\Delta S)^2} \right] = \left(\frac{\partial^2 U}{\partial S^2} \right)_{V,N} \geq 0.
\tag{16.11}
$$

Since we know that

$$
\left(\frac{\partial U}{\partial S} \right)_{V,N} = T,
\tag{16.12}
$$

eq. (16.11) can be expressed as a stability condition on a first derivative,

$$
\left(\frac{\partial T}{\partial S} \right)_{V,N} \geq 0.
\tag{16.13}
$$

This inequality can be rewritten in terms of the specific heat at constant volume,

$$
\left(\frac{\partial T}{\partial S} \right)_{V,N} = 1 \Big/ \left(\frac{\partial S}{\partial T} \right)_{V,N} = \frac{T}{Nc_V} \geq 0.
\tag{16.14}
$$

If the temperature is assumed to be positive, a condition for the system to be stable is

$$
c_V \geq 0.
\tag{16.15}
$$

This means that if heat is added to a system, its temperature will increase, which certainly agrees with experience. Eq. (16.15) is valid for all thermodynamic systems.

16.3   Stability Criteria based on the Helmholtz Free Energy Minimum Principle

Now we extend the stability criteria to cases in which the temperature of the composite system is held constant. This makes it natural to use the Helmholtz free energy minimization principle.

Recall that we do not (yet) have a stability condition involving derivatives with respect to temperature, so we will only look at the case of moving a piston to vary the volume.

The experiment we consider is the same as that illustrated in the lower picture in Fig. 15.1, except that instead of being thermally isolated, the entire cylinder is in contact with a heat bath at temperature T.

The Helmholtz free energy minimization principle now leads to the condition:

$$F(T, V + \Delta V, N) + F(T, V - \Delta V, N) - 2F(T, V, N) \geq 0. \tag{16.16}$$

This equation must also hold for arbitrary values of ΔV; it is not limited to small changes in the volumes of the subsystems.

Dividing eq. (16.16) by $(\Delta V)^2$ and taking the limit of $\Delta V \to 0$, we find a new stability condition,

$$\lim_{\Delta V \to 0} \left[\frac{F(T, V + \Delta V, N) + F(T, V - \Delta V, N) - 2F(T, V, N)}{(\Delta V)^2} \right] = \left(\frac{\partial^2 F}{\partial V^2} \right)_{T,N} \geq 0. \tag{16.17}$$

Since we know that

$$\left(\frac{\partial F}{\partial V} \right)_{T,N} = -P, \tag{16.18}$$

we can rewrite eq. (16.17) as

$$-\left(\frac{\partial P}{\partial V} \right)_{T,N} \geq 0. \tag{16.19}$$

Recalling the definition of the isothermal compressibility in eq. (14.14),

$$\kappa_T = \frac{-1}{V} \left(\frac{\partial V}{\partial P} \right)_{T,N}, \tag{16.20}$$

eq. (16.19) becomes

$$-\left(\frac{\partial P}{\partial V}\right)_{T,N} = \frac{-1}{\left(\frac{\partial V}{\partial P}\right)_{T,N}} = \frac{1}{V\kappa_T} \geq 0. \tag{16.21}$$

This tells us that the isothermal compressibility is also non-negative for all systems,

$$\kappa_T \geq 0. \tag{16.22}$$

16.4   Stability Criteria based on the Enthalpy Minimization Principle

Next we consider a consequence of the enthalpy minimization principle. Here, we are looking at a situation in which the total pressure on the system and on each of its subsystems is constant. We will consider an experiment in which heat is transferred between the identical subsystems. The derivation should look familiar, since it follows the same pattern used in previous derivations.

The enthalpy minimization principle gives us the condition:

$$H(S+\Delta S, P, N) + H(S-\Delta S, P, N) - 2H(S, P, N) \geq 0. \tag{16.23}$$

This equation must also hold for arbitrary values of ΔS; it is not limited to small changes.

Dividing eq. (16.23) by $(\Delta S)^2$ and taking the limit of $\Delta S \to 0$, we find yet another stability condition,

$$\lim_{\Delta S \to 0} \left[\frac{H(S+\Delta S, P, N) + H(S-\Delta S, P, N) - 2H(S, P, N)}{(\Delta S)^2}\right] = \left(\frac{\partial^2 H}{\partial S^2}\right)_{P,N} \geq 0. \tag{16.24}$$

Since we know that

$$\left(\frac{\partial H}{\partial S}\right)_{P,N} = T, \tag{16.25}$$

we can rewrite eq. (16.24) as

$$\left(\frac{\partial T}{\partial S}\right)_{P,N} \geq 0. \tag{16.26}$$

Recalling the definition of the specific heat at constant pressure in eq. (14.15),

$$c_P = \frac{T}{N}\left(\frac{\partial S}{\partial T}\right)_{P,N}, \tag{16.27}$$

eq. (16.26) becomes

$$\left(\frac{\partial T}{\partial S}\right)_{P,N} = \frac{1}{\left(\frac{\partial S}{\partial T}\right)_{P,N}} = \frac{T}{Nc_P} \geq 0. \tag{16.28}$$

This tells us that the specific heat at constant pressure is non-negative,

$$c_P \geq 0. \tag{16.29}$$

16.5 Inequalities for Compressibilities and Specific Heats

In Chapter 14 we derived a relationship between the specific heats at constant pressure and constant volume. According to eq. (14.73),

$$c_P = c_V + \frac{\alpha^2 TV}{N\kappa_T}. \tag{16.30}$$

Since all quantities in the second term on the right are non-negative, we immediately have another inequality,

$$c_P \geq c_V. \tag{16.31}$$

The equality would only occur when $\alpha = 0$, which can actually happen! Liquid water has $\alpha = 0$, and its maximum density at a temperature of $3.98°$ C and atmospheric pressure. Note that we have no stability condition on the coefficient of thermal expansion, α, even though we expect most things to expand when heated. This is as it should be, since there are several examples of materials that contract when heated, including water just above the freezing point. This does not affect the inequality in eq. (16.31), since only the square of α occurs in eq. (16.30).

There is an equation and an inequality linking κ_S and κ_T that are similar to those linking c_P and c_V. However, they are more fun to derive yourself, so they will be left as an exercise.

16.6 Other Stability Criteria

When we derived stability conditions for various second derivatives, we explicitly excluded derivatives with respect to T, P, or μ. Addressing this omission turns out to be quite easy, but requires a method that is different from that used in Chapters 15 and 16. One example will reveal the nature of the derivation.

We will derive a stability condition for the second derivative of the Helmholtz free energy with respect to temperature. We know that the first derivative gives the negative of the entropy,

$$\left(\frac{\partial F}{\partial T}\right)_{V,N} = -S. \tag{16.32}$$

Take the partial derivative of eq. (16.32) with respect to temperature to find an expression for the quantity in which we are interested,

$$\left(\frac{\partial^2 F}{\partial T^2}\right)_{V,N} = -\left(\frac{\partial S}{\partial T}\right)_{V,N}. \tag{16.33}$$

The right-hand side of eq. (16.33) is, of course, equal to $-Nc_V/T$, which must be negative. However, we are more concerned with relating it to the second derivative of the energy with respect to entropy, because that reveals a general property of thermodynamic potentials.

The first derivative of the energy with respect to entropy gives the temperature,

$$\left(\frac{\partial U}{\partial S}\right)_{V,N} = T. \tag{16.34}$$

Now take the second derivative of U with respect to S,

$$\left(\frac{\partial^2 U}{\partial S^2}\right)_{V,N} = \left(\frac{\partial T}{\partial S}\right)_{V,N} \geq 0. \tag{16.35}$$

The last inequality in eq. (16.35) is just the stability condition in eq. (16.11) that we derived earlier in this chapter.

Since we have the identity

$$\left(\frac{\partial S}{\partial T}\right)_{V,N} = 1 \bigg/ \left(\frac{\partial T}{\partial S}\right)_{V,N}, \tag{16.36}$$

we find that the second partial derivative of F with respect to T must be negative:

$$\begin{aligned}
\left(\frac{\partial^2 F}{\partial T^2}\right)_{V,N} &= -\left(\frac{\partial S}{\partial T}\right)_{V,N} \\
&= -1 \bigg/ \left(\frac{\partial T}{\partial S}\right)_{V,N} \\
&= -1 \bigg/ \left(\frac{\partial^2 U}{\partial S^2}\right)_{V,N} \\
&\leq 0.
\end{aligned} \tag{16.37}$$

Eq. (16.37) shows that the second partial derivative of F with respect to T must have the opposite sign from the second partial derivative of U with respect to S. It can be seen from the derivation that quite generally the second partial derivative of the Legendre transform (F in this case) with respect to the new variable (T) has the opposite sign from the second partial derivative of the original function (U) with respect to the old variable (S). It is also a general rule that the two second derivatives are negative reciprocals of each other.

Because of the simple relationship between the second derivatives of Legendre transforms, we can skip the explicit derivations of the remaining inequalities and simply summarize them in Table 16.1.

Table 16.1 *Summary of inequalities for second partial derivatives of thermodynamic potentials that are required for stability.*

$U(S,V,N)$	$\left(\dfrac{\partial^2 U}{\partial S^2}\right)_{V,N} \geq 0$	$\left(\dfrac{\partial^2 U}{\partial V^2}\right)_{S,N} \geq 0$
$F(T,V,N)$	$\left(\dfrac{\partial^2 F}{\partial T^2}\right)_{V,N} \leq 0$	$\left(\dfrac{\partial^2 F}{\partial V^2}\right)_{T,N} \geq 0$
$H(S,P,N)$	$\left(\dfrac{\partial^2 H}{\partial S^2}\right)_{P,N} \geq 0$	$\left(\dfrac{\partial^2 H}{\partial P^2}\right)_{S,N} \leq 0$
$G(T,P,N)$	$\left(\dfrac{\partial^2 G}{\partial T^2}\right)_{P,N} \leq 0$	$\left(\dfrac{\partial^2 G}{\partial P^2}\right)_{T,N} \leq 0$

I highly recommend memorizing the results shown in Table 16.1. Remembering them is easy because of the simplicity of their derivation, and they can save you from avoidable errors. It is astonishing how often these inequalities are violated in published data. Some prominent scientists have even made a hobby of collecting examples from the literature. Do not let them find any exhibits for their collections in your work!

16.7 Problems

PROBLEM 16.1
Thermodynamic stability

1. Starting with the stability conditions on the second derivatives of the Helmholtz free energy, prove that for the Gibbs free energy we have the general inequality:

$$\left(\frac{\partial^2 G}{\partial P^2}\right)_{T,N} \leq 0.$$

2. Use this inequality to prove that the isothermal compressibility is positive.

PROBLEM 16.2
A rubber band

The essential issues in the following questions concern whether the sign of some quantity is positive or negative. *Getting the signs right is essential!*

Consider an ordinary rubber band. If the length of the rubber band is L, and we apply a tension τ, a small change in length will change the energy of the rubber band by τdL. Assuming the number of molecules in the rubber band is fixed, the differential form of the fundamental relation in the energy representation is

$$dU = TdS + \tau dL.$$

Note that the sign of the τdL term is positive, in contrast to the more common $-PdV$ term.

1. Experimental question!

   Determine the sign of the quantity $\left(\dfrac{\partial T}{\partial L}\right)_S$ experimentally.

   Obtain a rubber band—preferably a clean rubber band.

   Stretch the rubber band quickly and, using your forehead as a thermometer, determine whether is becomes hotter or colder. If the rubber band is very clean, you might try using your lips, which are more sensitive.

   Since you are only interested in the sign of the derivative, you do not really have to carry out the experiment under true adiabatic (constant entropy) conditions. It will be sufficient if you simply do it quickly.

2. Now imagine that one end of the rubber band is attached to a hook in the ceiling, while the other end is attached to a weight. After the system is allowed to come to equilibrium with the weight hanging down, the rubber band is heated with a hair dryer.

   Using the result of your experiment with a rubber band and your knowledge of thermodynamic identities and stability conditions, predict whether the weight will rise or fall.

17

Phase Transitions

Life is pleasant. Death is peaceful. It's the transition that's troublesome.

Isaac Asimov (1920–1992), biochemist, and author of both
science fiction and non-fiction books.

One of the most interesting branches of thermal physics is the study of phase transitions. While most of thermodynamics is concerned with the consequences of analyticity in relating different measurable quantities, phase transitions occur at points where analyticity (Section 9.6.4) breaks down.

Examples of phase transitions abound. Water can freeze, going from a liquid to a solid state when the temperature is lowered. Water can also boil, going from a liquid state to a gaseous state when the temperature is raised. Iodine can sublimate, going directly from a solid to a gaseous state. In fact, almost all materials can exist in different states, with abrupt transitions between them.

Water boiling and freezing and iodine sublimating are examples of first-order transitions; that is, phase transitions in which the extensive variables change discontinuously.

An example of a second-order transition is given by the magnetization of iron, which goes to zero continuously at a 'critical' temperature of $1044\,K$. The partial derivative of the magnetization with respect to temperature is not only discontinuous, but it diverges as the critical temperature is approached from below.

The classification of phase transitions as first-order or second-order is a hold-over from an early classification scheme due to the Austrian physicist Paul Ehrenfest (1880–1933). Ehrenfest classified phase transitions at which the n-th partial derivative of the free energy was discontinuous as being n-th order. For first-order phase transitions this classification is very useful, but for higher-order transitions it is less so. Phase transitions have turned out to be much more complex and interesting than Ehrenfest thought. In current usage his definition of first-order transitions is retained, but 'second-order' generally refers to any transition with some sort of non-analytic behavior, but continuous first partial derivatives of the free energy.

An Introduction to Statistical Mechanics and Thermodynamics. Robert H. Swendsen, Oxford University Press (2020).
© Robert H. Swendsen. DOI: 10.1093/oso/9780198853237.001.0001

17.1 The van der Waals Fluid

To illustrate some of the basic ideas about phase transitions we will use a model for a fluid of interacting particles that was invented by the Dutch physicist, Johannes Diderik van der Waals (1837–1923), who was awarded the Nobel Prize in 1910.

17.2 Derivation of the van der Waals Equation

There are several ways of deriving the van der Waals equations, but they all begin with making approximations for the changes in the behavior of an ideal gas when the effects of interactions on the properties are included. As discussed in Chapter 7, a typical interaction between two particles has the form shown in Fig. 7.1. For very short distances the interactions between molecules are generally repulsive, while they are attractive at longer distances. In the van der Waals model the effects of the attractive and repulsive parts of the interactions between particles are considered separately. Rather than calculating the properties of the van der Waals fluid directly from a specific interaction potential of the form shown in Fig. 7.1, we will follow the usual procedure of introducing a parameter a for the overall strength of the attractive part of the interaction, and another parameter b for the strength of the repulsive part.

We will start from the Helmholtz free energy for the ideal gas. (X is the usual constant.)

$$F_{IG} = -Nk_BT\left[\ln\left(\frac{V}{N}\right) + \frac{3}{2}\ln(k_BT) + X\right]. \tag{17.1}$$

Since an interaction of the form shown in Fig. 7.1 implies that each particle is attracted to the particles in its neighborhood, we would expect the average energy of attraction to be proportional to the density of particles in the neighbor of any given particle, which we will approximate by the average density, N/V. The total energy of attraction for N particles can then be written as $-aN^2/V$, where $a > 0$ is a constant.

Assuming that the most important effect of the repulsive part of the interaction potential is a reduction in the available volume, we subtract a correction term proportional to the number of particles, so that $V \rightarrow V - bN$.

Naturally, the constants a and b will be different for different fluids.

With these two changes in eq. (17.1), we find the Helmholtz free energy for the van der Waals fluid,

$$F_{vdW} = -Nk_BT\left[\ln\left(\frac{V - bN}{N}\right) + \frac{3}{2}\ln(k_BT) + X\right] - a\left(\frac{N^2}{V}\right). \tag{17.2}$$

From eq. (17.2) we can use the usual partial derivatives of F_{vdW} with respect to T and V to find two equations of state,

$$P = \frac{Nk_BT}{V - bN} - \frac{aN^2}{V^2} \tag{17.3}$$

and

$$U = \frac{3}{2}Nk_BT - a\left(\frac{N^2}{V}\right). \tag{17.4}$$

Note that both van der Waals equations of state, eqs. (17.3) and (17.4), are unchanged (invariant) if U, V, and N are multiplied by an arbitrary constant λ, demonstrating that the van der Waals gas is extensive. This is because surfaces and interfaces are completely neglected in this model, and the fluid is assumed to be homogeneous.

17.3 Behavior of the van der Waals Fluid

At very high temperatures, the gas expands, the density is small, and V is large in comparison with the correction term bN. For low density the correction term for the pressure is also small, so that the predictions of the van der Waals equation, eq. (17.3), are very close to those of the equation for an ideal gas.

On the other hand, as the temperature is reduced, the deviations from ideal-gas behavior become pronounced. If the temperature is below a certain critical temperature, T_c, the deviations from the ideal-gas law are not only large, but the behavior is qualitatively different.

Below T_c, a plot of $P(V)$ vs. V looks qualitatively like Fig. 17.1, while above T_c the slope is negative everywhere. The value of the critical temperature, T_c, can be found from the condition that $P(V)$ has an inflection point at which both

$$\left(\frac{\partial P}{\partial V}\right)_{T,N} = 0, \tag{17.5}$$

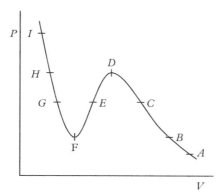

Fig. 17.1 *Schematic $P-V$ plot of an isotherm (constant temperature) for the van der Waals equation at a low temperature. The marked points are used to illustrate the calculation of the location of the phase transition. They are the same points as those marked in Fig. 17.2*

and

$$\left(\frac{\partial^2 P}{\partial V^2}\right)_{T,N} = 0. \tag{17.6}$$

With a little algebra, eqs. (17.5) and (17.6) can be solved for the critical values for the volume, pressure, and temperature,

$$V_c = 3bN \tag{17.7}$$

$$P_c = \frac{a}{27b^2} \tag{17.8}$$

$$k_B T_c = \frac{8a}{27b}. \tag{17.9}$$

A very interesting feature of the van der Waals equation is that although it contains only two arbitrary constants, a and b, it predicts the three critical values, V_c, P_c, and T_c. This means that the three values cannot be independent. Indeed, they can be combined to make a non-trivial prediction for the ratio

$$\frac{P_c V_c}{N k_B T_c} = \frac{3}{8} = 0.375. \tag{17.10}$$

The measured values of this ratio vary for different materials, but not as much as might be guessed from such a simple theory. For example, the experimental values of $P_c V_c / N k_B T_c$ for helium, water, and mercury are, respectively, 0.327, 0.233, and 0.909.

If we define the reduced values $\tilde{V} = V/V_c$, $\tilde{P} = P/P_c$, and $\tilde{T} = T/T_c$, we can write the van der Waals equation in dimensionless form as

$$\tilde{P} = \frac{8\tilde{T}}{3\tilde{V} - 1} - \frac{3}{\tilde{V}^2}, \tag{17.11}$$

or

$$\left(\tilde{P} + 3\tilde{V}^{-2}\right)\left(3\tilde{V} - 1\right) = 8\tilde{T}. \tag{17.12}$$

17.4   Instabilities

The qualitative differences between the PV plot for the low-temperature van der Waals gas and for an ideal gas are dramatic. The PV plot for an ideal gas is simply an hyperbola ($PV = N k_B T$). The plot of a low-temperature isotherm for the van der Waals gas, shown in Fig. 17.1, has the striking feature that the slope is positive in the region between points F and D,

$$\left(\frac{\partial P}{\partial V}\right)_{T,N} > 0. \tag{17.13}$$

This is surprising, because we know from Chapter 16 (see eq. (16.19)) that the derivative in eq. (17.13) must be negative for stability. Therefore, the region of the van der Waals plot from points F to D in Fig. 17.1 must be unstable.

We noted that the van der Waals equation was derived under the assumption that the system was homogeneous. The violation of the stability condition in the region between points F and D marks is where this assumption breaks down. This part of the curve must represent unphysical states.

More insight into the unusual features of the van der Waals equation can be seen if the axes of Fig. 17.1 are flipped, as shown in Fig. 17.2.

From Fig. 17.2 it can be seen that the van der Waals equation actually predicts that the volume is a triple-valued function of the pressure in the region between points B and H. The possibility of three solutions can also be seen directly from the form of the van der Waals equation. If eq. (17.3) is rewritten as

$$\left(PV^2 + aN^2\right)(V - bN) = Nk_BTV^2, \tag{17.14}$$

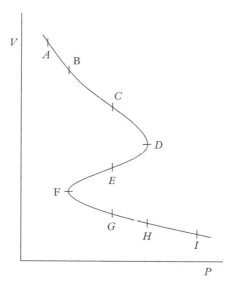

Fig. 17.2 *Schematic $V - P$ plot of an isotherm (constant temperature) for the van der Waals equation at a low temperature. This is a plot of the same function as in Fig. 17.1, but the axes have been flipped. The marked points are used to illustrate the calculation of the location of the phase transition. They are the same points as those marked in Fig. 17.1*

and the left side of eq. (17.14) is expanded, the van der Waals equation is seen to be a cubic equation in V for fixed P. It can therefore have either one or three real solutions, which is consistent with Fig. 17.2.

To find the stable states predicted by the van der Waals equation we must eliminate the section of the curve between points D and F on the basis of instability, but we still have to choose between the two remaining solutions. This will be done in the next section.

17.5 The Liquid–Gas Phase Transition

To resolve the problem of multiple solutions of the van der Waals equation, we need to go back to Chapter 16 and investigate the stability of the high- and low-density states in Fig. 17.2. For the stable state, the Gibbs free energy at constant temperature and pressure should be a minimum. Note that this criterion is not restricted to infinitesimal variations; it is also valid for the large differences in the volume between the states on the upper and lower branches in Fig. 17.2.

To use the stability condition on the Gibbs free energy, G, we must be able to calculate it. Since this is an extensive system, we know two very helpful things.

First, as shown in eq. (13.27) of Chapter 13,

$$G = U - TS + PV = \mu N, \qquad (17.15)$$

so that the condition that G be a minimum is the same as the condition that μ be a minimum.

Next, the Gibbs–Duhem equation, given in eq. (13.11),

$$d\mu = -\left(\frac{S}{N}\right) dT + \left(\frac{V}{N}\right) dP, \qquad (17.16)$$

is valid for the van der Waals fluid. Since Fig. 17.2 is an isotherm (constant temperature), we can set $dT = 0$ in eq. (17.16) and integrate it to find G or μ as a function of P,

$$G = \mu N = \int V dP. \qquad (17.17)$$

Starting with point A in Fig. 17.2 and integrating eq. (17.17) to point D, we find the increasing part of the curve shown in Fig. 17.3. Note that the curvature of this part of the plot is negative because the integrand (V) is decreasing. As we continue to integrate along the curve in Fig. 17.2 from point D to point F, we are moving in the negative P-direction, so the contributions of the integral are negative, and the value of $G = \mu N$ decreases. When we reach point F and continue to point I, we are again integrating in the positive P-direction and the value of $G = \mu N$ increases.

The integration of the van der Waals equation of state in Fig. 17.2, using eq. (17.17), is peculiar for at least two reasons. First, since the function $V = V(P)$ is multivalued, the integration must proceed in three stages: initially in the positive P-direction, then in the negative P-direction, and finally in the positive P-direction again. The more problematic feature is that we have already established that the states corresponding to the part of the function that lies between points D and F in Figs. 17.1 and 17.2 are unstable. They could only be regarded as equilibrium states if they somehow were subject to an added constraint that they remain homogeneous. However, since we only need the van der Waals equation in the unstable region for the formal integral, and that part of the curve is an analytic continuation of the van der Waals isotherm, the procedure is valid—strange, but valid.

The parts of the curve in Fig. 17.3 corresponding to equilibrium states are the two sections from point A to C and from point G to I, which have the minimum Gibbs free energy. Note that although the points C and G are distinct in Figs. 17.1 and 17.2, they have the same Gibbs free energy and occupy the same location in Fig. 17.3.

There are two kinds of instability that occur along the unstable parts of the curve. First, we have already noted that states corresponding to points between D and F are unstable to small perturbations because $\partial P / \partial V > 0$ in that region. On the other hand, $\partial P / \partial V < 0$ in the regions C to D and F to G, so that the corresponding states are stable to small perturbations. Nevertheless, they are still not equilibrium states, because there exist states with lower Gibbs free energy. They are unstable in a second sense: they are

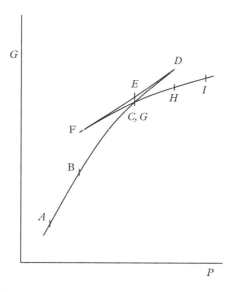

Fig. 17.3 *Schematic $G - P$ plot for the van der Waals fluid at a low temperature. The function $G = G(P)$ is obtained by performing the integral in eq. (17.17) on the van der Waals equation of state shown in Fig. 17.2*

'globally' unstable with respect to a large change. The term for such states is 'metastable'. Because they are stable to small perturbations, metastable states can exist for a long time. However, eventually a large fluctuation will occur and the system will make a transition to the true equilibrium state, which will lower the Gibbs free energy.

17.6 Maxwell Construction

The integration of the van der Waals equation of state in Fig. 17.2, using eq. (17.17), includes a positive contribution from point C to D, followed by a smaller negative contribution from D to E. The sum of these two contributions gives a net positive contribution to the Gibbs free energy. The magnitude of this net positive contribution is the shaded area under the curve EDC in Fig. 17.4.

The integral from E to F gives a negative contribution, and it is followed by a smaller positive contribution from F to G, giving a net negative contribution. The magnitude of this contribution is the shaded area above the curve EFG in Fig. 17.4.

Since the total integral from the point C to the point G must vanish (C and G have the same Gibbs free energy), the two shaded regions in Fig. 17.4 must have equal areas. The procedure of adjusting the position of the line CG until the two areas are equal is known as the Maxwell construction, first noted by the Scottish physicist James Clerk Maxwell (1831–1879).

17.7 Coexistent Phases

The points C and G in Figs. 17.1, 17.2, 17.3, or 17.4 indicate distinct phases that are in equilibrium with each other at the same temperature and pressure. The fact that they are in equilibrium with each other is demonstrated by their having the same Gibbs free

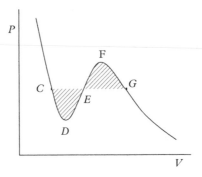

Fig. 17.4 *Schematic $P-V$ plot of an isotherm (constant temperature) for the van der Waals equation at a low temperature. The labeled points coincide with those in Fig. 17.1. By the Maxwell construction, the two shaded areas must have equal areas when the horizontal line bounding them is at the correct pressure to have equilibrium between the liquid and gas phases*

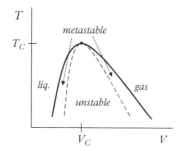

Fig. 17.5 *Schematic T − V plot of the coexistence diagram for the van der Waals fluid. The solid line indicates the liquid and gas phase volumes at each temperature. The dot at the top of the curve indicates the critical point. The dashed curves indicate the spinodals, which correspond to the points D and F in Figs. 17.1 through 17.4. Regions of liquid and gas phases are indicated, as well as regions of unstable and metastable states*

energy in Fig. 17.3. That they are distinct phases is clear because of the difference in their volumes.

The phase corresponding to point C in the figures corresponds to a small volume and a high density. We will call it the liquid phase to distinguish it from the gas phase corresponding to point G, which has a large volume and small density.

For each temperature, we can draw an isotherm and perform the Maxwell construction to determine the pressure at which the liquid and gas phases are in equilibrium with each other, as well as the volumes of the liquid and gas phases.

Fig. 17.5 shows a qualitative plot of the relationship between the liquid and gas volumes at coexistence and the temperature. The area under the solid curve is known as the coexistence region, because those temperatures and volumes can only be achieved by having distinct liquid and gas phases coexisting next to each other. The dot at the top of the coexistence curve indicates the critical point, with the critical temperature T_c and the critical volume V_c. As the temperature increases toward T_c from below, the volumes of the liquid and gas phases approach each other. Above T_c, there is no distinction between liquid and gas—only a single fluid phase.

The dashed lines in Fig. 17.5 correspond to the points D and F in Figs. 17.1 through 17.4. They are known as 'spinodals' and represent the boundaries of the metastable regions. All states under the dashed lines are unstable.

17.8 Phase Diagram

Equilibrium between liquid and gas phases occurs at a specific pressure for any given temperature, as determined by the Maxwell construction. The locus of the coexistence points on a $P − T$ diagram is plotted schematically in Fig. 17.6.

The curve in Fig. 17.6 is known as the 'coexistence curve' and separates regions of liquid and gas. At points along the curve, the liquid and gas phases coexist. The curve

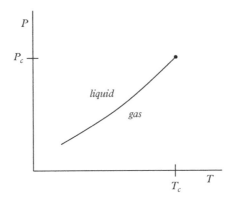

Fig. 17.6 *Schematic P − T plot of the phase diagram of the van der Waals fluid. The solid line separates the liquid and gas phase regions. The dot at the end of the curve indicates the critical point*

ends in a critical point with, logically enough, critical values of the temperature, T_c, and pressure, P_c.

At temperatures or pressures above the critical point, there is no distinction between liquid and gas; there is only a fluid. This has the curious consequence that, at least in some sense, liquid and gas phases are different aspects of the same phase.

Consider a process that starts with a gas on the coexistence curve (or infinitesimally below it) at a temperature T_o and pressure P_o. Raise the temperature of the gas at constant pressure. When the temperature is above T_c, the pressure is increased until it is above P_c. At this point, the temperature is again lowered until it is at the original value at T_o. Now lower the pressure until the original value of P_o is reached. The system has now returned to the original temperature and pressure, *but on the other side of the coexistence curve*. It is now in the liquid phase.

Since the coexistence curve was not crossed anywhere in the entire process, the fluid was taken smoothly from the gas to the liquid phase without undergoing a phase transition. The initial (gas) and final (liquid) phases must, in some sense, represent the same phase. Considering how obvious it seems that water and steam are different, it might be regarded as rather surprising to discover that they are not fundamentally different after all.

17.9   Helmholtz Free Energy

The graph of the Helmholtz free energy F as a function of the volume V for constant temperature T is qualitatively different than the graph of G as a function of P shown in Fig. 17.3. The essential difference is that in Fig. 17.3 the independent variable, P, was intensive, so that the phase transition occurred at a point. In Fig. 17.7 the independent variable is extensive, so that it takes on different values in the two phases at a first-order transition.

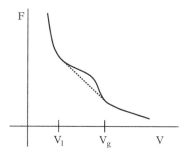

Fig. 17.7 *Schematic F − V plot for the van der Waals fluid at a fixed temperature T < T_c. The straight dashed line indicates the coexistence region in which part of the system is liquid in equilibrium with the rest of the system in the gas phase. (The true plot has the same features as shown here, but it is more difficult to see them clearly without magnification due to their relative magnitudes. Generating the true plot is left to the exercises.)*

The first partial derivative of F with respect to V is the negative of the pressure, so it is negative:

$$\left(\frac{\partial F}{\partial V}\right)_{T,N} = -P < 0. \tag{17.18}$$

The second partial derivative of F with respect to V is related to the isothermal compressibility, so it is positive:

$$\left(\frac{\partial^2 F}{\partial V^2}\right)_{T,N} = -\left(\frac{\partial P}{\partial V}\right)_{T,N} = \frac{V}{\kappa_T} > 0. \tag{17.19}$$

We know from the previous discussion of the $P - V$ diagram of the van der Waals gas that the model predicts an unstable region. In Fig. 17.1 this was found between the points D and F and identified by the positive slope in this region, which would imply a negative compressibility. This violates a stability condition.

In terms of the Helmholtz free energy, the unstable region is signaled by a negative curvature, which violates the inequality in eq. (17.19). As a consequence, the schematic $F - V$ plot exhibits a region of negative curvature between two inflection points, as shown in Fig. 17.7.

As we saw earlier in this chapter, below T_c the homogeneous solution to the van der Waals equations is not stable between two values of the volume corresponding to the liquid (V_l) and the gas (V_g). In this region, the Helmholtz free energy is lower for coexisting liquid and gas states. The values of F in this coexistence region are indicated by the straight dashed line in Fig. 17.7. The line is found as the common tangent to the van der Waals curve at the values V_l and V_g.

It is useful to compare the plot of G vs. P in Fig. 17.3 and the plot of F vs. V in Fig. 17.7. The former has an *intensive* variable on the horizontal axis, and the phase transition occurs at a single value of P. The latter has an *extensive* variable on the horizontal axis, and the phase transition occurs over a range of values of V. There are many diagrams of thermodynamic potentials as functions of various quantities, but they all fall into one of these two categories.

17.10   Latent Heat

We all know that it takes heat to boil water. More generally, to cross the coexistence curve in Fig. 17.6 requires adding energy to the liquid. If we add a small amount of heat $đQ$ to a system, its entropy changes by an amount $dS = đQ/T$. Therefore, there will be a difference in the entropy between the liquid state corresponding to point G and the gas state corresponding to the point C in Fig. 17.1.

To calculate the difference in entropy, ΔS, between the liquid and gas phases, consider a process by which we integrate the equation

$$dS = \left(\frac{\partial S}{\partial V}\right)_{T,N} dV \tag{17.20}$$

along the isotherm in Fig. 17.1 from point G to point C. Since the process is isothermal, the total heat needed to go from the liquid to the gas phase will be $T\Delta S$.

To find the partial derivative in eq. (17.20) we can use a Maxwell relation associated with the differential form of the fundamental relation $dF = -SdT - PdV + \mu dN$,

$$\left(\frac{\partial S}{\partial V}\right)_{T,N} = \left(\frac{\partial P}{\partial T}\right)_{V,N}. \tag{17.21}$$

With this Maxwell relation, the change in entropy is given by the integral

$$\Delta S = \int_G^C \left(\frac{\partial P}{\partial T}\right)_{V,N} dV. \tag{17.22}$$

Since the partial derivative in eq. (17.22) can be calculated from the van der Waals equation of state in eq. (17.3), evaluating the integral is straightforward.

The energy per particle needed to make a first-order phase transition is called the latent heat, denoted by ℓ,

$$\ell = \frac{T\Delta S}{N}. \tag{17.23}$$

It is also quite common to denote the energy needed to make a first-order phase transition for a complete system of N particles as the heat capacity, L,

$$L = N\ell = T\Delta S. \tag{17.24}$$

17.11 The Clausius–Clapeyron Equation

It might be supposed that the coexistence curve in the phase diagram in Fig. 17.6 merely separates the phases and is of no particular significance in itself. However, the German physicist Rudolf Clausius, whom we have already encountered as a founder of thermodynamics, and the French physicist Benoît Paul Émile Clapeyron (1799–1864) found a remarkable equation that links the slope of the coexistence curve with the change in volume across the transition and the value of the latent heat.

Consider two points along the coexistence curve. Denote the liquid phase at the first point by X and the gas phase at the same temperature and pressure by X'. Similarly denote the liquid and gas phases by Y and Y' at the other point. At each of these points, let the sample contain N particles. Assume that the points are very close together, so that the difference in temperature, $dT = T_Y - T_X = T_{Y'} - T_{X'}$, and the difference in pressure, $dP = P_Y - P_X = P_{Y'} - P_{X'}$, are both small.

Since the liquid and gas phases are in equilibrium, we must have

$$\mu_X = \mu_{X'}, \tag{17.25}$$

and

$$\mu_Y = \mu_{Y'}, \tag{17.26}$$

which implies that

$$\mu_Y - \mu_X = \mu_{Y'} - \mu_{X'}. \tag{17.27}$$

On the other hand, we know from the Gibbs–Duhem equation from eq. (13.11) in Section 13.2, that if we have the same number of particles on both sides of the transition, then,

$$\mu_Y - \mu_X = -\left(\frac{S}{N}\right)dT + \left(\frac{V}{N}\right)dP, \tag{17.28}$$

and

$$\mu_{Y'} - \mu_{X'} = -\left(\frac{S'}{N}\right)dT + \left(\frac{V'}{N}\right)dP. \tag{17.29}$$

Combining these equations, we see that

$$\frac{dP}{dT} = \frac{S' - S}{V' - V} = \frac{\Delta S}{\Delta V}. \tag{17.30}$$

Comparing eq. (17.30) with the definition of the latent heat in eq. (17.23), we find the Clausius–Clapeyron equation,

$$\frac{dP}{dT} = \frac{\ell N}{T \Delta V} = \frac{L}{T \Delta V}. \tag{17.31}$$

Note that the latent heat is positive by definition, and most materials expand when they melt or boil, $\Delta V > 0$. The slopes of *almost* all coexistence curves are positive, as is the coexistence curve for the van der Waals gas, shown schematically in Fig. 17.6.

However, ice contracts when it melts, so that $\Delta V < 0$ and $dP/dT < 0$ along the ice–water coexistence curve. The negative slope required by the thermodynamics analysis presented here is confirmed qualitatively and quantitatively by experiment.

17.12   Gibbs' Phase Rule

Phase diagrams in a general system can be considerably more complicated than Fig. (17.6). However, there is a limitation on the structure of these diagrams that is called 'Gibbs' phase rule'.

To derive Gibbs' phase rule, consider a general thermodynamic system with K components; that is, K different kinds of molecules. Let the number of distinct phases in equilibrium with each other be denoted by ϕ. The variables describing this system are T and P, plus the concentrations of each type of particle in each phase. For the j-th phase, $\{x_k^{(j)} = N_k^{(j)}/N^{(j)} | k = 1, \ldots, K\}$, where $N_k^{(j)}$ gives the number of type k particles in the j-th phase. Since the total number of particles in the j-th phase is $N^{(j)} = \sum_{k=1}^{K} N_k^{(j)}$, the sum of the concentrations is unity, $1 = \sum_{k=1}^{K} x_k^{(j)}$, and there are only $K - 1$ independent concentration variables for each of the ϕ phases. The total number of variables is then $2 + \phi(K - 1)$.

However, the equilibrium conditions on the chemical potentials limit the number of independent variables, since the chemical potential of a given component must take on the same value for every phase. Letting $\mu_k^{(j)}$ denote the chemical potential of the k-th component in the j-th phase, we have

$$\mu_k^{(1)} = \mu_k^{(2)} = \cdots = \mu_k^{(\phi)}. \tag{17.32}$$

Since there are K components, this gives a total of $K(\phi - 1)$ conditions on the variables describing the system.

Putting these two results together, the total number F of independent variables for a general system with K components and ϕ phases is

$$F = 2 + \phi(K-1) - K(\phi - 1), \qquad (17.33)$$

or

$$F = 2 + K - \phi. \qquad (17.34)$$

This is Gibbs' phase rule.

As an example, consider the case of a simple system containing only a single kind of molecule, so that $K = 1$. If we consider the boundary between normal ice and liquid water ($\phi = 2$), we see that the number of independent variables is $F = 2 + 1 - 2 = 1$. This means that we have one free parameter to describe the boundary between solid and liquid. In other words, the two phases will be separated by a line in a phase diagram (a plot of P vs. T), and the one parameter will tell us where the system is on that line. The same is true for the boundaries between the liquid and vapor phases and the solid and vapor phases.

As a second example, consider the triple point of water, at which three phases come together ($\phi = 3$). In this case, the number of independent variables is $F = 2 + 1 - 3 = 0$, so that there are no free parameters; coexistence of the three phases can occur only at a single point.

The phase diagram of ice is actually quite complicated, and there are at least ten distinct crystal structures at various temperatures and pressures. Gibbs' phase rule tells us that two different forms of ice can be separated in the phase diagram by a line and three phases can coexist at a point, but we cannot have *four* phases coexisting at a point. And, in fact, four phases do not coexist at any point in the experimental phase diagram.

17.13 Problems

PROBLEM 17.1
Properties of the van der Waals gas

1. Just for practice, starting from the expression for the Helmholtz free energy, derive the pressure, $P = P(T, V, N)$.
2. Again starting from the expression for the Helmholtz free energy for the van der Waals fluid, derive the energy as a function of temperature, volume, and number of particles,

$$U = U(T, V, N).$$

3. Using the conditions for the critical point,

$$\left(\frac{\partial P}{\partial V}\right)_{T,N} = 0$$

and

$$\left(\frac{\partial^2 P}{\partial V^2}\right)_{T,N} = 0$$

derive expressions for P_c, T_c, and V_c as functions of a and b.
4. Derive the value of the ratio

$$\frac{P_c V_c}{N k_B T_c}$$

as a pure number, independent of the values of a and b.

PROBLEM 17.2
Properties of the van der Waals gas by computation

I suggest doing the following calculations as functions of the van der Waals parameters a and b, However, make the plots in terms of the reduced quantities T/T_c, V/V_c, and P/P_C.

1. Write a program to compute and plot P as a function of V at constant T for the van der Waals gas.
2. Modify your program to include a comparison curve for the ideal gas law at the same temperature.
3. Make plots for values of T that are at, above, and below T_c.
4. For some values of temperature, the plots will have a peculiar feature. What is it?
5. Make plots of T as a function of V for values of P above, below, and at P_c.
6. Make a plot of the chemical potential μ as a function of P. You might want to include a plot of V vs. P for comparison.
   What does the plot of μ vs. P look like at, above, and below T_c?
   The data you need to do a numerical integration of the Gibbs-Duhem relation is the same as you've already calculated for the first plot in this assignment. You just have to multiply the volume times the change in P and add it to the running sum to calculate μ vs. P.

PROBLEM 17.3
Helmholtz free energy near a phase transition

Consider a material described by the phase diagram. A quantity of this material is placed in a cylinder that is in contact with a thermal reservoir at temperature T_a. The temperature is held constant, and the pressure is increased as shown on the phase diagram, through the point b (on the coexistence curve) to the point c.

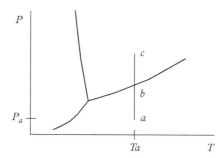

1. Make a qualitative sketch of the pressure P as following a function of volume V during the process that takes the system from the state a, through the state b, to the state c. Label the points on the phase diagram corresponding to states a, b, and c. Label the phases. Explain the signs and relative magnitudes of the slopes of the lines you draw. Explain any discontinuities in the function or its slope.
2. Sketch the Helmholtz free energy F as a function of the volume V for this process. Again label the phases and the points corresponding to states a, b, and c. Explain the relative magnitudes of the slopes and curvatures of the lines you draw. Explain any discontinuities in the function or its slope.

PROBLEM 17.4
Numerical evaluation of κ_T, α, and c_P as functions of P for the van der Waals gas – PART 1

As we'll see in this problem and the next one, the numerical evaluation of the properties of the van der Waals gas is much easier than finding an analytic solution. In this assignment you will derive the equations needed to write a computer program, which will be your next assignment.

We will use the volume V to parameterize the functions $P(V,T)$, etc. This will enable us to make plots of κ_T, α, and c_P as function of P, without actually solving the equations for $c_P(P,T)$ and $\kappa_T(P,T)$.

For the van der Waals gas:

1. Find an equation for κ_T as a function of P, T, and V by taking the derivative of the van der Waals equation of state with respect to V. You may find it convenient to express your result as $1/\kappa_T$.
2. Find an equation for α as a function of P, T, and V by taking the derivative of the van der Waals equation of state with respect to T.

   Hint: it is much easier to find an equation for α if you use a thermodynamic identity for

$$\left(\frac{\partial P}{\partial T} \right)_{V,N}.$$

3. Find c_V as a function of P, T, and V from the van der Waals equation for the energy U.
4. Find equations for c_P from a cleverly chosen thermodynamic identity, as a function of P, T, and V, AND α, AND κ_T, which you have already calculated.

PROBLEM 17.5
Numerical evaluation of κ_T, α, and c_P as functions of P for the van der Waals gas – PART 2

Now use the results of the previous assignment to do a numerical evaluation of the properties of the van der Waals gas.

Use the volume V to parameterize the functions $P(V,T)$, etc. Make plots of κ_T, α, and c_P as function of P. Note that you do not actually have to solve the equations for $\kappa_T(P,T), \alpha(P,T)$, and $c_P(P,T)$.

1. Write a program to plot κ_T, α, and c_P as functions of P AND their reciprocals as functions of P.
   Use V to first calculate P, and then use the values of V, P, and T to find $1/\kappa_T, 1/\alpha$, and $1/c_P$ from the equations derived in the previous assignment.
   Make plots for $T > T_c$ and values of P that range from below P_c to above P_c.
2. What happens when you use different values of T above T_c?
3. What happens at $T = T_c$?
4. What happens when $T < T_c$?

PROBLEM 17.6
Clausius–Clapeyron relation

The slope of a phase boundary in a $P - T$ phase diagram is not arbitrary. The Clausius–Clapeyron equation connects it directly to the latent heat and the change in volume at the transition.

The change in volume between the liquid and the gas, $V_{gas} - V_{liquid}$ can be approximated by noting that the liquid is much denser than the gas if you are not too close to the critical point. We can also approximate the latent heat by a constant under the same assumption.

Find an analytic approximation for the function $P(T)$ along the liquid–gas phase boundary. Sketch your answer and compare it to a phase diagram of water that you can find on the Web.

18

The Nernst Postulate: The Third Law of Thermodynamics

For every complex problem there is an answer that is clear, simple, and wrong.

H. L. Mencken, American journalist (1880–1956)

In the early part of the twentieth century the German chemist and physicist Walther Hermann Nernst (1884–1941) formulated the 'Nernst Postulate'. It has also become known as the Third Law of Thermodynamics, due to its importance.

> The Nernst Postulate: The entropy of a thermodynamic system goes to a constant as the temperature goes to zero.

This postulate has far-reaching consequences. Most directly, it puts important constraints on the behavior of real systems at low temperatures. Even more importantly, in its universal applicability, it reveals the pervasive influence of quantum mechanics on macroscopic phenomena.

18.1 Classical Ideal Gas Violates the Nernst Postulate

Since we have only discussed classical statistical mechanics so far in this book, the Nernst Postulate might seem rather startling. The results for the classical ideal gas that we derived in Part I are completely inconsistent with it. To see this, consider eq. (7.2) for the entropy of the classical ideal gas,

$$S(U,V,N) = kN\left[\frac{3}{2}\ln\left(\frac{U}{N}\right) + \ln\left(\frac{V}{N}\right) + X\right].$$
(18.1)

If we insert the equation of state for the energy,

$$U = \frac{3}{2}Nk_B T,$$
(18.2)

An Introduction to Statistical Mechanics and Thermodynamics. Robert H. Swendsen, Oxford University Press (2020).
© Robert H. Swendsen. DOI: 10.1093/oso/9780198853237.001.0001

we find an equation for the entropy as a function of temperature,

$$S(T,V,N) = kN \left[\frac{3}{2} \ln \left(\frac{3k_B T}{2} \right) + \ln \left(\frac{V}{N} \right) + X \right].$$

(18.3)

Since $\ln T \to -\infty$ as $T \to 0$, the entropy of the classical ideal gas does not go to a constant at zero temperature. In fact, it will be shown in Chapter 19 that the entropies of all classical systems go to $-\infty$ at zero temperature.

The reason for the validity of the Nernst Postulate for real systems lies entirely in quantum statistical mechanics, and will be derived in Chapter 23.

18.2   Planck's Form of the Nernst Postulate

Extending the Nernst Postulate, Max Planck made the stronger statement that the entropy should not only go to a constant, but that the constant must be zero. This is, indeed, the form in which the Nernst Postulate is usually remembered—which is unfortunate, since it is not entirely true. Although Nernst's formulation of his postulate is true for all quantum mechanical models and all real systems, Planck's version is not always valid.

We will return to a justification of the Nernst Postulate and a discussion of Planck's alternative in Chapter 23 on quantum statistical mechanics.

18.3   Consequences of the Nernst Postulate

Why is the Nernst Postulate important? One of the main reasons is that it places severe limits on the low-temperature behavior of the specific heat and the coefficient of thermal expansion.

18.3.1   Specific Heat at Low Temperatures

The first consequence of the Nernst Postulate is that the specific heat of anything must go to zero at zero temperature.

To prove this statement, use the connection between a small amount of heat, $đQ$, added to a system and the corresponding change in entropy, dS, in eq. (10.23),

$$dS = \frac{đQ}{T}.$$

(18.4)

We can relate the change of temperature to $đQ$ through the specific heat, which is defined in Section 14.4 as

$$c_X(T) = \frac{T}{N} \left(\frac{\partial S}{\partial T} \right)_{X,N},$$

(18.5)

where X stands for either V or P, depending on which is held constant for the experiment in question. The heat added is therefore

$$đQ = Nc_X(T)dT, \tag{18.6}$$

where we have explicitly indicated the temperature-dependence of the specific heat. Putting eqs. (18.4) and (18.6) together and integrating, we find an expression for the change in entropy between the temperatures T_1 and T_2,

$$S(T_2) - S(T_1) = \int_1^2 \frac{đQ}{T} = \int_{T_1}^{T_2} \frac{c_X(T)}{T} dT. \tag{18.7}$$

Now suppose that the specific heat went to a constant in the limit of zero temperature,

$$\lim_{T \to 0} c_X(T) = c_o. \tag{18.8}$$

Then, for sufficiently low temperatures, we could approximate the specific heat by c_o to calculate the change in entropy,

$$S(T_2) - S(T_1) \approx \int_{T_1}^{T_2} \frac{c_o}{T} dT = c_o(\ln T_2 - \ln T_1). \tag{18.9}$$

In the limit that T_1 goes to zero, eq. (18.9) implies that $S(T_2) - S(T_1) = \infty$, which contradicts the Nernst Postulate. Therefore, the Nernst Postulate requires that the specific heat of anything go to zero at zero temperature.

We have already seen $\lim_{T \to 0} S(T) = -\infty$ for the classical ideal gas, so that it violates the Nernst Postulate. This can also be seen from the fact that the specific heat at constant volume does not go to zero at zero temperature, but instead has the constant value $c_V = (3/2)k_B$.

18.4 Coefficient of Thermal Expansion at Low Temperatures

Another consequence of the Nernst Postulate is that the coefficient of thermal expansion, defined in Section 14.4 as

$$\alpha(T) = \frac{1}{V} \left(\frac{\partial V}{\partial T} \right)_{P,N}, \tag{18.10}$$

must also vanish at zero temperature. To see why this is true, note that a Maxwell relation that can be derived from

$$dG = -SdT + VdP + \mu dN, \tag{18.11}$$

gives us a relationship between a derivative of the entropy and the coefficient of thermal expansion,

$$\left(\frac{\partial S}{\partial P}\right)_{T,N} = -\left(\frac{\partial V}{\partial T}\right)_{P,N} = -V\alpha(T). \tag{18.12}$$

If the entropy goes to a constant that is independent of pressure, this gives us

$$\lim_{T\to 0} \alpha(T) = 0. \tag{18.13}$$

18.5   The Impossibility of Attaining a Temperature of Absolute Zero

It is often stated that a temperature of absolute zero is unattainable *because* of the Third Law of Thermodynamics. This is incorrect.

Classical statistical mechanics prohibits the attainment of $T = 0$ because it would require an infinite amount of energy. Quantum statistical mechanics actually makes it easier to achieve a low temperature, so that it would only require a finite amount of energy to achieve $T = 0$.

In both quantum and classical statistical mechanics, any process that would achieve $T = 0$ would require an infinite number of steps, so that it could not be completed in a finite amount of time.

The Nernst Postulate makes it easier to achieve low temperatures by making specific heats go to zero, as discussed in Section 18.3.1. Because of the equation

$$C = \frac{\partial Q}{\partial T}, \tag{18.14}$$

if we wish to lower the temperature by dT, it requires removing from the system of interest an amount of energy

$$dQ = C\,dT. \tag{18.15}$$

The larger the specific heat, C, the more energy must be removed from the system to lower the temperature by a given value of dT. Since the Nernst postulate requires $C \to 0$, as $T \to 0$, lowering the temperature is easier with the Nernst Postulate (quantum statistical mechanics) than it is without the Nernst Postulate (classical statistical mechanics).

These ideas will be explored further in the problems at the end of this chapter.

18.6   Summary and Signposts

In Part I of this book we developed the foundations of classical statistical mechanics based on the existence of atoms and some assumptions about model probabilities. We followed Boltzmann in defining the entropy in terms of the logarithm of the probability distribution of macroscopic observables in a composite system, and we derived a general formula for the entropy of a classical system of interacting particles. For the special case of non-interacting particles (the classical ideal gas), we explicitly evaluated the entropy as a function of energy, volume, and number of particles.

In Part II we developed the theory of thermodynamics on the basis of the general properties of the entropy that we found in Part I. This led us to develop powerful methods for generating exact relationships between measurable thermodynamic quantities.

With this chapter we come to the end of our introduction to thermodynamics, and with it the end of Part II of the book.

We have seen the power of thermodynamics in relating physical properties that, at first glance, might seem entirely unrelated to each other. On the other hand, we have seen that thermodynamics makes almost no predictions about the actual value of any quantity. Indeed, the main exceptions are the predictions for the vanishing of the specific heat and the coefficient of thermal expansion at zero temperature.

Up to this point, the emphasis has been on the concepts in thermal physics and their consequences for relationships between measurable quantities. In the remainder of the book we will develop more powerful methods for carrying out practical calculations in statistical mechanics.

Part III will discuss classical statistical mechanics, introducing the canonical ensemble, which is the workhorse of the field. It will also return to the question of irreversibility and the uniqueness of the model probability assumptions we made in setting up the foundations of statistical mechanics.

Part IV will discuss quantum statistical mechanics, including black-body radiation, and Bose–Einstein and Fermi–Dirac statistics. It will also show how quantum mechanics leads naturally to the Nernst Postulate. The book will end with an introduction to the theory of phase transitions, using the Ising model as an illustration.

By the end of the book you should have a good foundation in both thermodynamics and statistical mechanics—but it will still be only a foundation. The entire field of condensed-matter physics will be open to you, with all its complexity and surprises. May you enjoy your further explorations in this fascinating area.

18.7   Problems

PROBLEM 18.1
Is a temperature of absolute zero attainable?

The Third Law of Thermodynamics (the Nernst postulate) requires the heat capacity of any object to go to zero as the temperature goes to absolute zero. It is often stated in thermodynamics textbooks that the third law of thermodynamics also prohibits the attainment of a temperature of absolute zero. In this problem we will consider to what extent that claim is correct. If it is correct, that would mean that a temperature of absolute zero could be obtained if the Nernst postulate were not valid.

We will examine the consequences of a heat capacity that does not go to zero as the temperature goes to zero.

To simplify the argument, we will assume that we have two objects, A and B, with constant, non-zero heat capacities C_A and C_B. We will also assume that we have an ideal heat engine at our disposal.

1. Assume that the initial temperatures of the two objects are $T_{A,i} > 0$ and $T_{B,i} > 0$. Calculate the total work required to reduce the temperature of object A to absolute zero.
2. Now let the object B become infinite, so that it acts as a heat reservoir, and its temperature does not increase during the process. Again calculate the total work required to reduce the temperature of object A to absolute zero.
3. For the third part of this problem, consider again two *finite* objects, A and B. Assume that $C_A(T) \propto T^\alpha$, and $C_B(T) \propto T^\alpha$, where $\alpha > 0$ is a positive constant. Since $C_A(T) \to 0$ as $T \to 0$, this is consistent with the third law. We will see later in the course that materials with $\alpha = 1$ and $\alpha = 3$ really do exist. Again calculate the total work required to reduce the temperature of object A to absolute zero.

Part III

Classical Statistical Mechanics

19

Ensembles in Classical Statistical Mechanics

In theory, there is no difference between theory and practice. In practice, there is.

Yogi Berra

This chapter begins Part III, in which we go more deeply into classical statistical mechanics. In Chapter 7, eq. (7.45) provided a formal expression for the entropy of a general classical system of interacting particles as a function of the energy, volume, and number of particles. In principle, this completes the subject of classical statistical mechanics, since it is an explicit formula that determines the fundamental thermodynamic relation for any classical system of interacting particles. In practice, it does not work that way.

Although eq. (7.45) is *almost* correct (we will have to modify it for first-order phase transitions in Chapter 21), carrying out a 10^{20}-dimensional integral is non-trivial. It can be done for the ideal gas—which is why that example was chosen for Part I—but if the particles interact with each other, the whole business becomes much more difficult.

To make further progress we must develop more powerful formulations of the theory. The new expressions we derive will enable us to calculate some properties of interacting systems exactly. More importantly, they will enable us to make systematic approximations when we cannot find an exact solution.

The development of the new methods in classical statistical mechanics runs parallel to the development of the various representations of the fundamental relation in thermodynamics, which was discussed in Chapter 12. For every Legendre transform used in finding a new thermodynamic potential, there is a corresponding Laplace transform, named after Pierre-Simon, marquis de Laplace, French mathematician and scientist (1749–1827), that will take us to a new formulation of classical statistical mechanics. Just as different Legendre transforms are useful for different problems in thermodynamics, different Laplace transforms are useful for different problems in statistical mechanics.

With a slightly different interpretation, the canonical formalism will also allow us to refine the definition of the entropy. The idea is that we can use the canonical distribution as an improvement over the assumed delta-function distribution of the energy in defining

An Introduction to Statistical Mechanics and Thermodynamics. Robert H. Swendsen, Oxford University Press (2020).
© Robert H. Swendsen. DOI: 10.1093/oso/9780198853237.001.0001

the Boltzmann entropy. This will prove especially useful when we discuss quantum mechanical models that exhibit negative temperatures.

In Part IV we will see that the correspondence between Legendre transforms in thermodynamics and ensembles in statistical mechanics carries over into quantum statistical mechanics.

19.1 Microcanonical Ensemble

In Chapter 7 we defined the classical microcanonical ensemble by a uniform probability distribution in phase space, subject to the constraints that the particles are all in a particular volume and that the total energy is constant. The constraint on the positions was realized mathematically by limiting the integrals over the positions to the volume containing the particles. The constraint on the energy was expressed by a Dirac delta function. The Hamiltonian (energy as a function of momentum and position) for a system of particles with pair-wise interactions is given by

$$H(p,q) = \sum_{j=1}^{N} \frac{|\vec{p}_j|^2}{2m} + \sum_{j=1}^{N}\sum_{i>j}^{N} \phi(\vec{r}_i, \vec{r}_j). \tag{19.1}$$

The energy constraint is expressed by a delta function,

$$\delta(E - H(p,q)). \tag{19.2}$$

As in eq. (7.45), the entropy of a classical system of interacting particles is given by

$$S(E,V,N) = k \ln \Omega(E,V,N) \tag{19.3}$$

$$= k \ln \left[\frac{1}{h^{3N} N!} \int dq \int dp \, \delta(E - H(p,q)) \right].$$

The integrals $\int dq \int dp$ in eq. (19.3) are over the $6N$-dimensional phase space, which, unfortunately, makes them difficult or impossible to carry out explicitly for interacting systems. In this chapter we will investigate several approaches to the problem that enable us to make accurate and efficient approximations for many cases of interest. The first approach, which we will consider in the next section, is the use of computer simulations.

19.2 Molecular Dynamics: Computer Simulations

Computer simulations are remarkably useful for providing us with a good intuitive feeling for the properties of ensembles in statistical mechanics, as well as evaluating measurable quantities. The most natural method of simulating the microcanonical ensemble is known as Molecular Dynamics (MD). It consists of discretizing time, and then iterating discretized versions of Newton's equations of motion. Since Newton's equations conserve energy, MD is designed to explore a surface of constant energy in phase space, which is exactly what we want to do for the microcanonical ensemble.

For simplicity we will restrict the discussion to a single particle in one dimension. This avoids unnecessary notational complexity (indices) while demonstrating most of the important features of the method. The extension to more dimensions and more particles involves more programming, but no new ideas.

Consider a single particle in one dimension, moving under the influence of a potential energy $V(x)$. The equations of motion can be written in terms of the derivatives of the position and momentum

$$\frac{dx}{dt} = \frac{p}{m} \tag{19.4}$$

$$\frac{dp}{dt} = -\frac{dV(x)}{dx}. \tag{19.5}$$

The paths in phase space traced out by the solutions to these equations are called 'trajectories'.

The simplest way to discretize these equations is to define a time step δt and update the positions and momentum at every step.

$$x(t + \delta t) = x(t) + \frac{p}{m} \delta t \tag{19.6}$$

$$p(t + \delta t) = p(t) - \frac{dV(x)}{dx} \delta t. \tag{19.7}$$

While this is not the most efficient method, it is the easiest to program, and it is more than sufficient for the simulations in the assignments.

There are a quite a few ways of implementing discretized versions of Newton's equations—all of which go under the name of Molecular Dynamics. However, since our purpose is to understand physics rather than to write the most efficient computer program, we will only discuss the simplest method, which is given in eqs. (19.6) and (19.7).

There are two caveats:

1. Although solutions of the differential equations of motion (19.4) and (19.5) conserve energy exactly, the discretized equations (19.6) and (19.7) do not. As long as the time step δt is sufficiently short, the error is not significant. However, results should always be checked by changing the time step to determine whether it makes a difference.

2. Even if energy conservation is not a problem, it is not always true that an MD simulation will cover the entire surface of constant energy in phase space. If the trajectories do come arbitrarily close to any point on the constant energy surface, *after a very long time*, the system is said to be ergodic. Some examples in the assigned problems are ergodic, and some are not.

The MD simulations in the assignments will show, among other things, the distribution of positions in the microcanonical ensemble. The results will become particularly interesting when we compare them to the position distributions in the canonical ensemble (constant temperature, rather than constant energy), which we will explore with the Monte Carlo method in Section 19.11.

19.3 Canonical Ensemble

In this section we will show how to reduce the mathematical difficulties by calculating the properties of a system at fixed temperature rather than fixed energy. The result is known as the canonical ensemble, and it is an extremely useful reformulation of the essential calculations in statistical mechanics.

The canonical ensemble is the probability distribution in phase space for a system in contact with a thermal reservoir at a known temperature. The derivation of the canonical probability distribution is essentially the same as that of the Maxwell–Boltzmann probability distribution for the momentum of a particle, which was done in Subsection 8.3.1.

We will carry out the calculation in two different ways. First, we will calculate the energy distribution between the reservoir and the system of interest, based on the results summarized in Chapter 7. Next, we will calculate the probability density for the microstates (points in phase space) of the system of interest and show that the two calculations are consistent.

19.3.1 Canonical Distribution of the Energy

Assume that the system of interest is in thermal contact with a reservoir at temperature T. The entropy of the reservoir is given by an expression like that given in eq. (19.3),

$$S_R = k_B \ln \Omega_R(E_R). \tag{19.8}$$

We have suppressed the dependence on V_R and N_R in eq. (19.8) to make the notation more compact. Assume that the composite system of the thermal reservoir and the system of interest is isolated from the rest of the universe, so that the total energy of the composite system is fixed,

$$E_T = E + E_R. \tag{19.9}$$

As shown in Chapter 7, eq. (7.31), the probability distribution for the energy in the system of interest is given by

$$P(E) = \frac{\Omega(E)\Omega_R(E_T - E)}{\Omega_T(E_T)}. \tag{19.10}$$

Recall that the most important characteristic of a reservoir is that it is much bigger than the system of interest, so that $E_R \gg E$ and $E_T \gg E$. We can use these inequalities to make an extremely good approximation to eq. (19.10).

Since Ω_R is a rapidly varying function of energy, it is convenient to take the logarithm of both sides of eq. (19.10),

$$\ln P(E) = \ln \Omega(E) + \ln \Omega_R(E_T - E) - \ln \Omega_T(E_T). \tag{19.11}$$

Expand $\ln \Omega_R$ in powers of E,

$$\ln P(E) = \ln \Omega(E) + \ln \Omega_R(E_T) - E\left(\frac{\partial \ln \Omega_R(E_T)}{\partial E_T}\right)$$
$$- \ln \Omega_T(E_T) + \mathcal{O}\left((E/E_T)^2\right). \tag{19.12}$$

Since the system and the reservoir are at the same temperature in equilibrium,

$$T = T_R, \tag{19.13}$$

and

$$\beta = \frac{1}{k_B T} = \beta_R, \tag{19.14}$$

we have

$$\frac{\partial \ln \Omega_R(E_T)}{\partial E_T} = \beta_R = \beta = \frac{1}{k_R T}. \tag{19.15}$$

Since $\ln \Omega_R(E_T)$ and $\ln \Omega_T(E_T)$ are constants, we can replace them in eq. (19.12) with a single constant that we will write as $\ln Z$. With this change of notation, insert eq. (19.15) in eq. (19.12),

$$\ln P(E) = \ln \Omega(E) - \beta E - \ln Z. \tag{19.16}$$

Exponentiating eq. (19.16), we find a general expression for the canonical probability distribution of the energy,

$$P(E) = \frac{1}{Z}\Omega(E)\exp(-\beta E). \tag{19.17}$$

In eq. (19.17), Z is simply a normalization 'constant'. It is constant in the sense that it does not depend on the energy E, although it does depend on T, V, and N, which are held constant during this calculation. Using the fact that $P(E)$ must be normalized, an expression for Z is easy to obtain,

$$Z(T,V,N) = \int \Omega(E,V,N)\exp(-\beta E)dE. \qquad (19.18)$$

The integral is over all possible values of the energy E.

The kind of integral transformation shown in eq. (19.18) is known as a Laplace transform. It is similar to a Fourier transform, but the factor in the exponent is real, rather than imaginary.

Z turns out to be extremely important in statistical mechanics. It even has a special name: the 'partition function'. The reason for its importance will become evident in the rest of the book.

> The universal use of the letter Z to denote the partition function derives from its German name, *Zustandssumme*: the sum over states. (German *Zustand* = English *state*.) Eq. (19.18) shows that Z is actually defined by an integral in classical mechanics, but in quantum mechanics the partition function can be written as a sum over eigenstates, as we will see in Chapter 23.

19.3.2 Canonical Distribution in Phase Space

This subsection derives the probability density of points in phase space for the canonical ensemble. This is more fundamental than the distribution of the energy since $P(E)$ can be derived from the probability density in phase space by integrating over the delta function $\delta(E - H(p,q))$.

Denote the set of momentum and position variables describing the system of interest as $\{p,q\}$ and the corresponding set for the reservoir as $\{p_R, q_R\}$. The usual assumption of a uniform probability distribution then applies to the total phase space described by all momentum and position variables from both systems, subject to conservation of the total energy, E_T.

To find the probability distribution in the phase space of the system of interest, we simply integrate out the variables $\{p_R, q_R\}$ to obtain the marginal distribution $P(p,q)$. This gives us the canonical probability distribution at the temperature of the reservoir.

As shown in Chapter 7, eq. (7.31), the integral over $\{p_R, q_R\}$ gives us Ω_R. However, since the system of interest has an energy $H(p,q)$, the energy in the reservoir is $E_T - H(p,q)$. The resultant equation for the probability distribution in the phase space of the system of interest can be written as

$$P(p,q) = \frac{\Omega_R(E_T - H(p,q))}{\Omega_T(E_T)}. \qquad (19.19)$$

Since the reservoir is much larger than the system of interest, $E_T >> H(p,q)$. Taking the logarithm of both sides of eq. (19.19) and expanding $\ln \Omega_R$ in powers of $H(p,q)/E_T$, we find

$$\ln P(p,q) = \ln \Omega_R(E_T) - H(p,q)\left(\frac{\partial (\ln \Omega_R(E_T))}{\partial E_T}\right) - \ln \Omega_T(E_T) + \cdots . \qquad (19.20)$$

Recalling that

$$\beta = \beta_R \equiv \frac{\partial (\ln \Omega_R(E_T))}{\partial E_T} \qquad (19.21)$$

and noting that only the second term on the right in eq. (19.20) depends on p or q, we can write

$$\ln P(p,q) = -\beta H(p,q) - \ln \tilde{Z}(T,V,N) \qquad (19.22)$$

or

$$P(p,q) = \frac{1}{\tilde{Z}(T,V,N)} \exp[-\beta H(p,q)] \qquad (19.23)$$

The function $\tilde{Z}(T,V,N)$ is given by the normalization condition

$$\tilde{Z}(T,V,N) = \int dq \int dp \exp[-\beta H(q,p)]. \qquad (19.24)$$

The integral in eq. (19.24) is over all of phase space.

The only approximation in eq. (19.23) is the assumption that the reservoir is very large in comparison with the system of interest, which is usually an excellent approximation.

We still have to relate $\tilde{Z}(T,V,N)$ to the partition function $Z(T,V,N)$, which we will do in the next section.

19.4   The Partition Function as an Integral over Phase Space

If we combine the general equation for Ω, eq. (7.30),

$$\Omega(E,V,N) = \frac{1}{h^{3N}N!} \int dq \int dp \, \delta(E - H(q,p)), \qquad (19.25)$$

with the definition of the partition function in eq. (19.18),

$$Z(T,V,N) = \int \Omega(E) \exp(-\beta E) dE, \qquad (19.26)$$

and carry out the integral over the delta function, we find a very useful expression for the partition function in terms of an integral over phase space,

$$
\begin{aligned}
Z &= \int dE \frac{1}{h^{3N}N!} \int dq \int dp\, \delta(E - H(q,p)) \exp(-\beta E) \\
&= \frac{1}{h^{3N}N!} \int dq \int dp \int dE\, \delta(E - H(q,p)) \exp(-\beta E) \\
&= \frac{1}{h^{3N}N!} \int dq \int dp\, \exp[-\beta H(q,p)].
\end{aligned}
\tag{19.27}
$$

The final expression in eq. (19.27) is probably the most important equation for practical calculations in classical statistical mechanics. You should be able to derive it, but it is so important that you should also memorize it.

Comparing eq. (19.27) to eq. (19.24), we see that $\tilde{Z}(T,V,N)$ must be given by,

$$
\tilde{Z}(T,V,N) = h^{3N}N!Z.
\tag{19.28}
$$

The canonical probability density in phase space is then

$$
P(p,q) = \frac{1}{h^{3N}N!Z} \exp[-\beta H(p,q)].
\tag{19.29}
$$

19.5 The Liouville Theorem

Points in phase space move as functions of time to trace out trajectories. This rather obvious fact raises disturbing questions about the stability of the canonical distribution as a function of time. If a system is described by eq. (19.29) at a given time, will it still be described by the same distribution at a later time?

It is certainly not the case that an arbitrary, non-equilibrium probability distribution in phase space remains unchanged as time passes. We can see macroscopic changes in non-equilibrium systems, so the probability distribution must change.

Fortunately, for the canonical ensemble, the probability distribution is stable and does not change with time. This statement is based on the Liouville theorem, named after the French mathematician Joseph Liouville (1809–1882).

To prove the Liouville theorem, consider the probability density in phase space $P(p,q)$. Since points in phase space represent microscopic states, they can neither be created nor destroyed along a trajectory. We can then regard the points as abstract 'particles' moving in a 6N-dimensional space, with $P(p,q)$ corresponding to the density of such points. As for all gases of conserved particles, the points in phase space must obey a continuity equation,

$$\frac{\partial P(p,q)}{\partial t} + \nabla \cdot (P(p,q)\,\vec{v}) = 0. \tag{19.30}$$

The gradient in eq. (19.30) is defined by the vector,

$$\nabla = \left\{ \frac{\partial}{\partial q_j}, \frac{\partial}{\partial p_j} \,\middle|\, j = 1,\ldots,3N \right\}, \tag{19.31}$$

and the 6N-dimensional 'velocity' is defined by

$$\vec{v} = \left\{ \frac{\partial q_j}{\partial t}, \frac{\partial p_j}{\partial t} \,\middle|\, j = 1,\ldots,3N \right\} = \left\{ \dot{q}_j, \dot{p}_j \,\middle|\, j = 1,\ldots,3N \right\}, \tag{19.32}$$

where we have used a dot to indicate a partial derivative with respect to time.

The continuity equation then becomes

$$\frac{\partial P}{\partial t} + \sum_{j=1}^{3N} \left[\frac{\partial}{\partial q_j} \left(P(p,q)\,\dot{q}_j \right) + \frac{\partial}{\partial p_j} \left(P(p,q)\,\dot{p}_j \right) \right] = 0 \tag{19.33}$$

or

$$\frac{\partial P}{\partial t} + P \sum_{j=1}^{3N} \left[\frac{\partial \dot{q}_j}{\partial q_j} + \frac{\partial \dot{p}_j}{\partial p_j} \right] + \sum_{j=1}^{3N} \left[\frac{\partial P}{\partial q_j} \dot{q}_j + \frac{\partial P}{\partial p_j} \dot{p}_j \right] = 0. \tag{19.34}$$

Recall in classical mechanics that the time development of a point in phase space is given by Hamilton's equations, first derived by the Irish physicist and mathematician Sir William Rowan Hamilton (1805–1865)

$$\dot{q}_j = \frac{\partial H}{\partial p_j} \tag{19.35}$$

$$\dot{p}_j = -\frac{\partial H}{\partial q_j}. \tag{19.36}$$

$H = H(p,q)$ is, of course, the Hamiltonian. These two equations give us an interesting identity,

$$\frac{\partial \dot{q}_j}{\partial q_j} = \frac{\partial^2 H}{\partial q_j \partial p_j} = \frac{\partial^2 H}{\partial p_j \partial q_j} = -\frac{\partial \dot{p}_j}{\partial p_j}. \tag{19.37}$$

Inserting eq. (19.37) into eq. (19.34), we find that it simplifies to

$$\frac{\partial P}{\partial t} + \sum_{j=1}^{3N}\left[\frac{\partial P}{\partial q_j}\dot{q}_j + \frac{\partial P}{\partial p_j}\dot{p}_j\right] = 0. \tag{19.38}$$

To complete the proof of Liouville's theorem, we must only insert eq. (19.38) into the equation for the total derivative of $P(p, q)$ with respect to time.

$$\frac{dP}{dt} = \frac{\partial P}{\partial t} + \sum_{j=1}^{3N}\left[\frac{\partial P}{\partial q_j}\dot{q}_j + \frac{\partial P}{\partial p_j}\dot{p}_j\right] = 0 \tag{19.39}$$

The interpretation of eq. (19.39) is that the probability density in the neighborhood of a moving point remains constant throughout the trajectory.

The application of Liouville's theorem to the canonical ensemble rests on the property that the canonical probability density depends only on the total energy. Since the trajectory of a point in phase space conserves energy, the canonical probability distribution does not change with time.

19.6   Consequences of the Canonical Distribution

The probability density, $P(E)$, for the energy in the canonical ensemble is extremely sharply peaked. The reason is that $P(E)$ is the product of two functions that both vary extremely rapidly with energy: $\Omega(E)$ increases rapidly with increasing E, while $\exp(-\beta E)$ goes rapidly to zero.

To see that this is true, first recall that $\Omega(E)$ for the ideal gas has an energy dependence of the form

$$\Omega(E) \propto E^f, \tag{19.40}$$

where $f = 3N/2$. For more general systems, f is usually a relatively slowly varying function of the energy, but remains of the same order of magnitude as the number of particles N—at least to within a factor of 1000 or so—which is sufficient for our purposes.

If we approximate f by a constant, we can find the location of the maximum of $P(E)$ by setting the derivative of its logarithm equal to zero,

$$\frac{\partial \ln P(E)}{\partial E} = \frac{\partial}{\partial E}[\ln \Omega(E) - \beta E - \ln Z] = \frac{f}{E} - \beta = 0. \tag{19.41}$$

The location of the maximum of $P(E)$ gives the equilibrium value of the energy:

$$E_{eq} = \frac{f}{\beta} = fk_BT. \tag{19.42}$$

The width of the probability distribution for the energy can be found from the second derivative if we approximate $P(E)$ by a Gaussian function, as discussed in connection with applications of Bayes' theorem in Section 5.5,

$$\frac{\partial^2 \ln P(E)}{\partial E^2} = \frac{\partial}{\partial E}\left[\frac{f}{E} - \beta\right] = -\frac{f}{E^2} = -\frac{1}{\sigma_E^2}. \tag{19.43}$$

Evaluating the expression for the second derivative in eq. (19.43) at $E = E_{eq}$, we find the width of $P(E)$

$$\sigma_E = \frac{E_{eq}}{\sqrt{f}}. \tag{19.44}$$

When f is of the order N, and $N \approx 10^{20}$ or larger, the relative width of the probability distribution is about 10^{-10} or smaller. This is much smaller than the accuracy of thermodynamic measurements. Note that the $1/\sqrt{N}$ dependence is the same as we found for the width of the probability distribution for the number of particles in Chapter 4.

19.7 The Helmholtz Free Energy

The fact that the width of the probability distribution for the energy is roughly proportional to $1/\sqrt{N}$ in the canonical ensemble has an extremely important consequence, which is the basis for the importance of the partition function.

Consider eq. (19.16) for the logarithm of the canonical probability density for the energy

$$\ln P(E) = -\beta E + \ln \Omega(E) - \ln Z. \tag{19.45}$$

From eq. (7.45), we know that $S = k_B \ln \Omega$, so we can rewrite eq. (19.45) as

$$\begin{aligned} \ln Z &= -\beta E + S/k_B - \ln P(E) \\ &= -\beta(E - TS) - \ln P(E) \\ &= -\beta F - \ln P(E). \end{aligned} \tag{19.46}$$

Since the probability distribution $P(E)$ is normalized and its width is roughly proportional to $1/\sqrt{N}$, its height must be roughly proportional to $\sqrt{N}$. If we evaluate eq. (19.46) at the maximum of $P(E)$, which is located at $E = E_{eq}$, the term $\ln P(E)$ must be of the order $(1/2)\ln N$. However, the energy E, the entropy S, and the free energy $F = E - TS$ are all of order N. If $N \approx 10^{20}$, then

$$\frac{\ln P(E_{eq})}{\beta F} \approx \frac{\ln N}{N} \approx 5 \times 10^{-19}. \qquad (19.47)$$

This means that the last term in eq. (19.46) is completely negligible. We are left with a simple relationship between the canonical partition function and the Helmholtz free energy

$$\ln Z(T,V,N) = -\beta F(T,V,N) \qquad (19.48)$$

or

$$F(T,V,N) = -k_B T \ln Z(T,V,N). \qquad (19.49)$$

To remember this extremely important equation, it might help to recast the familiar equation,

$$S = k_B \ln \Omega, \qquad (19.50)$$

in the form

$$\Omega = \exp[S/k_B]. \qquad (19.51)$$

In carrying out the Laplace transform in eq. (19.18), we multiplied by Ω the factor $\exp[-\beta E]$ before integrating over E. The integrand in eq. (19.18) can therefore be written as

$$\begin{aligned}
\Omega \exp[-\beta E] &= \exp[-\beta E + S/k_B] \\
&= \exp[-\beta(E - TS)] \\
&= \exp[-\beta F]. \qquad (19.52)
\end{aligned}$$

19.8 Thermodynamic Identities

Since the postulates of thermodynamics are based on the results of statistical mechanics, it should not be too surprising that we can derive thermodynamic identities directly from statistical mechanics.

For example, if we take the partial derivative of the logarithm of the partition function in eq. (19.27) with respect to β, we find that it is related to the average energy

$$\frac{\partial}{\partial\beta}\ln Z = \frac{1}{Z}\frac{\partial Z}{\partial\beta}$$

$$= \frac{1}{Z}\frac{\partial}{\partial\beta}\left[\frac{1}{h^{3N}N!}\int dq\int dp\,\exp[-\beta H(q,p)]\right]$$

$$= -\frac{1}{Z}\frac{1}{h^{3N}N!}\int dq\int dp\,H(q,p)\exp[-\beta H(q,p)]$$

$$= -\frac{\int dq\int dp\,H(q,p)\exp[-\beta H(q,p)]}{\int dq\int dp\,\exp[-\beta H(q,p)]}$$

$$= -\langle E\rangle. \tag{19.53}$$

Since we know that $\ln Z = -\beta F$, eq. (19.53) is equivalent to the thermodynamic identity

$$U = \langle E\rangle = \frac{\partial(\beta F)}{\partial\beta}, \tag{19.54}$$

which was proved as an exercise in Chapter 14.

19.9 Beyond Thermodynamic Identities

Because statistical mechanics is based on the microscopic structure of matter, rather than the phenomenology that led to thermodynamics, it is capable of deriving relationships that are inaccessible to thermodynamics. An important example is a general relationship between the fluctuations of the energy and the specific heat.

Begin with the expression for the average energy from eq. (19.53),

$$U = \langle E\rangle$$

$$= \frac{1}{Z}\frac{1}{h^{3N}N!}\int dq\int dp\,H(q,p)\exp[-\beta H(q,p)]$$

$$= \frac{\int dq\int dp\,H(q,p)\exp[-\beta H(q,p)]}{\int dq\int dp\,\exp[-\beta H(q,p)]}. \tag{19.55}$$

Note that the dependence on the temperature appears in only two places in eq. (19.55): once in the numerator and once in the denominator. In both cases, it appears in an exponent as the inverse temperature $\beta = 1/k_B T$. It is usually easier to work with β than with T in statistical mechanical calculations. To convert derivatives, either use the identity,

$$\frac{d\beta}{dT} = \frac{d}{dT}\left(\frac{1}{k_B T}\right) = \frac{-1}{k_B T^2}, \tag{19.56}$$

or the identity

$$\frac{dT}{d\beta} = \frac{d}{d\beta}\left(\frac{1}{k_B\beta}\right) = \frac{-1}{k_B\beta^2}. \tag{19.57}$$

The specific heat per particle at constant volume can then be written as

$$c_V = \frac{1}{N}\left(\frac{\partial U}{\partial T}\right)_{V,N} = \frac{1}{N}\frac{\partial}{\partial T}\langle E\rangle = \frac{1}{N}\frac{-1}{k_B T^2}\frac{\partial}{\partial \beta}\langle E\rangle. \tag{19.58}$$

We can now take the partial derivative of $\langle E\rangle$ in eq. (19.55),

$$
\begin{aligned}
\frac{\partial}{\partial \beta}\langle E\rangle &= \frac{\partial}{\partial \beta}\left[\frac{\int dq\,\int dp\,H(q,p)\exp[-\beta H(q,p)]}{\int dq\,\int dp\,\exp[-\beta H(q,p)]}\right] \\
&= \frac{\int dq\,\int dp\,H(q,p)(-H(q,p))\exp[-\beta H(q,p)]}{\int dq\,\int dp\,\exp[-\beta H(q,p)]} \\
&\quad - \frac{\int dq\,\int dp\,H(q,p)\exp[-\beta H(q,p)]\int dq\,\int dp\,(-H(q,p))\exp[-\beta H(q,p)]}{\left[\int dq\,\int dp\,\exp[-\beta H(q,p)]\right]^2} \\
&= -\langle E^2\rangle + \langle E\rangle^2, \tag{19.59}
\end{aligned}
$$

and use eq. (19.58) to find the specific heat:

$$c_V = \frac{1}{N}\frac{-1}{k_B T^2}\frac{\partial}{\partial \beta}\langle E\rangle = \frac{1}{N k_B T^2}\left(\langle E^2\rangle - \langle E\rangle^2\right). \tag{19.60}$$

In relating the fluctuations in the energy to the specific heat, eq. (19.60) goes beyond thermodynamics. It gives us a direct connection between the microscopic fluctuations and macroscopic behavior. This relationship turns out to be extremely useful in computational statistical mechanics, where it has been found to be the most accurate method for calculating the specific heat from a computer simulation.

It should be clear from the derivation of eq. (19.60) that it can be generalized to apply to the derivative of any quantity of interest with respect to any parameter in the Hamiltonian. The generality and flexibility of this technique has found many applications in computational statistical mechanics.

19.10 Integration over the Momenta

A great advantage of working with the canonical partition function is that the integrals over the momenta can be carried out exactly for any system in which the forces do not depend on the momenta. This, unfortunately, eliminates systems with moving particles in magnetic fields, but still leaves us with an important simplification for most systems of interest. As an example, we will consider the case of pairwise interactions for simplicity, although the result is still valid for many-particle interactions.

If $H(p,q)$ only depends on the momenta through the kinetic energy, the partition function in eq. (19.27) becomes

$$Z = \frac{1}{h^{3N}N!} \int dp \int dq \exp\left[-\beta\left(\sum_{j=1}^{N}\frac{|\vec{p}_j|^2}{2m} + \sum_{j=1}^{N}\sum_{i>j}^{N}\phi(\vec{r}_i,\vec{r}_j)\right)\right]$$

$$= \frac{1}{h^{3N}N!} \int dp\, \exp\left[-\beta\sum_{k=1}^{3N}\frac{p_k^2}{2m}\right] \int dq\, \exp\left[-\beta\sum_{j=1}^{N}\sum_{i>j}^{N}\phi(\vec{r}_i,\vec{r}_j)\right]$$

$$= \frac{1}{h^{3N}N!} (2\pi mk_BT)^{3N/2} \int dq\, \exp\left[-\beta\sum_{j=1}^{N}\sum_{i>j}^{N}\phi(\vec{r}_i,\vec{r}_j)\right], \tag{19.61}$$

where we have used $\beta = 1/k_BT$ in the last line of eq. (19.61). The momenta have been integrated out exactly.

19.10.1   The Classical Ideal Gas

For the ideal gas, the interactions between particles vanish $[\phi(\vec{r}_i,\vec{r}_j) = 0]$, and we can rederive the main results of Part I in a few lines. The integrals remaining in eq. (19.61) give a factor of V for each particle, so that,

$$Z = \frac{1}{h^{3N}N!} (2\pi mk_BT)^{3N/2} \int dq\, \exp(0)$$

$$= \frac{1}{h^{3N}N!} (2\pi mk_BT)^{3N/2} V^N. \tag{19.62}$$

The logarithm of eq. (19.62) gives the Helmholtz free energy, $F = -k_BT\ln Z$, of the classical ideal gas as a function of temperature, volume, and number of particles, which is a representation of the fundamental relation and contains all thermodynamic information. It is left as an exercise to confirm that the explicit expression for F agrees with that derived from the entropy of the classical ideal gas in Part I.

19.10.2   Computer Simulation in Configuration Space

Eq. (19.61) is used extensively in Monte Carlo computer simulations because it eliminates the need to simulate the momenta and therefore reduces the required computational effort. Monte Carlo computations of the thermal properties of models in statistical mechanics will be discussed in the next section.

19.11   Monte Carlo Computer Simulations

Just as the Molecular Dynamics method is the most natural way to simulate a microcanonical ensemble (constant energy), the Monte Carlo (MC) method is the most natural way to simulate a canonical ensemble (constant temperature). The MC method ignores the natural trajectories of the system in favor of a random sampling that reproduces the canonical probability distribution. In practice, applications of MC to classical statistical mechanics usually simulate the equilibrium canonical probability distribution for the configurations after the momenta have been integrated out,

$$P_{eq}(q) = \frac{1}{Q} \exp[-\beta V(q)]. \tag{19.63}$$

In eq. (19.63), which follows directly from eq. (19.29) upon integrating out the momenta, the normalization constant Q plays a role very similar to the partition function. It is defined by

$$Q \equiv \int \exp[-\beta V(q)]\, dq, \tag{19.64}$$

where the integral goes over the allowed volume in configuration space.

To implement an MC simulation, we first construct a stochastic process for probability distributions defined in configuration space. A stochastic process is a random sequence of states (configurations in our case) in which the probability of the next state depends on the previous states and the parameters of the model (temperature, interaction energies, and so on). Our goal is to find a stochastic process for which an arbitrary initial probability distribution will evolve into the desired equilibrium distribution

$$\lim_{t \to \infty} P(q,t) = P_{eq}(q) = \frac{1}{Q} \exp[-\beta V(q)]. \tag{19.65}$$

Once we have such a process we can use it to generate states with the equilibrium probability distribution and average over those states to calculate thermal properties.

If the next state in a stochastic process does not depend on states visited before the previous one, it is called a Markov process. Since we will see that they are particularly convenient for simulations, we will consider only Markov processes for MC simulations.

For a given Markov process, we can define a conditional probability for a particle to be in configuration q' at time $t + \delta t$ if it was in configuration q at time t. We will denote this conditional probability as $W(q'|q)$. (It is also called a transition probability and is often denoted by $W(q \to q')$ to indicate the direction of the transition.) Since the transition from one state to the next in a Monte Carlo simulation is discrete, δt is called the time step and is *not* an infinitesimal quantity.

The change in the probability density $P(q,t)$ for q between times t and $t + \delta t$ is then given by the 'Master Equation',

$$P(q, t + \delta t) - P(q, t) = \int \left[W(q|q')P(q',t) - W(q'|q)P(q,t) \right] dq', \tag{19.66}$$

where the integral in eq. (19.66) goes over all allowed configurations q'. The term $W(q|q')P(q',t)$ in the integrand of eq. (19.66) represents the increase in the probability of state q due to transitions from other states, while $W(q'|q)P(q,t)$ represents the decrease in the probability of state q due to transitions out of that state. Obviously, the transition probabilities must be normalized, so that,

$$\int W(q'|q)dq' = 1, \tag{19.67}$$

for all q.

A necessary condition for the simulation to go to equilibrium as indicated in eq. (19.65) is that once the probability distribution reaches $P_{eq}(q)$, it stays there. In equilibrium, $P(q,t) = P_{eq}(q)$ is independent of time, and the master equation becomes

$$P_{eq}(q) - P_{eq}(q) = 0 = \int \left[W(q|q')P_{eq}(q') - W(q'|q)P_{eq}(q) \right] dq'. \tag{19.68}$$

Although the Markov process will remain in equilibrium as long as the integral in eq. (19.68) vanishes, the more stringent condition that the integrand vanishes turns out to be more useful,

$$W(q|q')P_{eq}(q') - W(q'|q)P_{eq}(q) = 0. \tag{19.69}$$

Eq. (19.69) is called the condition of 'detailed balance', because it specifies that the number of transitions between any two configurations is the same in both directions.

It can be shown that if there is a finite sequence of transitions such that the system can go from any configuration to any other configuration with non-zero probability *and* that detailed balance holds, the probability distribution will go to equilibrium as defined in eq. (19.65). After the probability distribution has gone to equilibrium, we can continue the Markov process and use it to sample from $P_{eq}(q)$.

The condition that there is a finite sequence of transitions such that the system can go from any configuration to any other configuration with non-zero probability is called 'ergodicity'. This terminology is unfortunate, because the word 'ergodicity' is also used to refer to the property that dynamical trajectories come arbitrarily close to any given point on an energy surface, as discussed in connection with MD simulations. It is much too late in the history of the subject to produce words that clearly distinguish these two concepts, but we may take some consolation in the fact that one meaning is used only with MC, and the other only with MD.

The great thing about the condition of detailed balance in eq. (19.69) is that we are able to decide what the conditional probabilities $W(q'|q)$ will be. For simulating a canonical ensemble, we want to use this power to create a Markov process that produces the canonical distribution,

$$\lim_{t \to \infty} P(q,t) = P_{eq}(q) = \frac{1}{Q} \exp[-\beta V(q)]. \tag{19.70}$$

As long as the conditional (transition) probabilities do not vanish, the condition of detailed balance can be written as,

$$\frac{W(q'|q)}{W(q|q')} = \frac{P_{eq}(q')}{P_{eq}(q)} = \exp[-\beta(V(q') - V(q))], \tag{19.71}$$

or,

$$\frac{W(q'|q)}{W(q|q')} = \exp[-\beta \Delta E], \tag{19.72}$$

where $\Delta E = V(q') - V(q)$. Note that the 'partition function', Q, has canceled out of eq. (19.72)—which is very convenient, since we usually cannot evaluate it explicitly. There are many ways of choosing the $W(q'|q)$ to satisfy eq. (19.72). The oldest and simplest is known as the Metropolis algorithm.

Metropolis algorithm:

A trial configuration q_{trial} is chosen at random.

- If $\Delta E \leq 0$, the new state is $q' = q_{trial}$.
- If $\Delta E > 0$, the new state is $q' = q_{trial}$ with probability $\exp(-\beta \Delta E)$, and the same as the old state $q' = q$ with probability $1 - \exp(-\beta \Delta E)$.

It is easy to confirm that the Metropolis algorithm satisfies detailed balance.

There are many things that have been swept under the rug in this introduction to the Monte Carlo method. Perhaps the most serious is that there is nothing in eq. (19.70) that says that convergence will occur within your computer budget. Slow MC processes can also make acquiring data inefficient, which has prompted a great deal of effort to go into devising more efficient algorithms. There are also many methods for increasing the amount of information acquired from MC (and MD) computer simulations. Fortunately, there are a number of excellent books on computer simulation methods available for further study.

We will again restrict our examples in the assignments to a single particle in one dimension, as we did in Section 19.2 for Molecular Dynamics simulations. Remember to compare the results of the two simulation methods to see the different kinds of information each reveals about the thermal properties of the system.

19.12   Factorization of the Partition Function: The Best Trick in Statistical Mechanics

The key feature of the Hamiltonian that allowed us to carry out the integrals over the momenta in eq. (19.61) and the integrals over the coordinates in eq. (19.62) is that the integrand $\exp[-\beta H]$ can be written as a product of independent terms. Writing a $3N$-dimensional integral as a one-dimensional integral raised to the $3N$-th power takes a nearly impossible task and makes it easy.

We can extend this trick to any system in which the particles do not interact with each other, even when they do interact with a potential imposed from outside the system. Consider the following Hamiltonian:

$$H(p,q) = \sum_{j=1}^{N} \frac{|\vec{p}_j|^2}{2m} + \sum_{j=1}^{N} \tilde{\phi}_j(\vec{r}_j). \tag{19.73}$$

The partition function for the Hamiltonian in eq. (19.73) separates neatly into a product of three-dimensional integrals:

$$Z = \frac{1}{h^{3N}N!} (2\pi mk_B T)^{3N/2} \int dq \exp\left[-\beta \sum_{j=1}^{N} \tilde{\phi}_j(\vec{r}_j)\right]$$

$$= \frac{1}{h^{3N}N!} (2\pi mk_B T)^{3N/2} \int dq \prod_{j=1}^{N} \exp\left[-\beta\tilde{\phi}_j(\vec{r}_j)\right]$$

$$= \frac{1}{h^{3N}N!} (2\pi mk_B T)^{3N/2} \prod_{j=1}^{N} \int d\vec{r}_j \exp\left[-\beta\tilde{\phi}_j(\vec{r}_j)\right]. \tag{19.74}$$

Eq. (19.74) has reduced the problem from one involving an integral over a $3N$-dimensional space, as in eq. (19.61), to a product of N three-dimensional integrals.

When the three-dimensional integrals are all the same, eq. (19.74) simplifies further:

$$Z = \frac{1}{h^{3N}N!} (2\pi mk_B T)^{3N/2} \left(\int d\vec{r}_1 \exp\left[-\beta\tilde{\phi}_1(\vec{r}_1)\right]\right)^N. \tag{19.75}$$

Since a three-dimensional integral can always be evaluated, even if only numerically, all problems of this form can be regarded as solved.

This trick is used repeatedly in statistical mechanics. Even when it is not possible to make an exact factorization of the partition function, it is often possible to make a good approximation that does allow factorization.

Factorization of the partition function is the most important trick you need for success in statistical mechanics. Do not forget it!

19.13   Simple Harmonic Oscillator

The simple harmonic oscillator plays an enormous role in statistical mechanics. The reason might be that many physical systems behave as simple harmonic oscillators, or it might be just that it is one of the few problems we can solve exactly. It does provide a nice example of the use of factorization in the evaluation of the partition function.

19.13.1   A Single Simple Harmonic Oscillator

First consider a single, one-dimensional, simple harmonic oscillator (SHO). The Hamiltonian is given by the following equation, where the subscript 1 indicates that there is just a single SHO:

$$H_1 = \frac{1}{2}Kx^2 + \frac{p^2}{2m}. \tag{19.76}$$

The partition function is found by simplifying eq. (19.27). The integrals over x and p are both Gaussian, and can be carried out immediately,

$$\begin{aligned} Z_1 &= \frac{1}{h}\int_{-\infty}^{\infty} dx \int_{-\infty}^{\infty} dp \, \exp\left(-\beta\left(\frac{1}{2}Kx^2 + \frac{1}{2m}p^2\right)\right) \\ &= \frac{1}{h}(2\pi k_B T/K)^{1/2}(2\pi m k_B T)^{1/2} \\ &= \frac{1}{\beta\hbar}\left(\frac{m}{K}\right)^{1/2} \\ &= \frac{1}{\beta\hbar\omega} \end{aligned} \tag{19.77}$$

where

$$\omega = \sqrt{\frac{K}{m}} \tag{19.78}$$

is the angular frequency of the SHO.

As we have mentioned earlier, the factor of $\hbar$ has no true significance within classical mechanics. However, we will see in Section 24.10 that the result in eq. (19.77) does coincide with the classical limit of the quantum simple harmonic oscillator given in eq. (24.71).

19.13.2 *N* Simple Harmonic Oscillators

Now consider a macroscopic system consisting of N one-dimensional simple harmonic oscillators,

$$H_N = \sum_{j=1}^{N} \left(\frac{1}{2} K_j x_j^2 + \frac{p_j^2}{2m_j} \right). \tag{19.79}$$

Note that the spring constants K_j and the masses m_j might all be different.

Since the oscillating particles are localized by the harmonic potential, we do not need to consider exchanges of the particles with another system. Therefore, we do not include an extra factor of $1/N!$ to appear in the partition function. Actually, since N is never varied in this model, a factor of $1/N!$ would not have any physical consequences. However, excluding it produces extensive thermodynamic potentials that are esthetically pleasing.

After integrating out the momenta, the partition function can be written as,

$$
\begin{aligned}
Z_N &= \frac{1}{h^N} \prod_{j=1}^{N} (2\pi m_j k_B T)^{1/2} \int_{-\infty}^{\infty} d^N x \, \exp\left(-\beta \sum_{j=1}^{N} \frac{1}{2} K_j x_j^2 \right) \\
&= \frac{1}{h^N} \prod_{j=1}^{N} \left[(2\pi m_j k_B T)^{1/2} \int_{-\infty}^{\infty} dx_j \, \exp\left(-\beta \frac{1}{2} K_j x_j^2 \right) \right] \\
&= \frac{1}{h^N} \prod_{j=1}^{N} \left[(2\pi m_j k_B T)^{1/2} (2\pi k_B T / K_j)^{1/2} \right] \\
&= \left(\frac{2\pi k_B T}{h} \right)^N \prod_{j=1}^{N} \left(\frac{m_j}{K_j} \right)^{1/2} \\
&= \prod_{j=1}^{N} \left(\frac{1}{\beta \hbar \omega_j} \right),
\end{aligned}
\tag{19.80}
$$

where

$$\omega_j = \sqrt{\frac{K_j}{m_j}} \tag{19.81}$$

is the angular frequency of the j-th SHO.

If all N SHO's have the same frequency, the partition function simplifies further,

$$Z_N = \left(\frac{1}{\beta \hbar \omega_1} \right)^N. \tag{19.82}$$

19.14 Problems

....

PROBLEM 19.1
Two one-dimensional relativistic particles

Consider two one-dimensional **relativistic** ideal-gas particles with masses confined to a one-dimensional box of length L. Because they are relativistic, their energies are given by $E_A = |p_A|c$ and $E_A = |p_B|c$.

Assume that the particles are in thermal equilibrium with each other, and that the total kinetic energy is $E = E_A + E_B$. Use the usual assumption that the probability density is uniform in phase space, subject to the constraints.

Calculate the probability distribution $P(E_A)$ for the energy of one of the particles.

PROBLEM 19.2
Classical simple harmonic oscillator

Consider a one-dimensional, classical, simple harmonic oscillator with mass m and potential energy

$$\frac{1}{2}Kx^2.$$

The SHO is in contact with a thermal reservoir (heat bath) at temperature T.

Calculate the classical partition function, Helmholtz free energy, and energy.

PROBLEM 19.3
Canonical ensemble for anharmonic oscillators

We have shown that the classical canonical partition function is given by

$$Z = \frac{1}{h^{3N}N!} \int dp^{3N} \int dq^{3N} \exp\left[-\beta H(p,q)\right].$$

Consider N non-interacting (ideal gas) particles in an anharmonic potential,

$$\frac{K}{r}\left(|x|^r + |y|^r + |z|^r\right)$$

where r is a positive constant.

The full Hamiltonian can be written as:

$$H(p,q) = \sum_{n=1}^{N}\frac{p^2}{2m} + \frac{K}{r}\sum_{j=1}^{3N}|q_j|^r.$$

Calculate the energy and the specific heat using the canonical partition function.

Hint: The integrals over the momenta shouldn't cause any problems, since they are the same as for the ideal gas. However, the integrals over the positions require a trick that we used in deriving the value of the gaussian integral.

Another hint: You might find that there is an integral that you can't do, but don't need.

PROBLEM 19.4
Partition function of ideal gas

1. Calculate the partition function and the Helmholtz free energy of the classical ideal gas using the canonical ensemble.
2. Starting with the entropy of the classical ideal gas, calculate the Helmholtz free energy and compare it with the result from the canonical ensemble.

PROBLEM 19.5
Molecular Dynamics (MD) simulation of a simple harmonic oscillator

This assignment uses the Molecular Dynamics method, which simply consists of discretizing Newton's equations of motion, and then iterating them many times to calculate the trajectory.

Write a computer program to perform a Molecular Dynamics (MD) computer simulation of a one-dimensional simple harmonic oscillator (SHO),

$$H = \frac{1}{2} K x^2.$$

[You can take $K = 1$ in answering the questions.]

Include a plot of the potential energy, including a horizontal line indicating the initial energy. [You will also do Monte Carlo (MC) simulations of the same models later, so remember to make your plots easy to compare.]

The program should read in the starting position, the size of the time step, and the number of iterations. It should then print out a histogram of the positions visited during the simulation.

Your program should also print out the initial and final energy as a check on the accuracy of the algorithm. Don't forget to include the kinetic energy in your calculation!

1. For a starting value of $x = 1.0$. Try various values of the size of the time step to see the effect on the results.

   What happens when the time step is very small or very large?
2. Choose $x = 1.0$ and run the program for an appropriate value of dt. Print out histograms of the positions visited. [Later, you will compare them with you results for MC.]

PROBLEM 19.6
Molecular Dynamics (MD) simulation of an anharmonic oscillator

This assignment again uses the Molecular Dynamics method. Only minor modifications of your code should be necessary, although I would recommend using functions to simplify the programming.

Write a computer program to perform a Molecular Dynamics (MD) computer simulation of a one-dimensional particle in the following potential:

$$H = Ax^2 + Bx^3 + Cx^4.$$

[The values of A, B, and C should be read in or set at the beginning of the program.]

Include a plot of the potential energy, including a horizontal line indicating the initial energy.

The program should read in the starting position, the size of the time step, and the number of iterations. It should then print out a histogram of the positions visited during the simulation.

Your program should also print out the initial and final energy as a check on the accuracy of the algorithm. Don't forget to include the kinetic energy in your calculation!

1. Carry out simulations with the values $A = -1.0$, $B = 0.0$, and $C = 1.0$.
   First try various starting value of the position x. Choose the starting positions to adjust the (constant) total energy to take on 'interesting' values. Part of the problem is to decide what might be interesting on the basis of the plot of the potential energy.
2. Now change the value of B to -1.0, while retaining $A = -1.0$ and $C = 1.0$. Again look for interesting regions.

PROBLEM 19.7

Molecular Dynamics (MD) simulation of a chain of simple harmonic oscillators

This assignment again uses the Molecular Dynamics method, but this time we will simulate a chain of particles connected by harmonic interactions.

Write a computer program to perform a Molecular Dynamics (MD) computer simulation of the following Hamiltonian:

$$H = \sum_{j=1}^{N} \frac{p_j^2}{2m} + \frac{1}{2}K\sum_{j=1}^{N}(x_j - x_{j-1})^2.$$

Take the particle at $x_0 = 0$ to be fixed.

1. For various (small) values of N, try different values of the time step while monitoring the change in total energy during the simulation. See if a good choice of time step agrees with what you found for a single particle.
2. For $N = 1$, make sure that you get the same answer as you did for HW2. [Just confirm that you did it; you don't have to hand it in.]
3. For $N = 2$ and 3, and random starting values of position and momentum, what distributions do you get for x_1 and $x_{N-1} - x_{N-2}$?
4. For $N = 10$, start particle 1 with an initial position of $x[1] = 1.0$, let the initial positions of all other particles be 0.0, and let the initial value of all momenta be 0.0. What distribution do you get for $x_j - x_{j-1}$ for whatever value of j you choose?
5. Which results resemble what you might expect if your chain were in equilibrium at some temperature? Can you estimate what that temperature might be?

PROBLEM 19.8
Molecular Dynamics (MD) simulation of a chain of simple harmonic oscillators - AGAIN

This assignment again uses the MD program you wrote to simulate a chain of particles connected by harmonic interactions:

$$H = \sum_{j=1}^{N} \frac{p_j^2}{2m} + \frac{1}{2}K\sum_{j=1}^{N}(x_j - x_{j-1})^2.$$

[The particle at $x_0 = 0$ is fixed.]

1. You know that the average energy of a single simple harmonic oscillator at temperature T is given by

$$U = \langle H \rangle = k_B T,$$

   so that the specific heat is

$$c = k_B.$$

   Demonstrate analytically that the energy per particle and the specific heat per particle are the same for a harmonic chain and a single SHO *without* using a Fourier transform of the Hamiltonian.
2. Modify your program to calculate the expected kinetic energy per particle from the initial (randomly chosen) energy. Then use the average kinetic energy to predict an effective temperature.
3. Modify your program to find the contributions to the specific heat from the fluctuations of the potential energy under the assumption that the kinetic degrees of freedom act as a thermal reservoir at the predicted effective temperature.
   [The contributions to the specific heat from the momenta are $k_B/2$ per particle.]
4. Calculate the effective temperature and the specific heat for various lengths of the harmonic chain. To what extent are the relationships between the energy and the effective temperature and the fluctuations and the specific heat confirmed by your data?

PROBLEM 19.9
Not-so-simple harmonic oscillators

Consider a one-dimensional, generalized oscillator with Hamiltonian

$$H = K|x|^\alpha$$

with the parameter $\alpha > 0$.

   Calculate the energy and specific heat as functions of the temperature for arbitrary values of α.

PROBLEM 19.10
Monte Carlo simulations of not-so-simple harmonic oscillators

Consider a one-dimensional, generalized oscillator with Hamiltonian

$$H = A|x|^\alpha .$$

1. Modify your program for MC simulations to simulate the potential energy given in this problem. For values of $\alpha = 0.5, 1.0,$, abd 4.0, run MC simulations and compare the results to your theoretical results for the heat capacity from an earlier assignment.

PROBLEM 19.11
Relativistic particles in a gravitational field

Consider N relativistic ideal-gas particles at temperature T. The particles are confined to a three-dimensional box of area A and $0 < z < \infty$. Because they are relativistic, the Hamiltonian is given by

$$H(p,q) = \sum_{j=1}^{N} |\vec{p}_j|c + \sum_{j=1}^{N} mgz_j .$$

1. Find the probability distribution for the height of a single particle, i.e.: $P(z_1)$.
2. Find the average height of a single particle.
3. In what way did your answer for the average height depend on the relativistic nature of the particles?
4. Find the probability distribution for the energy of a single particle, $E_1 = |\vec{p}_1|c + mgz_1$, including the normalization constant.

PROBLEM 19.12
A classical model of a rubber band

The rubber band is modeled as a one-dimensional polymer, that is, a chain of N monomers. Each monomer has a fixed length δ, and the monomers are connected end-to-end to form the polymer. Each monomer is completely free to rotate in *two* dimensions (there are no steric hinderances from the other monomers). A microscopic state of the rubber band is therefore described by the set of angles

$$\theta \equiv \{\theta_n | n = 1, 2, \ldots, N\}$$

describing the orientations of the monomers. [You can decide whether to define the angles with respect to the orientation of the previous monomer, or according to a fixed frame of reference. However, you must say which you've chosen!]

The rubber band (polymer) is used to suspend a mass M above the floor. One end of the rubber band is attached to the ceiling, and the other end is attached to the mass, so that the mass is hanging above the floor by the rubber band. For simplicity, assume that the rubber band is weightless. You can ignore the mass of the monomers and their kinetic energy.

The whole system is in thermal equilibrium at temperature T.

1. Calculate the energy of the system (polymer plus mass) for an arbitrary microscopic state θ of the monomers (ignoring the kinetic energy).
2. Find an expression for the canonical partition function in terms of a single, one-dimensional integral.
3. Find an expression for the average energy U.
4. For very high temperatures, find the leading term in an expansion of the average energy U.

20

Classical Ensembles: Grand and Otherwise

Life is nature's solution to the problem of preserving information despite the second law of thermodynamics.

Howard L. Resnikoff, *The Illusion of Reality*

Although the canonical ensemble is the work horse of classical statistical mechanics, several other ensembles often prove convenient. The most important of these is the 'grand canonical ensemble', which will be discussed in this chapter. This discussion will also serve as an introduction to the very important use of the quantum grand canonical ensemble, which will play a prominent role in the theories of boson and fermion fluids in Part IV.

20.1 Grand Canonical Ensemble

The physical situation described by the grand canonical ensemble is that of a system that can exchange both energy and particles with a reservoir. As usual, we assume that the reservoir is much larger than the system of interest, so that its properties are not significantly affected by relatively small changes in its energy or particle number. Note that the reservoir must have the same type (or types) of particle as the system of interest, which was not a requirement of the reservoir for the canonical ensemble.

Because the system of interest and the reservoir are in equilibrium with respect to both energy and particle number, they must have the same temperature $T = T_R$ and chemical potential $\mu = \mu_R$. In thermodynamics, this situation would correspond to the thermodynamic potential given by the Legendre transform with respect to temperature and chemical potential,

$$U[T, \mu] = U - TS - \mu N. \tag{20.1}$$

An Introduction to Statistical Mechanics and Thermodynamics. Robert H. Swendsen, Oxford University Press (2020).
© Robert H. Swendsen. DOI: 10.1093/oso/9780198853237.001.0001

The function $U[T,\mu]$ is sometimes referred to as the 'grand potential'. Since Legendre transforms conserve information, the grand potential is another representation of the fundamental relation and contains all thermodynamic information about the system.

For extensive systems, Euler's equation, eq. (13.5), holds, and $U[T,\mu]$ has a direct connection to the pressure

$$U[T,\mu] = U - TS - \mu N = -PV. \tag{20.2}$$

While the grand canonical ensemble is useful in classical statistical mechanics, it turns out to be essential in quantum statistical mechanics, as we will see in Chapters 27, 28, and 29. Since the basic idea is the same in both classical and quantum statistical mechanics, it is convenient to see how the grand canonical ensemble works in the classical case, without the added complexity of quantum mechanics.

20.2 Grand Canonical Probability Distribution

To describe a system that can exchange energy and particles with a reservoir, we must greatly expand our description of a microscopic state. Both the microcanonical and canonical ensembles for a system with N particles required only a $6N$-dimensional phase space to describe a microscopic state. When the particle number can vary, we need a different $6N$-dimensional phase space for each value of N. Since N can vary from zero to the total number of particles in the reservoir—which we can approximate by infinity—this is a significant change.

To find the probability of a state in this expanded ensemble of an infinite number of phase spaces, we return to the basic calculation of the probability distribution in a composite system. Assume that the reservoir and the system of interest can exchange both energy and particles with each other, but are completely isolated from the rest of the universe. The total number of particles is $N_T = N + N_R$ and the total energy is $E_T = E + E_R$, where the subscripted R indicates properties of the reservoir.

The probability of the system of interest having N particles and energy E is given by a generalization of eq. (19.10)

$$P(E,N) = \frac{\Omega(E,V,N)\Omega_R(E_T - E, V_R, N_T - N)}{\Omega_T(E_T, V_T, N_T)}. \tag{20.3}$$

Following the same procedure as in Section 19.3, we take the logarithm of eq. (20.3) and expand $\ln\Omega_R(E_T - E, V_R, N_T - N)$ in powers of E/E_T and N/N_T,

$$\ln P(E,N) = \ln \Omega(E,V,N) + \ln \Omega_R(E_T - E, V_R, N_T - N)$$
$$- \ln \Omega_T(E_T, V_T, N_T)$$
$$\approx \ln \Omega(E,V,N) + \ln \Omega_R(E_T, V_R, N_T)$$
$$- E \frac{\partial}{\partial E_T} \ln \Omega_R(E_T, V_R, N_T)$$
$$- N \frac{\partial}{\partial N_T} \ln \Omega_R(E_T, V_R, N_T)$$
$$- \ln \Omega_R(E_T, V_T, N_T). \tag{20.4}$$

In eq. (20.4) we have neglected terms of higher order in E/E_T and N/N_T, since the reservoir is assumed to be much bigger than the system of interest.

Recalling that

$$S_R = k_B \ln \Omega_R(E_T, V_R, N_T), \tag{20.5}$$

we see that we have already found the derivatives of $\ln \Omega$ in eq. (8.12). Therefore, $\beta_R = 1/k_B T_R$ is given by

$$\beta_R \equiv \frac{\partial}{\partial E_T} \ln \Omega_R(E_T, V_R, N_T). \tag{20.6}$$

Similarly, by comparison with eq. (8.34), we see that the chemical potential of the reservoir is given by the other partial derivative in eq. (20.4),

$$-\mu_R \beta_R \equiv \frac{\partial}{\partial N_T} \ln \Omega_R(E_T, V_R, N_T). \tag{20.7}$$

Since $\beta = \beta_R$ and $\mu = \mu_R$, and the quantities E_T, N_T, V_R do not depend on either E or N, we can combine $\ln \Omega_R(E_T, V_R, N_T)$ and $\ln \Omega_R(E_T, V_T, N_T)$ into a single value denoted by $-\ln \mathcal{Z}$,

$$\ln P(E,N) \approx \ln \Omega(E,V,N) - \beta E + \beta \mu N - \ln \mathcal{Z}. \tag{20.8}$$

Exponentiating this equation gives us the grand canonical probability distribution for E and N,

$$P(E,N) = \frac{1}{\mathcal{Z}} \Omega(E,V,N) \exp[-\beta E + \beta \mu N]. \tag{20.9}$$

The normalization condition determines $\mathcal{Z}$, which depends on both $T = 1/k_B \beta$ and μ (and, of course, V),

$$\mathcal{Z} = \sum_{N=0}^{\infty} \int_0^{\infty} dE\, \Omega(E,V,N) \exp\left[-\beta E + \beta \mu N\right]. \tag{20.10}$$

The lower limit in the integral over the energy in eq. (20.10) has been taken to be zero, which is the most common value. More generally it should be the lowest allowed energy.

The normalization 'constant' $\mathcal{Z}$ is called the grand canonical partition function. It is constant in the sense that it does not depend on E or N, although it does depend on β and μ.

The grand canonical partition function can also be expressed as a Laplace transform of the canonical partition function. Since eq. (19.18) gave the canonical partition function as,

$$Z(T,V,N) = \int_0^{\infty} dE\, \Omega(E,V,N) \exp(-\beta E)dE, \tag{20.11}$$

we can rewrite eq. (20.10) as,

$$\mathcal{Z}(T,V,\mu) = \sum_{N=0}^{\infty} Z(T,V,N) \exp\left[\beta \mu N\right]. \tag{20.12}$$

20.3   Importance of the Grand Canonical Partition Function

The grand canonical partition function plays much the same role as the canonical partition function. It is directly related to the grand canonical thermodynamic potential $U[T,\mu]$. To see this, consider the value of $\ln P(E,N)$ at its maximum. The location of the maximum, as before, gives the equilibrium values of $E = E_{eq} = U$ and $N = N_{eq}$. We can rewrite eq. (20.8) to solve for $\ln \mathcal{Z}$,

$$\ln \mathcal{Z} = \ln \Omega(E_{eq}, V, N_{eq}) - \beta E_{eq} + \beta \mu N_{eq} - \ln P(E_{eq}, N_{eq}). \tag{20.13}$$

We already know that the first three terms on the right of eq. (20.13) are proportional to the size of the system. However, $P(E_{eq}, N_{eq})$ will be proportional to $\sqrt{\langle E_{eq}\rangle}$ and $\sqrt{\langle N_{eq}\rangle}$, so that $\ln P(E_{eq}, N_{eq})$ will only be of order $\ln E_{eq}$ and $\ln N_{eq}$. For $N_{eq} \approx 10^{20}$, $\ln P(E_{eq}, N_{eq})$ is completely negligible, and we can discard it in eq. (20.13),

$$\ln \mathcal{Z} = \ln \Omega(E, V, N) - \beta E + \beta \mu N. \tag{20.14}$$

Since we know that the microcanonical entropy is given by

$$S = k_B \ln \Omega, \tag{20.15}$$

we can rewrite eq. (20.14) as

$$\ln \mathcal{Z} = S/k_B - \beta E + \beta \mu N, \tag{20.16}$$

or,

$$\ln \mathcal{Z} = -\beta (E - TS - \mu N) = -\beta U[T,\mu]. \tag{20.17}$$

If the system is extensive, Euler's equation, eq. (13.5), holds, and,

$$\ln \mathcal{Z} = -\beta U[T,\mu] = \beta PV. \tag{20.18}$$

This equation gives us a convenient way to calculate the pressure as a function of T and μ, as illustrated next for the classical ideal gas in eq. (20.21).

Eq. (20.18) is directly analogous to eq. (19.48). It provides a way to calculate the fundamental relation, and we can immediately invoke everything we learned in Part II about how to use thermodynamic potentials to calculate thermodynamic properties.

20.4 $\mathcal{Z}(T,V,\mu)$ for the Ideal Gas

In Section 19.10.1 we derived the partition function for the classical ideal gas, with the result given in eq. (19.62),

$$Z = \frac{1}{h^{3N} N!} (2\pi m k_B T)^{3N/2} V^N.$$

Inserting this equation into eq. (20.12) for the grand canonical partition function, we find

$$\mathcal{Z}(T,V,\mu) = \sum_{N=0}^{\infty} \frac{1}{h^{3N} N!} (2\pi m k_B T)^{3N/2} V^N \exp [\beta \mu N] \tag{20.19}$$

$$= \sum_{N=0}^{\infty} \frac{1}{N!} \left(\frac{(2\pi m k_B T)^{3/2}}{h^3} V e^{\beta \mu} \right)^N$$

$$= \exp \left((2\pi m k_B T)^{3/2} h^{-3} V e^{\beta \mu} \right).$$

Taking the logarithm of eq. (20.19) gives us the grand canonical thermodynamic potential for the classical ideal gas,

$$-\beta U[T,\mu] = (2\pi m k_B T)^{3/2} h^{-3} V e^{\beta \mu} = \beta PV. \tag{20.20}$$

The last equality is, of course, true only because the ideal gas is extensive. Note that the pressure depends only on β and μ, and can be obtained by dividing out the product βV in eq. (20.20),

$$P = k_B T (2\pi m k_B T)^{3/2} h^{-3} e^{\beta \mu}. \tag{20.21}$$

20.5 Summary of the Most Important Ensembles

The three ensembles that we have discussed so far are the most important.

The microcanonical ensemble fixes the energy, volume, and number of particles. The formula for the entropy of an isolated system is,

$$S = k_B \ln \Omega(E, V, N), \tag{20.22}$$

where

$$\Omega(E, V, N) = \frac{1}{h^{3N} N!} \int dq \int dp \, \delta(E - H(p, q)). \tag{20.23}$$

The origin of these equations is Boltzmann's definition of the entropy of a composite system in terms of the logarithm of the probability distribution. Eq. (20.22) can also be written in the form

$$\Omega(E, V, N) = \exp(S(E, V, N)/k_B). \tag{20.24}$$

The width of both the energy distribution and the particle-number distribution is zero in the microcanonical ensemble. This is clearly an approximation.

The canonical ensemble describes a system in equilibrium with a thermal reservoir at temperature T. The canonical partition function is given by

$$Z = \int dE \, \Omega(E, V, N) \exp(-\beta E), \tag{20.25}$$

and is related to the Helmholtz free energy F by

$$Z = \exp(-\beta F) = \exp(S[\beta]/k_B). \tag{20.26}$$

The width of the particle-number distribution is zero in the canonical ensemble. The energy distribution has a relative width of the order of $1/\sqrt{N}$, which is not exactly correct, but is a much better approximation than zero. We will return to this point in Chapter 21.

The grand canonical ensemble describes a system that can exchange both energy and particles with a reservoir, which is assumed to be characterized by temperature T and chemical potential μ. The grand canonical partition function is given by

$$\mathcal{Z} = \sum_{N=0}^{\infty} \int_{o}^{\infty} dE\,\Omega(E,V,N)\exp\left(-\beta E + \beta\mu N\right), \qquad (20.27)$$

and its relationship to $U[T,\mu]$ is

$$\mathcal{Z} = \exp\left(-\beta U[T,\mu]\right) = \exp\left(S[\beta,(\beta\mu)]/k_B\right) = \exp(\beta PV), \qquad (20.28)$$

where the last equality is valid only for extensive systems.

Both the energy distribution and the particle-number distribution have non-zero widths in the grand canonical ensemble.

20.6   Other Classical Ensembles

For every thermodynamic potential there is a corresponding ensemble in statistical mechanics. Every Legendre transform in thermodynamics corresponds to a Laplace transform in statistical mechanics. In each case, the logarithm of a partition function produces a thermodynamic potential. The only feature that is slightly different is found in the grand canonical partition function, for which the Laplace transform takes the form of a sum instead of an integral. None of these transforms should present any difficulties, since they are all derived using the same principles.

20.7   Problems

....

PROBLEM 20.1
A non-extensive thermodynamic system

Consider a classical gas of N weakly interacting atoms of mass m, enclosed in a container of volume V and surface area A.

The interactions between the atoms are so weak that they may be neglected (ideal gas approximation). However, the interaction of the atoms with the walls of the container may NOT be neglected. Some fraction of the atoms are adsorbed onto the walls of the container in equilibrium. Your job will be to calculate the average number N' of adsorbed atoms on the walls.

A simple model for the atoms adsorbed onto the walls is to treat them as a two-dimensional ideal gas. The energy of an adsorbed atom is taken to be

$$\epsilon(\vec{p}) = \frac{|\vec{p}|^2}{2m} - \epsilon_o$$

where $\vec{p}$ is the two-dimensional momentum and ϵ_o is a known parameter that describes the energy of adsorption.

The entire system is in equilibrium and in thermal contact with a heat reservoir at temperature T.

1. What is the classical partition function of the adsorbed atoms if N' of them are adsorbed onto the walls of the container?
2. What is the chemical potential μ_S of the atoms adsorbed on the surface?
3. What is the chemical potential μ_V of the $N - N'$ atoms in the volume of the container?
4. When the atoms in the volume and those adsorbed on the walls are in equilibrium, what is the average number of atoms adsorbed as a function of the temperature?
5. What are the high- and low-temperature limits of the ratio of the number of adsorbed atoms to the total number of atoms?

21

Refining the Definition of Entropy

$$S = k \log W$$

Inscription on Boltzmann's tombstone
(first written in this form by Max Planck in 1900)

In this section we will refine the definition of the entropy beyond that given in Part I. Specifically, we will be concerned with the distribution of the energy, which had been taken to be a delta function. This is not correct if the system of interest has ever been in contact with another macroscopic system.[1]

21.1 The Canonical Entropy

In the canonical distribution, the energy distribution is given by eq. (19.17), which has a non-zero probability for all energies. Although this distribution is not exact for the interaction of a system with another finite system, it is a reasonable approximation and far better than a delta function. More importantly, the entropy calculated from it is a significant improvement for any macroscopic system that has been in contact with any other macroscopic system.[2] We will denote this quantity as the canonical entropy.

21.1.1 The entropy of a system is independent of the size of the system with which it has exchanged energy

Consider three macroscopic systems labeled A, B, and C. Assume that systems A and B are the same size and composition, so that the entropies of systems A and B are equal by symmetry. All three systems are initially in equilibrium with each other.

Next, remove system C from thermal contact with the other two systems. A and B must still be in equilibrium with each other at the same temperature as before C was removed. They must also have the same average energy as they did when they were is contact with C.

---

[1] R. H. Swendsen, 'Continuity of the entropy of macroscopic quantum systems,' *Phys. Rev. E*, 92: 052110 (2015).

[2] R. H. Swendsen, 'Thermodynamics, Statistical Mechanics, and Entropy,' *Entropy*, 19: 603 (2017).

An Introduction to Statistical Mechanics and Thermodynamics. Robert H. Swendsen, Oxford University Press (2020).
© Robert H. Swendsen. DOI: 10.1093/oso/9780198853237.001.0001

The entropies must be unchanged. If the total entropy were to decrease, it would be a violation of the second law of thermodynamics (and the second essential postulate). If the entropy were to increase upon separation, then bringing system C back into thermal contact with systems A and B would decrease the entropy, which would also be a violation of the second law. The only possibility consistent with the second law is that the entropy is unchanged.

Note that the fluctuations of the energies in A and B are narrower when they are separated from the larger system C, but that this small change in the width has no effect on the entropy. Therefore, the entropy of an isolated system is correctly given by the canonical entropy.

It is impossible for thermodynamic measurements to determine the size of the system with which a given system has been in contact.

21.1.2   The calculation of the canonical entropy using a Massieu function

Massieu functions do not require the inversion of the fundamental relation $S = S(U, V, N)$ to find $U = U(S, V, N)$, which will make them particularly useful when we consider quantum systems with a non-monotonic density of states. We will use a dimensionless entropy $\tilde{S} = S/k_B$, which was introduced in Section 12.6 along with the differential form of the fundamental relation in the $\tilde{S}$-representation,

$$d\tilde{S} = \beta\, dU + (\beta P)dV - (\beta\mu)dN, \tag{21.1}$$

where $\beta = 1/k_B T$, P is the pressure, V is the volume, μ is the chemical potential, and N is the number of particles. From eq. (21.1) we immediately have

$$\left(\frac{\partial \tilde{S}}{\partial U}\right)_{V,N} = \beta. \tag{21.2}$$

We will need the Massieu function $\tilde{S}[\beta]$,

$$\tilde{S}[\beta] = \tilde{S} - \beta U = -\beta\,(U - TS) = -\beta F, \tag{21.3}$$

and the differential form of the fundamental relation in this representation,

$$d\tilde{S}[\beta] = -Ud\beta + (\beta P)dV - (\beta\mu)dN. \tag{21.4}$$

Since

$$\tilde{S}[\beta] = -\beta F = \ln Z(\beta, V, N), \tag{21.5}$$

this enables us to calculate the Massieu function directly from the canonical partition function.

Given the Massieu function, we can obtain the energy U from

$$\left(\frac{\partial \tilde{S}[\beta]}{\partial \beta}\right)_{V,N} = -U. \tag{21.6}$$

To obtain $\tilde{S}$ from $\tilde{S}[\beta]$, use

$$\tilde{S} = \tilde{S}[\beta] + \beta U, \tag{21.7}$$

and substitute $\beta = \beta(U)$.

21.1.3   The canonical entropy of the classical ideal gas

To calculate the canonical and grand canonical forms of the entropy, the density of states of the classical ideal gas must first be evaluated as

$$\Omega_{CIG}(E,V,N) = \frac{\pi^{3N/2}}{(3N/2-1)!} \frac{1}{h^{3N}N!} V^N (2m)^{3N/2} E^{3N/2-1}. \tag{21.8}$$

This can be done by carrying out the integrals as shown in Chapters 4 and 6.

Finding the explicit expression for the canonical entropy involves a bit of algebra, which we will leave as a problem at the end of the chapter. The result is

$$S_C = k_B \left[N\left(\frac{3}{2}\right)\ln\left(\frac{U}{N}\right) + \ln\left(\frac{V^N}{N!}\right) + N\left(\frac{3}{2}\right)\ln\left(\frac{4\pi m}{3h^2}\right) + \frac{3N}{2} \right]. \tag{21.9}$$

Note that Eq. (21.9) uses $U = \langle E \rangle$. The factorials associated with the energy integral over the surface of a sphere in momentum space, appear in the Boltzmann entropy, but not in the canonical entropy. Only the term associated with the volume involves a factorial $(1/N!)$.

21.2   The Grand Canonical Entropy

The microcanonical and canonical ensembles make the approximation that the value of N is specified exactly. This approximation is very good, but not quite correct if the system has ever exchanged particles with another system. The situation is the same as for the energy dependence, and it results in a small error in the predictions of the entropy.

For the classical ideal gas, this error leads to an expression for the entropy that is not exactly extensive. Systems with short-range interactions, such as those shown schematically in Fig. 7.1, are expected to have small deviations from extensivity that only go to zero as N goes to infinity. However, the classical ideal gas is a model that has no interactions between particles. Every particle is expected to contribute separately to the macroscopic properties of the system. The small deviations from extensivity are surprising from this point of view.

The resolution is to include the width of the N-distribution in the calculation of the entropy. This leads us to the grand canonical distribution as providing an approximation to the true N-distribution. The same kind of argument that we used in Section 21.1.1 to justify the canonical distribution also applies here.

For the grand canonical ensemble, we will use $\tilde{S}[\beta, (\beta\mu)]$ to derive the grand canonical entropy. As earlier, parentheses around the second variable indicate that the product of β and μ is treated as a single variable.

The Legendre transform of $\tilde{S}$ with respect to both β and $(\beta\mu)$ uses the equation

$$-(\beta\mu) = \left(\frac{\partial \tilde{S}}{\partial N}\right)_{U,N} \tag{21.10}$$

in addition to eq. (21.2).

The Legendre transform of $\tilde{S}$ with respect to both β and $(\beta\mu)$ is then given by the Massieu function

$$\tilde{S}[\beta, (\beta\mu)] = \tilde{S} - \beta U + (\beta\mu)N, \tag{21.11}$$

so that

$$\tilde{S}[\beta, (\beta\mu)] = \ln \mathcal{Z}(\beta, V, (\beta\mu)), \tag{21.12}$$

where $\mathcal{Z}$ is the grand canonical partition function.

The differential of the Massieu function $\tilde{S}[\beta, (\beta\mu)]$ is

$$d\tilde{S}[\beta, (\beta\mu)] = -Ud\beta + (\beta P)dV + Nd(\beta\mu). \tag{21.13}$$

This immediately gives

$$\left(\frac{\partial \tilde{S}[\beta, (\beta\mu)]}{\partial \beta}\right)_{V,(\beta\mu)} = -U, \tag{21.14}$$

and

$$\left(\frac{\tilde{S}[\beta, (\beta\mu)]}{\partial (\beta\mu)}\right)_{\beta,V} = N. \tag{21.15}$$

To obtain $\tilde{S}$ from $\tilde{S}[\beta,(\beta\mu)]$, use

$$\tilde{S} = \tilde{S}[\beta,(\beta\mu)] + \beta U - (\beta\mu)N, \tag{21.16}$$

and replace the β and $(\beta\mu)$ dependence by U and $\langle N \rangle$.

The calculation involves some algebra, which we have left to a problem at the end of the chapter. The answer is

$$S_{GC} = \langle N \rangle k_B \left[\frac{3}{2}\ln\left(\frac{U}{\langle N \rangle}\right) + \ln\left(\frac{V}{\langle N \rangle}\right) + \ln\left(\frac{4\pi m}{3h^2}\right)^{3/2} + \frac{5}{2} \right]. \tag{21.17}$$

This expression for the grand canonical entropy of a classical ideal gas is exactly extensive, and no use has been made of Stirling's approximation.

21.3   Problems

PROBLEM 21.1
The canonical entropy.
A different view of the entropy of the ideal gas.

Consider a classical ideal gas of N particles in three dimensions. The particles are contained in a box of volume V. Denote the dimensionless entropy by $\tilde{S} = S/k_B$. Do not use Stirling's approximation.

1. Write the partition function.
2. Evaluate the partition function in closed form.
3. Denote the Legendre transform of the dimensionless entropy with respect to β as $\tilde{S}[\beta]$. Evaluate $\tilde{S}[\beta]$.
4. Do an inverse Legendre transform of $\tilde{S}[\beta]$ to find $\tilde{S}$.
5. Find S.

PROBLEM 21.2
The grand canonical entropy.
A different view of the entropy of the ideal gas.

Consider a classical ideal gas of N particles in three dimensions. The particles are contained in a box of volume V. Denote the dimensionless entropy by $\tilde{S} = S/k_B$. Do not use Stirling's approximation.

1. Calculate the grand canonical partition function for the classical ideal gas. You can use the results of the previous assignment in your solution.
2. Denote the Legendre transform of the dimensionless entropy with respect to β and $(\beta\mu)$ as $\tilde{S}[\beta,(\beta\mu)]$.
   Evaluate $\tilde{S}[\beta,(\beta\mu)]$.
3. Do an inverse Legendre transform of $\tilde{S}[\beta,(\beta\mu)]$ to find $\tilde{S}$.
4. Find S.

22

Irreversibility

It would be much more impressive if it flowed the other way.

Oscar Wilde, on seeing the Niagara Falls

22.1 What Needs to be Explained?

The first thing we must establish is the meaning of the term 'irreversibility'. This is not quite as trivial as it might seem.[1] The irreversible behavior I will try to explain is that which is observed. Every day we see that time runs in only one direction in the real world. If I drop my keys, they fall to the floor and stay there; keys lying on the floor do not suddenly jump into my hand. This asymmetry of time, or 'arrow of time', has seemed to many as being incompatible with the time-reversal invariance of the fundamental equations of both classical and quantum physics.[2] This is the issue addressed in this chapter.

As with the development of statistical mechanics in general, the explanation of irreversibility given here is based on a large number of particles in a macroscopic system for which we have very limited information about the microscopic state. As in the rest of the book, this leads us to describe a macroscopic system on the basis of probability theory.

We will present an explanation of irreversibility using the example of the free expansion of a classical ideal gas. The microscopic equations of motion for this problem are time reversal invariant, but the macroscopic behavior will nevertheless turn out to be irreversible. Because of the simplicity of the example we will be able to carry out every mathematical step exactly, so that the argument can be analyzed completely.[3]

---

[1] This chapter is based on an earlier paper by the author: R. H. Swendsen, 'Explaining irreversibility', *Am. J. Phys.,* 76, 643–48 (2008).

[2] The history of the debate on the origins of irreversibility is fascinating (and continuing), but we will not have space to delve into it. There are many books on this topic that the interested student might consult.

[3] The mathematical analysis we use was first derived by Harry Lloyd Frisch, an American chemist, who was born in Vienna, Austria (1928–2007). H. L. Frisch, 'An approach to equilibrium,' *Phys. Rev.,* 109, 22–9 (1958).

An Introduction to Statistical Mechanics and Thermodynamics. Robert H. Swendsen, Oxford University Press (2020).
© Robert H. Swendsen. DOI: 10.1093/oso/9780198853237.001.0001

22.2   Trivial Form of Irreversibility

The irreversibility observed in daily life must be distinguished from a trivial form of irreversibility that appears only in infinite systems.[4]

Consider a particle moving in empty space. At some initial time it is observed to be located in some finite region of space. Since the particle is moving, it will eventually leave this region. If space is infinite, it will not return. This is, technically, irreversible behavior, but of a trivial kind.

This trivial form of irreversibility is real, and occurs in the radiation of light from a star. It is quite general in infinite systems, whether classical or quantum. An open system also displays this trivial form of irreversibility, since it is really just a piece of an infinite system.

However, we would like to separate this trivial irreversibility from the non-trivial irreversibility that we experience every day. Therefore, we will restrict the discussion to an isolated, finite system.

22.3   Boltzmann's H-Theorem

The history of the debate on irreversibility has been closely associated with responses to Boltzmann's famous H-theorem, in which he claimed to explain irreversibility. Although objections were aimed at Boltzmann's equations rather than the apparent paradox itself, discussing them can help clarify the problem.

Boltzmann derived an equation for the time-derivative of the distribution of atoms in a six-dimensional space of positions and momenta by approximating the number of collisions between atoms (*Stosszahlansatz*). This approximation had the effect of replacing the true macroscopic dynamics by a process that included random perturbations of the particle trajectories. Within the limits of his approximation, Boltzmann showed that a particular quantity that he called 'H' could not increase with time.

The fact that Boltzmann used an approximation—and an essential one—in deriving his result meant that his derivation could not be regarded as a proof. The arguments about its validity began soon after his publication appeared, and have continued to the present day.

22.4   Loschmidt's *Umkehreinwand*

The first objection came from Boltzmann's friend, Johann Josef Loschmidt (Austrian physicist, 1821–1895), who noted that if all momenta in an isolated system were reversed, the system would retrace its trajectory. If Boltzmann's H-function had been decreasing at the moment of reversal, its value after time-reversal must increase. Loschmidt argued on this basis that Boltzmann's conclusions could not be correct, and his argument is usually referred to by the German term *Umkehreinwand*.

---

[4] R. H. Swendsen, 'Irreversibility and the thermodynamic limit,' *J. of Stat. Phys.*, 10: 175–7 (1974).

Loschmidt's argument is very close to the central paradox. If every microscopic state that approaches equilibrium corresponds to a time-reversed state that moves away from equilibrium, should they not be equally probable?

22.5   Zermelo's *Wiederkehreinwand*

Ernst Zermelo (German mathematician, 1871–1953) was a prominent mathematician, who raised a different objection. He cited a recently derived theorem by Jules Henri Poincaré (French mathematician and physicist, 1854–1912) that proved that any isolated classical system must exhibit quasi-periodic behavior; that is, the system must return repeatedly to points in phase space that are arbitrarily close to its starting point. Zermelo claimed that Poincaré's theorem is incompatible with Boltzmann's H-theorem, which predicts that the system will never leave equilibrium. Zermelo's argument is referred to by the German term *Wiederkehreinwand*.

22.6   Free Expansion of a Classical Ideal Gas

In this section we will present an exact analysis of a simple model that exhibits irreversible behavior. Since the model is governed by time-reversal-invariant equations of motion, this will serve as a demonstration of the compatibility of irreversible macroscopic behavior and time-reversal invariance.

Consider an ideal gas of N particles in a volume V. Isolate the gas from the rest of the universe and assume that the walls of the box are perfectly reflecting. Initially, the gas is confined to a smaller volume V_o by an inner wall. At time $t = 0$, the inner wall is removed.

For convenience, let the box be rectangular and align the coordinate axes with its sides. The inner wall that initially confines the gas to a subvolume V_o is assumed to be perpendicular to the x-direction and located at $x = L_o$. The length of the box in the x-direction is L. The dependence of the probability distribution on the y- and z-coordinates does not change with time, so that we can treat it as a one-dimensional problem.

At time $t = 0$, the confining wall at L_o is removed and the particles are free to move throughout the box. Following Frisch's 1958 paper, we will eliminate the difficulties involved in describing collisions with the walls at $x = 0$ and $x = L$ by mapping the problem onto a box of length $2L$ with periodic boundary conditions. When a particle bounces off a wall in the original system, it corresponds to a particle passing between the positive and negative sides of the box without change of momentum. The mapping from the periodic system to the original system is then simply $x_j \rightarrow |x_j|$.

The key assumption is that the initial microscopic state must be described by a probability distribution. Assume that before the inner wall is removed, the initial positions of the particles are uniformly distributed in V_o, and the momenta are independent of each other and have a probability distribution $h(p_j)$. This probability distribution is assumed to be time-reversal invariant, $h(-p_j) = h(p_j)$, so that there is no time-asymmetry in the initial conditions. Since we are assuming that the positions and momenta of different particles are initially independent, we can write the total probability distribution as

$$f_N\left(\{x_j,p_j|j=1\ldots N\},t=0\right)=\prod_{j=1}^{N}f\left(x_j,p_j,t=0\right)=\prod_{j=1}^{N}g\left(x_j\right)h\left(p_j\right) \qquad (22.1)$$

where

$$g\left(x\right)=\begin{cases}1/L_o & -L_0<x<L_0.\\0 & |x|\geq L_0.\end{cases} \qquad (22.2)$$

This initial probability distribution for the periodic system is shown schematically in Fig. 22.1.

Since the particle probabilities are independent, we can restrict our attention to the distribution function of a single particle. The periodic boundary conditions then allow us to Fourier transform the initial conditions to obtain

$$g(x)=g_0+\sum_{n=1}^{\infty}g_n\cos\left(\frac{\pi n}{L}x\right) \qquad (22.3)$$

where no sine terms enter because of the symmetry of the initial conditions in the expanded representation with period $2L$.

Using the standard procedure of multiplying by $\cos\left(\pi n'x/L\right)$ and integrating to obtain the coefficients in eq.(22.3), we find

$$\int_{-L}^{L}g(x)\cos\left(\frac{\pi n'}{L}x\right)dx$$
$$=\int_{-L}^{L}g_0\cos\left(\frac{\pi n'}{L}x\right)dx+\sum_{n=1}^{\infty}g_n\int_{-L}^{L}\cos\left(\frac{\pi n}{L}x\right)\cos\left(\frac{\pi n'}{L}x\right)dx. \qquad (22.4)$$

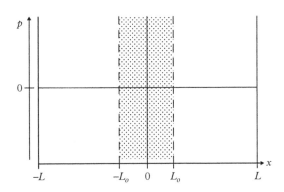

Fig. 22.1 *The initial probability distribution at $t=0$.*

For $n' = 0$, this gives us

$$g_0 = \frac{1}{L}, \tag{22.5}$$

and for $n' \geq 1$

$$g_{n'} = \frac{2}{n' \pi L_o} \sin\left(\frac{n' \pi}{L} L_o\right). \tag{22.6}$$

The time development for a given momentum is now simple:

$$f(x, p, t) = f\left(x - \left(\frac{p}{m}\right) t, p, 0\right)$$

$$= h(p) \left[g_0 + \sum_{n=1}^{\infty} g_n \cos\left(\frac{n\pi}{L}\left(x - \left(\frac{p}{m}\right) t\right)\right)\right]. \tag{22.7}$$

Fig. 22.2 shows the probability distribution at a time $t > 0$. The shaded areas, representing non-zero probability density, tilt to the right. As time goes on, the shaded areas become increasingly flatter and closer together.

To find an explicit solution for a special case, assume that the initial probability distribution for the momentum is given by the Maxwell–Boltzmann distribution,

$$h(p) = \left(\frac{\beta}{2\pi m}\right)^{1/2} \exp\left[-\beta\left(\frac{p^2}{2m}\right)\right] \tag{22.8}$$

where T is the temperature, and $\beta = 1/k_B T$. The assumption of a Maxwell–Boltzmann distribution is not essential, but it provides an explicit example of how the approach to equilibrium comes about.

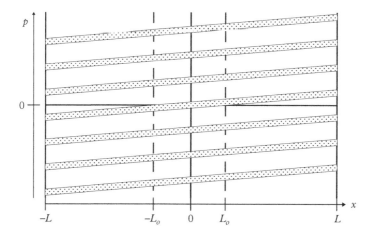

Fig. 22.2 *The probability distribution for $t > 0$.*

Inserting eq. (22.8) in eq. (22.7), we can integrate over the momenta to calculate the local probability density as a function of time,

$$
\begin{aligned}
g(x,t) &= \int_{-\infty}^{\infty} f(x,t)\, dp \\
&= \int_{-\infty}^{\infty} h(p) L^{-1} dp \\
&\quad + \sum_{n=1}^{\infty} g_n \int_{-\infty}^{\infty} h(p) \cos\left(\frac{n\pi}{L}\left(x - \left(\frac{p}{m}\right)t\right)\right) dp.
\end{aligned}
\tag{22.9}
$$

The first integral is trivial, since $h(p)$ is normalized,

$$
\int_{-\infty}^{\infty} h(p)\, dp = 1.
\tag{22.10}
$$

The integrals in the sum are rather tedious, but not particularly difficult when you use the identity $\cos\theta = \frac{1}{2}\left(e^{i\theta} + e^{-i\theta}\right)$ to write them in the form

$$
\int_{-\infty}^{\infty} \exp\left[-\beta \frac{p^2}{2m} \pm \frac{n\pi}{L} i \frac{p}{m} t\right] dp,
\tag{22.11}
$$

and complete the square in the exponent. The result is

$$
g(x,t) = L^{-1} + \frac{2}{\pi L_o} \sum_{n=1}^{\infty} \frac{1}{n} \sin\left(\frac{n\pi}{L}L_o\right) \cos\left(\frac{n\pi}{L}x\right) \exp\left[-\lambda_n^2 t^2\right],
\tag{22.12}
$$

where

$$
\lambda_n^2 = \frac{n^2\pi^2}{2mL^2\beta} = \frac{n^2\pi^2}{2mL^2} k_B T.
\tag{22.13}
$$

Note that since the temperature is related to the one-dimensional root-mean-square velocity by

$$
\frac{1}{2}k_B T = \frac{1}{2m}\langle p^2\rangle = \frac{m}{2}\langle v^2\rangle = \frac{m}{2}v_{rms}^2.
\tag{22.14}
$$

The coefficients λ_n can be written as

$$
\lambda_n^2 = \frac{n^2\pi^2}{2L^2} v_{rms}^2.
\tag{22.15}
$$

Eq. (22.12) can also be expressed in terms of the characteristic time for a particle traveling with the speed v_{rms} to cross the box,

$$\tau = \frac{L}{v_{rms}}. \tag{22.16}$$

In terms of τ, the time-dependent local probability density becomes

$$g(x,t) = L^{-1}\left[1 + \frac{L}{L_o}\sum_{n=1}^{\infty}\left(\frac{2}{n\pi}\right)\sin\left(\frac{n\pi}{L}L_o\right)\cos\left(\frac{n\pi}{L}x\right)\exp\left[-\left(\frac{\pi^2 n^2}{2}\right)\left(\frac{t^2}{\tau^2}\right)\right]\right]. \tag{22.17}$$

Since

$$\left|\left(\frac{2}{n\pi}\right)\sin\left(\frac{n\pi}{L}L_o\right)\cos\left(\frac{n\pi}{L}x\right)\right| \le \frac{2}{n\pi} \tag{22.18}$$

for all x, the sum converges rapidly for $t > 0$. At $t = \tau$, the n-th term in the sum is less than $(2/n\pi)\exp\left[-\left(\frac{\pi^2}{2}\right)n^2\right]$, and even the leading non-constant term has a relative size of less than $0.005\,(L/L_o)$. The rapidity of convergence is striking. The inclusion of interactions in the model would actually slow down the approach to equilibrium.

It is also possible to calculate the energy density as a function of position and time by including a factor of $p^2/2m$ in the integral over the momentum,

$$U(x,t) = \int_{-\infty}^{\infty} f(x,p,t)\left(\frac{p^2}{2m}\right) dp$$

$$= \int_{-\infty}^{\infty} h(p)\left(\frac{p^2}{2m}\right) L^{-1} dp$$

$$+ \sum_{n=1}^{\infty} g_n \int_{-\infty}^{\infty} h(p)\left(\frac{p^2}{2m}\right)\cos\left(\frac{n\pi}{L}\left(x - \left(\frac{p}{m}\right)t\right)\right) dp. \tag{22.19}$$

The integrals are even more tedious than those in eq. (22.9), but again not particularly difficult:

$$U(x,t) = \frac{1}{2L\beta} - \sum_{n=1}^{\infty}\left(\frac{n^2\pi^2 t^2 - L^2 m\beta}{L^2 L_o mn\pi\beta^2}\right)$$

$$\sin\left(\frac{n\pi}{L}L_o\right)\cos\left(\frac{n\pi}{L}x\right)\exp\left[-\left(\frac{\pi^2 n^2 t^2}{2\tau^2}\right)\right]. \tag{22.20}$$

After the internal wall is removed, the energy distribution is not uniform and not proportional to the particle density, since the faster particles move more rapidly into the region that was originally vacuum. However, the energy density converges rapidly to the expected constant, $1/2L\beta = \frac{1}{2}k_B T/L$, as $t \to \infty$.

22.7   Zermelo's *Wiederkehreinwand* Revisited

We have now established that that the particle density and energy density both go to equilibrium. This leaves the question of reconciling the approach to equilibrium with Poincaré recurrences. Fortunately, it is easy to see how quasi-periodic behavior can arise in the ideal gas model, without contradicting the approach to equilibrium. To do this, it helps to have an intuitive picture of how Poincaré cycles occur.

First consider two particles with speeds v_1 and v_2. They will each return to their original states with periods $\tau_1 = 2L/v_1$ and $\tau_2 = 2L/v_2$, respectively. In general, the ratio τ_1/τ_2 will be irrational, but it can be approximated to arbitrary accuracy by a rational number, $\tau_1/\tau_2 \approx n_1/n_2$, where n_1 and n_2 are sufficiently large integers. Therefore, after a time $\tau_{1,2} = n_2\tau_1 \approx n_1\tau_2$, both particles will return to positions and velocities arbitrarily close their initial states .

Now add a third particle with speed v_3 and period $\tau_3 = 2L/v_3$. A rational approximation $\tau_{1,2}/\tau_3 \approx n_{1,2}/n_3$, will give us an approximate recurrence after a time $\tau_{1,2,3} = n_{1,2}\tau_3 \approx n_3\tau_{1,2} \approx n_3 n_2 \tau_1$. Since n_2 and n_3 will usually be large numbers, $\tau_{1,2,3}$ will usually be a long time.

If we repeat this procedure for 10^{23} particles, we will arrive (with probability one) at a recurrence time that would be enormous even in comparison to the age of the universe. It might naturally be said that we are not interested in such extremely long times, but it is interesting to note that eqs. (22.17) and (22.20) do not exhibit the quasi-periodic behavior that Poincaré recurrence might seem to require.

The resolution of the apparent contradiction lies in our lack of knowledge of the exact initial velocities and the extreme sensitivity of the Poincaré recurrence time to tiny changes in initial velocities. Even with far more detailed information than we are ever going to obtain in a real experiment, we would not be able predict a Poincaré recurrence with an uncertainty of less than many ages of the universe.

Poincaré recurrences are included in the exact solution for the free expansion of an ideal gas. However, since we cannot predict them, they appear as extremely rare large fluctuations. Since such large fluctuations are always possible in equilibrium, there is no contradiction between observed irreversible behavior and Poincaré recurrences.

22.8   Loschmidt's *Umkehreinwand* Revisited

Loschmidt's *Umkehreinwand* was directed against Boltzmann's equation, which was not time-reversal invariant. However, the exact solution to the free expansion of an ideal gas retains the time-reversal properties of the microscopic equations. If we reverse all velocities in the model some time after the inner wall is removed, the particles return to their original positions.

Under normal conditions, the reversal of molecular velocities is experimentally difficult, to say the least. However, the reversal of spin precession can be accomplished and is fundamental to magnetic resonance imaging. It is known as the spin-echo effect. After an initial magnetic pulse aligns the spins in a sample, interactions between the spins and inhomogeneities in the sample lead to decoherence and a decay of this magnetization. If no further action is taken, the signal will not return. However, if a second magnetic pulse is used to rotate the spins by 180° at a time t after the first pulse, it effectively reverses the direction of precession of the spins. The spins realign (or 'refocus') after a total time $2t$ after the initial pulse, and the magnetization appears again.

The reversal of the time-development of an 'irreversible' process is therefore not a flaw in the free-expansion model, but a reflection of the reality that such experiments can be carried out.

22.9   What is 'Equilibrium'?

An interesting feature of the exact solution to the free-expansion model can be seen in Fig. 22.2. Even though the shaded regions, indicating non-zero probabilities, become progressively thinner and closer together, the local probability density at any point along the trajectory of the system remains constant. This is consistent with Liouville's theorem that the total time derivative of the probability distribution vanishes.[5]

This property of all isolated Hamiltonian systems has caused difficulties for those who would like to define equilibrium in terms of a smooth distribution in phase space. Although we assumed that our model experiment started with a canonical probability distribution, it will certainly never evolve to one.

I believe that the difficulty is a confusion about the direction of inference. It has been amply demonstrated that a canonical distribution accurately describes an equilibrium state. However, that does not imply that a system in equilibrium can only be described by a canonical distribution. The probability distribution for our model at long times will give the same macroscopic predictions as the canonical distribution, even though the two distributions are different.

22.10   Entropy

Liouville's theorem has also caused difficulties for the traditional textbook definition of the entropy as the logarithm of a volume in phase space. The theorem requires that this volume remain constant, so the traditional expression for the entropy cannot increase in an isolated system.

This seems to violate the Second Law, but it is correct—in a certain sense. The traditional definition of entropy is related to the total information we have about the

---

[5] See Section 19.5 for a derivation of the Liouville theorem.

system—not the thermodynamic information about the current and future behavior of the system. The information that the system was initially confined to a smaller volume is contained in the layered structure of the probability density shown in Fig. 22.2. Since that information does not change, the traditional entropy does not change. The apparent violation of the Second Law arises because the traditional entropy does not correspond to the thermodynamic entropy.

If we use Boltzmann's definition of the thermodynamic entropy in terms of the probability of a macroscopic state, we obtain an expression that increases with time as expected for the thermodynamic entropy.

The specific form of the entropy depends on what we are measuring in the experiment. Since we are interested in the time development of the system as it approaches a uniform state, it would be reasonable to observe its properties as a function of position. Let us divide the system into M subsystems, each with length $\Delta L = L/M$, and measure the energy and density in each subsystem. For convenience, assume that the first m subsystems were inside the inner wall before the free expansion began, so that $L_o = mL/M$.

If M is large, we can assume that the energy and density are uniform across a subsystem. The number of particles in the j-th subsystem, $N_j(t)$, is given by ΔL times the expression in eq. (22.12), and the energy in the j-th subsystem, $E_j(t)$, is given by ΔL times the expression in eq. (22.20).

If the subsystems are large enough that $1/\sqrt{N_j(t)}$ is much smaller than the relative error in experimental measurements, they can be regarded as macroscopic systems. Their individual entropies might then be well approximated by the equilibrium entropy of a one-dimensional, classical ideal gas,

$$S(E_j(t), L/M, N_j(t)) = N_j(t)k_B T \left[\ln\left(\frac{L/M}{N_j(t)}\right) + \frac{1}{2}\ln\left(\frac{E_j(t)}{N_j(t)}\right) + X \right], \qquad (22.21)$$

where X is the usual constant. The total time-dependent entropy of the whole system is then given by

$$S(\{E_j(t)\}, L, N_j(t), /M) = \sum_{j=1}^{M} S\left(E_j(t), L/M, N_j(t)\right). \qquad (22.22)$$

This expression has the properties that at $t = 0$, it takes on the value of the initial entropy before the inner wall was removed,

$$S(\{E_j(0)\}, L, N_j(0), /M) = S(E, L_o, N), \qquad (22.23)$$

and as $t \to \infty$, it goes rapidly to the equilibrium entropy of the full system,

$$\lim_{t \to \infty} S(\{E_j(t)\}, L, N_j(t), /M) = S(E, L, N). \qquad (22.24)$$

These two properties are independent of the number M of subsystems that are observed.

22.11   Interacting Particles

The explanation of irreversible phenomena given here is, of course, not complete. The apparent conflict between microscopic and macroscopic laws has been resolved, but we have ignored the effects of interactions. However, now that it is clear that there is no real conflict between time-reversal-invariant microscopic laws and macroscopic irreversibility, it should be sufficient to demonstrate the equilibration of the momentum distribution by molecular dynamics computer simulations that are generalizations of those carried out in the assignments in Chapter 19.

Part IV

Quantum Statistical Mechanics

23

Quantum Ensembles

If you are not completely confused by quantum mechanics, you do not understand it.

John Archibald Wheeler, American theoretical physicist (1911–2008)

In Part IV of this book we will extend the application of statistical ideas to quantum mechanical systems. Most of the results we will derive have counterparts in classical statistical mechanics, but some important features will be quite new.

The differences between classical and quantum statistical mechanics are all based on the differing concepts of a microscopic 'state'. While the classical microscopic state (specified by a point in phase space) determines the exact position and momentum of every particle, the quantum mechanical state determines neither; quantum states can only provide probability distributions for observable quantities. This feature has the important consequence that quantum statistical mechanics involves two different kinds of probability, while classical statistical mechanics involves only one.

Classical and quantum statistical mechanics both require the assignment of probabilities to the possible microscopic states of a system. The calculation of all quantities involves averages over such probability distributions in both theories.

Quantum statistical mechanics further requires averages over the probability distributions that are obtained from each individual microscopic state.

We will begin this chapter by recalling the basic equations of quantum mechanics, assuming that the reader is already familiar with the material. If any part of the discussion seems mysterious, it would be advisable to consult a textbook on quantum mechanics.

After a review of the basic ideas we will discuss the special features that distinguish quantum statistical mechanics from the classical theory.

Subsequent chapters will develop the theory of quantum statistical mechanics and apply it to models that demonstrate the significant differences between the predictions of quantum and classical statistical mechanics. We will show that even simple harmonic oscillators and ideal gases have dramatically different properties in quantum and classical theories. Since the real world ultimately obeys quantum mechanics, quantum statistical mechanics is essential to obtain agreement with experimental results.

An Introduction to Statistical Mechanics and Thermodynamics. Robert H. Swendsen, Oxford University Press (2020).
© Robert H. Swendsen. DOI: 10.1093/oso/9780198853237.001.0001

23.1 Basic Quantum Mechanics

The most fundamental characteristic of quantum mechanics is that the microscopic state of a system is described by a wave function, rather than a point in phase space. For a simple, one-particle system, the wave function can be represented by a complex function of space and time, $\psi\,(\vec{r},t)$, which contains all possible information about the particle.

Even if the wave function is known exactly, the result of a measurement of the position of the particle can only be predicted in terms of probabilities. For example, the absolute value of the square of the wave function, $|\psi\,(\vec{r},t)|^2$, gives the probability density for the result of a measurement of the position of a particle. (After the measurement, of course, the wave function of the particle would change.) Only the average position of the particle and the various moments of its probability distribution can be calculated from the wave function.

The momentum, $\vec{p}$, is represented by a vector operator rather than a real vector,

$$\vec{p} = -i\hbar\vec{\nabla} = -i\hbar\left(\frac{\partial}{\partial x}, \frac{\partial}{\partial y}, \frac{\partial}{\partial z}\right). \tag{23.1}$$

As is the case for the position of the particle, the momentum does not have a fixed value. The expectation value (average value) of the momentum is given by an integral over the wave function. This can be denoted as

$$\langle\psi|\vec{p}|\psi\rangle = -i\hbar\int \psi^*\,(\vec{r},t)\,\vec{\nabla}\psi\,(\vec{r},t)\,d^3r. \tag{23.2}$$

In eq. (23.2), ψ^* denotes the complex conjugate of the function ψ, and we have introduced the 'bra', $\langle\psi| = \psi^*\,(\vec{r},t)$, and 'ket', $|\psi\rangle = \psi\,(\vec{r},t)$ notation. The integral is over the allowed volume.

The kinetic and potential energy are represented by a Hamiltonian,

$$H\,(\vec{r},\vec{p}) = H\left(\vec{r},-i\hbar\vec{\nabla}\right) = \frac{\vec{p}\cdot\vec{p}}{2m} + V\,(\vec{r}) = \frac{-\hbar^2}{2m}\nabla^2 + V\,(\vec{r}). \tag{23.3}$$

In this equation,

$$\nabla^2 = \vec{\nabla}\cdot\vec{\nabla} = \frac{\partial^2}{\partial x^2} + \frac{\partial^2}{\partial y^2} + \frac{\partial^2}{\partial z^2}. \tag{23.4}$$

The wave function must satisfy the Schrödinger equation, named after the Austrian physicist Erwin Schrödinger (1887–1961, Nobel Prize 1933),

$$i\hbar\frac{\partial}{\partial t}\psi\,(\vec{r},t) = H\left(\vec{r},-i\hbar\vec{\nabla}\right)\psi\,(\vec{r},t). \tag{23.5}$$

23.2   Energy Eigenstates

There are certain special quantum states that can be written as a product of a function of position times a function of time,

$$\psi\,(\vec{r},t) = \psi\,(\vec{r})f(t). \tag{23.6}$$

For these states, the Schrödinger equation can be written in a form that separates the functions of time and space:

$$i\hbar \frac{1}{f(t)}\frac{\partial f(t)}{\partial t} = \frac{1}{\psi\,(\vec{r})}H\!\left(\vec{r},-i\hbar\vec{\nabla}\right)\psi\,(\vec{r}). \tag{23.7}$$

Since the left side of eq. (23.7) depends only on time and the right side depends only on space, they must both be equal to a constant, which we will denote as E. For these special states, the Schrödinger equation separates into two equations:

$$\frac{\partial f(t)}{\partial t} = \frac{-iE}{\hbar}f(t), \tag{23.8}$$

and

$$H\!\left(\vec{r},-i\hbar\vec{\nabla}\right)\psi\,(\vec{r}) = E\psi\,(\vec{r}). \tag{23.9}$$

Since the expectation value of H is just

$$\langle H\rangle = E, \tag{23.10}$$

we can identify E as the energy of the state.

Eq. (23.8) can be easily solved as a function of time,

$$f(t) = \exp\left(-i\frac{E}{\hbar}t\right). \tag{23.11}$$

In general, eq. (23.9) only has solutions for particular values of E, which are the only observable values of the energy. These special values are known as energy eigenvalues, and the corresponding wave functions are known as eigenfunctions. Eq. (23.9) is called an eigenvalue equation. The eigenfunctions and eigenvalues can be identified by a quantum 'number' n, which is really a set of d numbers that describe the eigenstate for a wave function in a d-dimensional system.

Including the quantum number explicitly, the eigenvalue equation can be written as

$$H\psi_n\,(\vec{r}) = E_n\psi_n\,(\vec{r}), \tag{23.12}$$

or

$$H|n\rangle = E_n|n\rangle, \tag{23.13}$$

where we have suppressed the explicit dependence on $\vec{r}$ in the second form of the eigenvalue equation. E_n is called the energy eigenvalue. It is possible for distinct eigenstates to have the same energy eigenvalue, in which case the states are called 'degenerate'.

The time-dependence of an eigenstate therefore has the form

$$\psi_n(\vec{r}, t) = \psi_n(\vec{r}) \exp\left(-\frac{iE}{\hbar}t\right). \tag{23.14}$$

Since the expectation value of any operator is constant in time for eigenstates, they are also called 'stationary states'.

It is standard procedure to normalize the eigenfunctions

$$\langle n|n\rangle = \int \psi_n^*(\vec{r}, t)\,\psi_n(\vec{r}, t)\,d^3r = 1. \tag{23.15}$$

In this equation, we have again used the 'bra-ket' notation ($\psi_n(\vec{r}, t) = |n\rangle$).

It is easily proved that for $E_n \neq E_m$, $\langle n|m\rangle = 0$. It is also straightforward to choose eigenfunctions in such a way that for $n \neq m$, $\langle n|m\rangle = 0$, even when $E_n = E_m$. These properties can be summarized using the Kronecker delta function

$$\langle n|m\rangle = \delta_{n,m}. \tag{23.16}$$

23.2.1   Expansion of a General Wave Function

A very important theorem is that any wave function can be expanded in the set of all eigenfunctions,

$$\psi(\vec{r}) = \sum_n c_n \psi_n(\vec{r}). \tag{23.17}$$

This can also be expressed more compactly using the ket notation

$$|\psi\rangle = \sum_n c_n|n\rangle. \tag{23.18}$$

The coefficients c_n are complex numbers that can be calculated from the wave functions,

$$\langle m|\psi\rangle = \sum_n c_n \langle m|n\rangle = \sum_n c_n \delta_{m,n} = c_m. \tag{23.19}$$

23.2.2   Magnitudes of Coefficients

Since wave functions are assumed to be normalized, there is a sum rule for the set of coefficients $\{c_n\}$,

$$1 = \langle \psi | \psi \rangle = \sum_n \sum_m c_n^* c_m \langle \psi_n | \psi_m \rangle = \sum_n \sum_m c_n^* c_m \delta_{n,m} = \sum_n |c_n|^2. \tag{23.20}$$

The expectation value of the energy can also be expressed in terms of the coefficients $\{c_n\}$ and the energy eigenvalues

$$\langle \psi | H | \psi \rangle = \sum_m \sum_n c_m^* c_n \langle m | H | n \rangle \tag{23.21}$$

$$= \sum_m \sum_n c_m^* c_n \langle m | E_n | n \rangle$$

$$= \sum_m \sum_n E_n c_m^* c_n \delta_{m,n}$$

$$= \sum_n E_n |c_n|^2.$$

Since the sum of $|c_n|^2$ is 1, and the expectation value of the energy is given by the weighted sum in eq. (23.21), it is natural to interpret $|c_n|^2$ as the probability that a measurement of the energy will put the system into the eigenstate $|n\rangle$.

It is important to note that if a system is in a general state, $|\psi\rangle$, given by eq. (23.18), it is *not* in an eigenstate unless $|c_n| = 1$ for some value of n. $|c_n|^2$ is the probability that the system would be found in the state $|n\rangle$ only after a measurement has been made that put the system into an eigenstate. Without such a measurement, the probability that a general system is in an eigenstate is zero.

23.2.3   Phases of Coefficients

Since the time-dependence of the eigenfunctions is known, the time-dependence of an arbitrary wave function can be written in terms of the expansion in eigenfunctions,

$$\psi(\vec{r}, t) = \sum_n c_n \exp\left(-i \frac{E_n}{\hbar} t\right) \psi_n(\vec{r}), \tag{23.22}$$

or more compactly,

$$|\psi, t\rangle = \sum_n c_n \exp\left(-i \frac{E_n}{\hbar} t\right) |n\rangle. \tag{23.23}$$

Eqs. (23.22) or (23.23) show that the time development of an arbitrary quantum state can be described by the changing phases of the expansion coefficients. They also show that eigenstates with different energies have phases that change at different rates, the relative phases between any two states, given by $(E_n - E_m)t/\hbar$, sweep uniformly through all angles. This property suggests that a model probability distribution in equilibrium should be uniform in the phases. Indeed, if a probability distribution is not uniform in the phases, it will not be time-independent.

23.3   Many-Body Systems

Quantum systems that contain many particles ('many-body systems' in common terminology) are also described by a single wave function, but one that is a function of the coordinates of every particle in the system: $\psi\left(\{\vec{r}_j | j = 1, \ldots, N\}\right)$. The Hamiltonian naturally depends on all the coordinates and all the gradients.

$$H = H\left(\left\{\vec{r}_j, -i\hbar\vec{\nabla}_j | j = 1, \ldots, N\right\}\right) \tag{23.24}$$

We can again find eigenstates of the Hamiltonian (at least in principle), but the quantum 'number' n is now a set of $3N$ numbers that are needed to describe the many-body wave function.

A general many-body wave function can also be expanded as a linear combination of eigenstates, as in eq. (23.18), and the time development of the many-body wave function can still be represented by eq. (23.23), with the appropriate interpretation of n as the set of many-body quantum numbers.

As in the single-particle case, the square of the absolute value of a coefficient in the expansion of a many-body wave function, $|c_n|^2$, can be interpreted as the probability that a measurement that put the system in an eigenstate would put it in eigenstate n. However, such a measurement is *never* carried out on a macroscopic system, which is never in an energy eigenstate.

23.4   Two Types of Probability

In Part I of this book we introduced the use of probabilities to describe our ignorance of the exact microscopic state of a many-body system. Switching from classical to quantum mechanics, we find a corresponding ignorance of the exact microscopic wave function. Again we need to construct a model probability distribution for the microscopic states to describe the behavior of a macroscopic system.

However, as we have seen previously, the properties of a quantum system are still only given by probability distributions even if we know the wave function. Therefore, in quantum statistical mechanics we need to deal with two kinds of probability distributions: one for the microscopic states, and a second for the observable properties.

23.4.1 Model Probabilities for Quantum Systems

We will denote a model probability distribution for the many-body wave functions as P_ψ. The calculation of the expectation value (average value) of any operator A must include averages over probability distribution of the wave functions *and* the probability distribution of each quantum state,

$$\langle A \rangle = \int_\psi P_\psi \langle \psi | A | \psi \rangle. \tag{23.25}$$

We have written the average over P_ψ as an integral because there is a continuum of wave functions.

We can write eq. (23.25) in a more convenient form by expressing the wave function in terms of an expansion in eigenfunctions, as in eq. (23.18). Since the expansion coefficients completely specify the wave function, we can express the probability distribution in terms of the coefficients

$$P_\psi = P_{\{c_n\}}. \tag{23.26}$$

The average of A then becomes

$$\langle A \rangle = \int_{\{c_n\}} \sum_n \sum_m P_{\{c_n\}} c_m^* c_n \langle m | A | n \rangle. \tag{23.27}$$

The integral in eq. (23.27) is over all values of all coefficients, subject to the normalization condition in eq. (23.20).

Since the phases play an important role in specifying the time dependence in quantum mechanics, it will be useful to modify eq. (23.27) to exhibit them explicitly. If we write the coefficients as

$$c_n = |c_n| \exp(i\phi_n), \tag{23.28}$$

we can rewrite eq. (23.27) as

$$\langle A \rangle = \int_{\{|c_n|\}} \int_{\{\phi_n\}} \sum_n \sum_m P_{\{c_n\}} |c_m||c_n| \exp(-i\phi_m + i\phi_n) \langle m | A | n \rangle. \tag{23.29}$$

Since we have introduced the phases explicitly into eq. (23.29), we have also separated the integrals over the coefficients into integrals over their magnitudes and their phases.

23.4.2   Phase Symmetry in Equilibrium

Eq. (23.29) gives a formal expression for the macroscopic average of any property in any macroscopic system. However, since we are primarily interested in macroscopic equilibrium states, we can greatly simplify the problem by introducing the model probability for the phases mentioned in Subsection 23.2.3. Since the macroscopic equilibrium state is time-independent, it would seem reasonable to assume that the equilibrium probability distribution of phase angles should be uniform and the phases independent. With this assumption, $P_{\{c_n\}} = P_{\{|c_n|\}}$, and we can integrate over the phase angles when $n \neq m$

$$\int_0^{2\pi} d\phi_m \int_0^{2\pi} d\phi_n \exp(-i\phi_m + i\phi_n) = 0, \tag{23.30}$$

or for $n = m$,

$$\int_0^{2\pi} d\phi_n \exp(-i\phi_n + i\phi_n) = \int_0^{2\pi} d\phi_n = 2\pi. \tag{23.31}$$

Inserting eqs. (23.30) and (23.31) into eq. (23.29) we find

$$\langle A \rangle = 2\pi \int_{\{|c_n|\}} \sum_n P_{\{|c_n|\}} |c_n|^2 \langle n|A|n \rangle, \tag{23.32}$$

or

$$\langle A \rangle = \sum_n \left[2\pi \int_{\{|c_n|\}} P_{\{|c_n|\}} |c_n|^2 \right] \langle n|A|n \rangle. \tag{23.33}$$

Eq. (23.33) is particularly interesting. If we define a set of values $\{P_n\}$, where

$$P_n = 2\pi \int_{\{|c_n|\}} P_{\{|c_n|\}} |c_n|^2, \tag{23.34}$$

we can rewrite the equation for $\langle A \rangle$ in a very useful form,

$$\langle A \rangle = \sum_n P_n \langle n|A|n \rangle. \tag{23.35}$$

The only quantum averages that appear in eq. (23.35) are of the form $\langle n|A|n \rangle$, so every term in the sum is an average over a single quantum eigenstate. For the purposes of calculating the macroscopic properties of a macroscopic system in equilibrium, each eigenstate contributes separately.

The quantities P_n have two useful properties. First, since P_ψ is a probability distribution, $P_\psi \geq 0$ for all ψ. From eq. (23.34), it follows that $P_n \geq 0$ for all n. Next, since the average of a constant must be the value of the constant, eq. (23.35) gives us a normalization condition on the set of quantities $\{P_n\}$,

$$1 = \langle 1 \rangle = \sum_n P_n \langle n|1|n \rangle = \sum_n P_n. \tag{23.36}$$

This equation also implies that $P_n \leq 1$.

Since $0 \leq P_n \leq 1$ for all n, and from eq. (23.36) the sum of all P_ns is unity, the P_n's fairly beg to be regarded as probabilities. The question is: what are they probabilities of? P_n can be interpreted as the probability that a measurement of the eigenstate of a system will result in the system being in eigenstate $|n\rangle$. However, it must be recognized that such a measurement is never made on a macroscopic system. Nevertheless, calculations using eq. (23.35) are carried out formally exactly *as if* the P_ns were probabilities of something physically relevant. There is no real harm in referring to P_n as the 'probability of being in eigenstate $|n\rangle$' (which is often found in textbooks), as long as you are aware of its true meaning. A more complete description of why this is a useful—if not entirely correct—way of thinking about quantum statistical mechanics is presented in Sections 23.5 and 23.6.

23.5 The Density Matrix

A very common representation of averages in a quantum ensemble is provided by the density operator, which is defined by a sum or integral over the microscopic states in an ensemble,

$$\rho = \int_\psi P_\psi |\psi\rangle \langle\psi|. \tag{23.37}$$

The density matrix $\rho_{n,m}$ is found by taking the matrix element between the states $\langle m|$ and $|n\rangle$,

$$\rho_{m,n} = \int_\psi P_\psi \langle m|\psi\rangle \langle\psi|n\rangle. \tag{23.38}$$

The density matrix is useful because it incorporates all the information needed to calculate ensemble averages. In particular, the ensemble average of an operator $\mathcal{A}$ is given by

$$\langle \mathcal{A} \rangle = Tr[\rho \mathcal{A}] \tag{23.39}$$

where Tr indicates the trace of the matrix. To prove eq. (23.39), simply evaluate the trace on the right-hand side of the equation,

$$Tr[\rho A] = \sum_n \int_\psi P_\psi \langle n|\psi\rangle\langle\psi|A|n\rangle \tag{23.40}$$

$$= \int_\psi \sum_n P_\psi \langle n|\psi\rangle\langle\psi|A|n\rangle$$

$$= \int_{\{c_n\}} \sum_n \sum_\ell \sum_m P_{\{c_n\}} c_\ell c_m^* \langle n|\ell\rangle\langle m|A|n\rangle$$

$$= \int_{\{c_n\}} \sum_n \sum_m P_{\{c_n\}} c_m^* c_n \langle m|A|n\rangle = \langle A\rangle.$$

The last equality is found by comparison with eq. (23.27).

If we write the coefficients as $c_n = |c_n|\exp(-i\phi_n)$, assume that the system is in equilibrium, and average over the phases, we find

$$Tr[\rho A] = \int_{\{c_n\}} \sum_n P_{\{c_n\}} |c_n|^2 \langle n|A|n\rangle, \tag{23.41}$$

or,

$$Tr[\rho A] = \sum_n P_n \langle n|A|n\rangle = \langle A\rangle. \tag{23.42}$$

This agrees with eq. (23.35).

23.6 Uniqueness of the Ensemble

A peculiar feature of the density matrix is that although eq. (23.40) shows that it contains the information to calculate any ensemble average, it does not uniquely specify the quantum ensemble. This can be seen from a simple example that compares the following two different quantum mechanical ensembles for a simple harmonic oscillator.

1. Ensemble 1: The SHO is either in the ground state $|0\rangle$ with probability $1/2$, or it is in the first excited state $|1\rangle$ with probability $1/2$. In either case, the system is in an eigenstate of the Hamiltonian. The density matrix operator corresponding to this ensemble is then

$$\rho_1 = \frac{1}{2}(|0\rangle\langle 0| + |1\rangle\langle 1|). \tag{23.43}$$

2. Ensemble 2: The system is in a state of the form

$$|\phi\rangle = \frac{1}{\sqrt{2}}\left(|0\rangle + e^{i\phi}|1\rangle\right), \tag{23.44}$$

with the values of ϕ being uniformly distributed between 0 and 2π. No member of the ensemble is an eigenstate of the Hamiltonian. The density matrix operator corresponding to this ensemble is then

$$
\begin{aligned}
\rho_2 &= \frac{1}{2\pi} \int_0^{2\pi} d\phi \frac{1}{\sqrt{2}} \left(|0\rangle + e^{i\phi} |1\rangle \right) \frac{1}{\sqrt{2}} \left(\langle 0| + e^{-i\phi} \langle 1| \right) \\
&= \frac{1}{4\pi} \int_0^{2\pi} d\phi \left(|0\rangle \langle 0| + |1\rangle \langle 1| + e^{-i\phi} |0\rangle \langle 1| + e^{i\phi} |1\rangle \langle 0| \right) \\
&= \frac{1}{2} \left(|0\rangle \langle 0| + |1\rangle \langle 1| \right) = \rho_1.
\end{aligned}
\tag{23.45}
$$

This shows that the same density matrix operator describes both an ensemble that contains no eigenstates and an ensemble that contains only eigenstates. If we combine this with the result in eq. (23.40), we see that the expectation value of any operator is exactly the same for different ensembles as long as the density matrix operators are the same—even if the two ensembles do not have a single quantum state in common!

It is usual to express the density matrix operator in terms of the eigenstates of a system, which gives the impression that the ensemble also consists entirely of eigenstates. Even though we know that this is not the case, the previous demonstration shows that all predictions based on the (erroneous) assumption that a macroscopic system is in an eigenstate will be consistent with experiment.

23.7   The Planck Entropy

In the early days of quantum mechanics, Max Planck suggested defining the entropy of a quantum system as

$$
S_P = k_B \sum_\ell \ln g_\ell,
\tag{23.46}
$$

where g_ℓ is the number of eigenstates associated with energy level ℓ and energy E_ℓ, known as the degeneracy of the energy level. This form is analogous to the classical microcanonical entropy. Eq. (23.46) has a number of advantages, and is still quite often presented in textbooks as the correct definition of the entropy of a quantum system. It also has a number of disadvantages, starting with the discreteness of the set of energies for which it is defined.

It has been mainly applied to simple Hamiltonians, for which the degeneracies allow a straightforward approximation of the energy dependence of S_P by a continuum. For more complicated Hamiltonians, the energy levels can be split, reducing the degeneracy of each level. If all the states become non-degenerate, $g_\ell = 1$ for all ℓ, and $S_P = 0$, which is clearly not correct.

In Chapter 24, we will compare the predictions of the Planck entropy with those of the canonical entropy.

23.8   The Quantum Microcanonical Ensemble

Now that we have reduced the problem of calculating averages in equilibrium quantum statistical mechanics to averages over individual eigenstates, we need only evaluate the set of 'probabilities', $\{P_n\}$, where n is the quantum number (or numbers) indexing the eigenstates. In principle this is straightforward, and we might try to follow the same procedure as in classical statistical mechanics. Consider a composite system, isolated from the rest of the universe, with fixed total energy $E_T = E_A + E_B$, assume all states are equally 'probable' (subject to the constraints), and calculate the probability distributions for observable quantities.

Unfortunately, we run into a technical difficulty that prevents us from carrying out this procedure for anything other than a particularly simple model.

If we assume that each subsystem is in an eigenstate—a nontrivial assumption that we return to in Chapter 24—the total energy of the composite system must be the sum of an energy eigenvalue from each subsystem, $E = E_{n,A} + E_{m,B}$. To calculate the probabilities for distributing the total energy among the two subsystems, we have to be able to take some amount of energy from subsystem A and transfer it to subsystem B. Unfortunately, the energy levels in a general macroscopic quantum system are not distributed in a regular fashion; if we change the energy in subsystem A to $E_{n',A} = E_{n,A} + \Delta E$, there will generally not be an eigenstate in subsystem B with an energy $E_{m,B} - \Delta E$. Even though the energy eigenvalues in a macroscopic system are very closely spaced, we will not always be able to transfer energy between the subsystem and maintain a constant total energy.

It is important to recognize that this is a problem that has arisen because we are assuming that each subsystem is in an eigenstate; all linear combinations of eigenstates, as in eq. (23.18), have been neglected.

Because macroscopic systems are never in an eigenstate, they do not have precisely specified energy. The typical spread in energy is far greater than the tiny gap between energy levels. Beyond that, the interactions between the two subsystems that are necessary to transfer energy between them will result in a single set of eigenvalues for the full composite system.

We must find a different way to do calculations. The simplest method is to make one of the subsystems extremely large and treat it as a thermal reservoir. As a quantum system becomes larger, its energy levels move closer together; in the limit of an infinite system, the energy spectrum becomes a continuum, and it is always possible to find a state with any desired energy. For this reason, we will abandon the microcanonical ensemble and turn to the canonical ensemble for most calculations in quantum statistical mechanics.

24

Quantum Canonical Ensemble

I do not like it, and I am sorry I ever had anything to do with it.

Erwin Schrödinger, Austrian physicist (1887–1961), Nobel Prize 1933,
speaking about quantum mechanics.

The quantum canonical ensemble, like its classical counterpart, which was discussed in Chapter 19, describes the behavior of a system in contact with a thermal reservoir. As in the classical case, the detailed nature of the reservoir is not important; it must only be big. This is partly to allow us to expand the entropy in powers of the ratio of the size of the system to that of the reservoir, and also to allow us to treat the eigenvalue spectrum of the reservoir as continuous.

We have already seen in Chapter 21 that the canonical entropy gives an improved expression for the entropy of a classical system. This is also true of the quantum canonical entropy, which has the advantage of being defined on a continuum of energy values, instead of just being defined on the eigenvalues of the Hamiltonian.

An important thing to notice about quantum statistical mechanics as you go through the following chapters is that the same ideas and equations keep showing up to solve different problems. In particular, the solution to the quantum simple harmonic oscillator (QSHO) turns out to be the key to understanding crystal vibrations, black-body radiation, and Bose–Einstein statistics. The basic calculation for the QSHO is discussed in Section 24.10; if you understand it completely, the topics in Chapters 25 through 28 will be much easier. The other kind of calculation that will appear repeatedly is one in which there are only a small number of quantum energy levels. Indeed, in the most important case there are only two energy levels. This might seem like a very limited example, but it provides the essential equations for Chapters 27, 29, and 31.

24.1 Derivation of the QM Canonical Ensemble

The derivation of the quantum canonical ensemble follows the same pattern as the derivation of the classical canonical ensemble in Section 19.3.

An Introduction to Statistical Mechanics and Thermodynamics. Robert H. Swendsen, Oxford University Press (2020).
© Robert H. Swendsen. DOI: 10.1093/oso/9780198853237.001.0001

Let an eigenstate $|n\rangle$ for the system of interest have energy E_n and the reservoir have energy E_R, so that the total energy is $E_T = E_n + E_R$. Let $\Omega_R(E_R) = \Omega_R(E_T - E_n)$ be the degeneracy (number of eigenstates) of the energy level E_R in the reservoir. The 'probability' P_n of an eigenstate with energy E_n in the system of interest is then given by:

$$P_n = \frac{\Omega_R(E_T - E_n)}{\Omega_T(E_T)}. \tag{24.1}$$

As we did in Section 19.3, take the logarithm of both sides and use the fact that $E_T \gg E_n$ to expand $\ln \Omega_R(E_T - E_n)$ in powers of E_n/E_T,

$$\ln P_n = \ln \Omega_R(E_T - E_n) - \ln \Omega_T(E_T)$$

$$\approx \ln \Omega_R(E_T) - E_n \frac{\partial}{\partial E_T} \ln \Omega_R(E_T) - \ln \Omega_T(E_T). \tag{24.2}$$

The approximation becomes exact in the limit of an infinite reservoir. In practice, the error due to using finite reservoirs is so small, that we will ignore the higher-order terms and use an equal sign for simplicity.

In analogy to the classical derivation in Section 19.3, we identify

$$\beta = \frac{1}{k_B T} = \frac{\partial}{\partial E_T} \ln \Omega_R(E_T), \tag{24.3}$$

so that we can write eq. (24.2) as

$$\ln P_n = -\ln Z - \beta E_n, \tag{24.4}$$

or

$$P_n = \frac{1}{Z} \exp(-\beta E_n), \tag{24.5}$$

where

$$-\ln Z = \ln \Omega_R(E_T) - \ln \Omega_T(E_T). \tag{24.6}$$

The normalization factor Z is known as the quantum mechanical partition function (in analogy to the classical partition function) and can be evaluated by summation over all eigenstates,

$$Z = \sum_n \exp(-\beta E_n). \tag{24.7}$$

Note that the quantum Boltzmann factor $\exp(-\beta E_n)$ plays the same role as the classical Boltzmann factor in eq. (19.17). The quantum mechanical partition function plays the same role as the classical partition function.

Eq. (24.7) is often used as a starting point in textbooks on quantum statistical mechanics because of its simplicity. Unfortunately, its validity and significance are not completely obvious without an explanation. It *is* a good starting point for many calculations in quantum statistical mechanics. However, a problem arises if we consider a quantum theory of distinguishable particles. A common procedure is to ignore this possibility, but since a quantum theory of colloidal particles (which are distinguishable) should make sense, we will include this case in a later chapter. It should not come as a surprise that a factor of $1/N!$ will play a role in the discussion, and eq. (24.7) will have to be modified to include the factor of $1/N!$ explicitly.

It is often convenient to reexpress eq. (24.7) in terms of the energy levels. In that case, we must include the degeneracy $\Omega(\ell)$ for each energy level ℓ,

$$Z = \sum_{\ell} \Omega(\ell) \exp(-\beta E_\ell). \tag{24.8}$$

It is essential to remember that in eq. (24.7) the index n runs over the quantum numbers for the eigenstates of the system—not the energy levels—and that you need to include the degeneracy as in eq. (24.8) if you are summing over energy levels.

24.2 Thermal Averages and the Average Energy

Combining eq. (24.7) with eq. (23.35), we find a very useful expression for the average of an operator A in the canonical ensemble, that is, at a temperature $T = 1/k_B\beta$,

$$\langle A \rangle = \sum_n P_n \langle n|A|n \rangle. \tag{24.9}$$

Inserting the expression for P_n from eq. (24.5), this becomes

$$\langle A \rangle = \frac{1}{Z} \sum_n \langle n|A|n \rangle \exp(-\beta E_n). \tag{24.10}$$

A particularly important case is the average energy

$$U = \langle H \rangle = \frac{1}{Z} \sum_n \langle n|H|n \rangle \exp(-\beta E_n) = \frac{1}{Z} \sum_n E_n \exp(-\beta E_n). \tag{24.11}$$

24.3 The Quantum Mechanical Partition Function

Although the partition function Z was introduced simply as a normalization factor in eq. (24.7), it turns out to be very useful in its own right in the same way that the classical partition function was found to be useful in Chapter 19. As in the classical case, the partition function in eq. (24.7) depends on the temperature as $\beta = 1/k_B T$. If we take the derivative of the logarithm of the partition function, we find

$$\left(\frac{\partial}{\partial\beta}\ln Z\right)_{V,N} = \frac{1}{Z}\left(\frac{\partial}{\partial\beta}\sum_n \exp(-\beta E_n)\right)_{V,N} = \frac{1}{Z}\sum_n(-E_n)\exp(-\beta E_n) = -U.$$

$$(24.12)$$

If we compare this equation to the thermodynamic identity

$$\left(\frac{\partial(\beta F)}{\partial\beta}\right)_{V,N} = U, \tag{24.13}$$

we see that we can integrate it to obtain

$$\ln Z = -\beta F + f(V,N), \tag{24.14}$$

in which the function $f(V,N)$ has not yet been determined.

To determine the function $f(V,N)$, first consider the derivative of

$$F = -k_b T\left[\ln Z - f(V,N)\right], \tag{24.15}$$

with respect to temperature,

$$\left(\frac{\partial F}{\partial T}\right)_{V,N} = -k_B[\ln Z - f(V,N)] - k_b T\frac{\partial\beta}{\partial T}\frac{\partial}{\partial\beta}\ln Z \tag{24.16}$$

$$= -k_B[-\beta F] - k_b T\left(\frac{-1}{k_B T^2}\right)(-U)$$

$$= \frac{1}{T}(F - U) = \frac{1}{T}(U - TS - U) = -S.$$

Clearly, the function $f(V,N)$ has no effect on either the energy or the entropy.

Next consider the derivative of F with respect to volume,

$$\left(\frac{\partial F}{\partial V}\right)_{T,N} = -k_B T\left[\frac{\partial}{\partial V}\ln Z - \frac{\partial}{\partial V}f(V,N)\right] \tag{24.17}$$

$$= -k_B T\frac{1}{Z}\sum_n\left(-\beta\frac{\partial E_n}{\partial V}\right)\exp(-\beta E_n) - k_B T\frac{\partial f}{\partial V}$$

$$= \frac{1}{Z}\sum_n\left(\frac{\partial E_n}{\partial V}\right)\exp(-\beta E_n) - k_B T\frac{\partial f}{\partial V}.$$

Now we need the relationship between the pressure the system would have in an eigenstate, and the partial derivative of the energy of that eigenstate with respect to the volume. To avoid confusion with P_n defined earlier in this section, we will denote the pressure of the system in an eigenstate by $P(|n\rangle)$,

$$P(|n\rangle) = -\left(\frac{\partial E_n}{\partial V}\right)_{T,N}. \tag{24.18}$$

Combining eqs. (24.17) and (24.18), we find

$$\left(\frac{\partial F}{\partial V}\right)_{T,N} = -\frac{1}{Z}\sum_n P(|n\rangle)\exp(-\beta E_n) - k_B T\frac{\partial f}{\partial V} = -P, \tag{24.19}$$

where P is the average pressure. Since P must be given by the weighted average of the pressures in the eigenstates,

$$P = \frac{1}{Z}\sum_n P(|n\rangle)\exp(-\beta E_n), \tag{24.20}$$

the partial derivative of $f(V,N)$ with respect to V must vanish; f can only be a function of N,

$$f(V,N) = f(N). \tag{24.21}$$

At this point, we would need to examine the N-dependence of the energy eigenvalues, but this is best done later in the context of the discussion of the quantum ideal gas and the exchange of particles with a reservoir. For the time being we will tentatively assign the value zero to the function $f(N)$. This will turn out to be a valid choice for fermions or bosons, but not for distinguishable particles.

The (tentative) choice of $f(N) = 0$ has the great advantage of simplifying eq. (24.14) and giving it the same form as in classical statistical mechanics:

$$\ln Z = -\beta F, \tag{24.22}$$

or

$$Z = \exp(-\beta F). \tag{24.23}$$

As noted in the first box in Section 24.1, the case of distinguishable particles has to be handled rather differently. We will see in Section 27.9 that the proper choice for a system of distinguishable particles is $f(N) = -\ln(N!)$.

24.4   The Quantum Mechanical Entropy

The expression for the entropy can also be written in another form, which is both useful and revealing.

As in eq. (24.16), take the derivative of the free energy with respect to temperature to find the (negative) entropy, but now set $f(V,N) = 0$,

$$\left(\frac{\partial F}{\partial T}\right)_{V,N} = -k_B \ln Z - k_b T \left(\frac{-1}{k_B T^2}\right) \frac{\partial}{\partial \beta} \ln Z$$

$$= -k_B \ln Z + \frac{1}{TZ} \sum_n (-E_n) \exp(-\beta E_n). \tag{24.24}$$

In the second line of this equation we have used eq. (24.7) to express the derivative of the partition function in terms of a sum. Now recall from eq. (24.5) that

$$P_n = \frac{1}{Z} \exp(-\beta E_n), \tag{24.25}$$

or

$$\ln P_n = -\ln Z - \beta E_n, \tag{24.26}$$

and, of course,

$$\sum_n P_n = \frac{1}{Z} \sum_n \exp(-\beta E_n) = 1. \tag{24.27}$$

Eq. (24.24) can now be written as

$$\left(\frac{\partial F}{\partial T}\right)_{V,N} = -k_B \left[\ln Z \sum_n P_n - \frac{\beta}{Z} \sum_n (-E_n) \exp(-\beta E_n) \right]$$

$$= -k_B \sum_n [P_n \ln Z + \beta E_n P_n]$$

$$= k_B \sum_n P_n \ln P_n = -S, \tag{24.28}$$

which gives us an alternative expression for the entropy

$$S = -k_B \sum_n P_n \ln P_n. \tag{24.29}$$

Eq. (24.29) is often taken as a fundamental starting point in books on statistical mechanics. Its simplicity is certainly an advantage of this choice. Nevertheless, the validity of this equation is not really obvious without a derivation such as the one given above.

24.5   The Origin of the Third Law of Thermodynamics

An important consequence of eq. (24.29) is that since $1 \geq P_n \geq 0$ and $\ln P_n \leq 0$, the entropy is always positive. This implies that the limit of the entropy as the temperature goes to zero is non-negative. Since the entropy at $T = 0$ must be a constant, and not minus infinity as it is in classical statistical mechanics, this establishes the Nernst Postulate, or Third Law of Thermodynamics, as a general consequence of quantum mechanics.

To investigate the value of the entropy at $T = 0$, we note that

$$P_n = \frac{1}{Z} \exp(-\beta E_n) = \frac{\exp(-\beta E_n)}{\sum_n \exp(-\beta E_n)}. \tag{24.30}$$

If we express the partition function in terms of energy levels indexed by ℓ, we can write

$$Z = \sum_\ell \Omega(\ell) \exp(-\beta E_\ell), \tag{24.31}$$

where $\Omega(\ell)$ is the degeneracy of the ℓth level.

Let $\ell = 0$ be the lowest energy level, so that $\Omega(0)$ is the degeneracy of the ground state. Eq. (24.31) can then be written as

$$Z = \Omega(0) \exp(-\beta E_0) \left[1 + \sum_{\ell > 0} \left(\frac{\Omega(\ell)}{\Omega(0)} \right) \exp(-\beta(E_\ell - E_0)) \right]. \tag{24.32}$$

Since $\ell = 0$ is the lowest energy level, $E_\ell - E_0 > 0$, for all $\ell > 0$. That implies that as $T \to 0$, the ground state probability, P_0, is given by

$$
\begin{aligned}
\lim_{T \to 0} P_0 &= \lim_{T \to 0} \left(\frac{\exp(-\beta E_0)}{\Omega(0) \exp(-\beta E_0) \left[1 + \sum_{\ell > 0} \left(\frac{\Omega(\ell)}{\Omega(0)} \right) \exp(-\beta(E_\ell - E_0)) \right]} \right) \\
&= \frac{1}{\Omega(0)} \lim_{T \to 0} \left[1 + \sum_{\ell > 0} \left(\frac{\Omega(\ell)}{\Omega(0)} \right) \exp(-\beta(E_\ell - E_0)) \right]^{-1} \\
&= \frac{1}{\Omega(0)},
\end{aligned}
\tag{24.33}
$$

and the probability of any higher level with $E_\ell > E_0$ is

$$\lim_{T \to 0} P_\ell = \lim_{T \to 0} \left(\frac{\exp(-\beta E_\ell)}{\Omega(0)\exp(-\beta E_0)\left[1 + \sum_{\ell' > 0} \left(\frac{\Omega(\ell')}{\Omega(0)}\right)\exp(-\beta(E_{\ell'} - E_0))\right]} \right)$$

$$= \frac{\exp(-\beta(E_\ell - E_0))}{\Omega(0)} \lim_{T \to 0} \left[1 + \sum_{\ell' > 0} \left(\frac{\Omega(\ell')}{\Omega(0)}\right)\exp(-\beta(E_{\ell'} - E_0)) \right]^{-1}$$

$$= 0. \tag{24.34}$$

Now return to eq. (24.29) for the entropy in terms of the set of 'probabilities' $\{P_n\}$. Since it is well known that $\lim_{x \to 0}(x \ln x) = 0$, we can see that

$$\lim_{T \to 0} S(T) = -k_B \Omega(0) \frac{1}{\Omega(0)} \ln \frac{1}{\Omega(0)} = k_B \ln \Omega(0). \tag{24.35}$$

If the ground state is non-degenerate, $\Omega(0) = 1$, and $S(T = 0) = 0$. Otherwise, it is a positive number.

We are often interested in the entropy per particle, rather than the total entropy. At zero temperature, $S(T = 0)/N$, is given by

$$\frac{S(T = 0)}{N} = \frac{k_B \ln \Omega(0)}{N}. \tag{24.36}$$

If $\Omega(0)$ does not depend on the size of the system, the entropy per particle vanishes as the system size diverges. For a finite but macroscopic system, the entropy per site at zero temperature is non-zero, but immeasurably small. This is the origin of the Planck formulation of the Nernst Postulate (or Third Law of Thermodynamics), which requires the entropy to be zero at $T = 0$.

However, there is another possibility. Suppose that the degeneracy of the ground state depends on the size of the system as,

$$\Omega(0) = a^N, \tag{24.37}$$

where $a > 1$ is some positive constant. Then the zero-temperature entropy per particle is given by

$$\frac{S(T = 0)}{N} = \frac{k_B \ln a^N}{N} = k_B \ln a > 0. \tag{24.38}$$

This would violate the Planck formulation of the Nernst Postulate, but Nernst's original formulation would still be valid.

The possibility suggested by eq. (24.38) actually occurs, both in model calculations and real experiments. The essential feature of a system that exhibits $S(T = 0)/N > 0$ is a disordered ground state. This is true of normal window glass, and has been confirmed by experiment. It is also true of an interesting class of materials called 'spin glasses', which are formed when a small fraction of a magnetic atom (such as iron) is dissolved in a non-magnetic material (such as copper). For reasons that go beyond the scope of this book, the ground state of this system is highly disordered, leading to a large number of very low-lying energy levels. Although the ground-state degeneracy for these materials does not strictly follow eq. (24.37), the predicted zero-temperature limit $S(T = 0)/N > 0$ is found experimentally. There are simple models of spin glasses for which the inequality in eq. (24.38) can be shown to be exactly correct.

24.6 Derivatives of Thermal Averages

Since the partition function depends on the temperature in a simple way, we can express derivatives of thermal averages with respect to T or β in terms of fluctuations. This extremely useful mathematical technique has already been discussed for the classical case in Section 19.8.

For simplicity, begin with derivatives with respect to β. Eq. (24.10) gives the formal expression for the thermal average $\langle A \rangle$,

$$\langle A \rangle = \frac{1}{Z} \sum_n \langle n|A|n \rangle \exp(-\beta E_n). \tag{24.39}$$

The partial derivative of $\langle A \rangle$ with respect to β gives the equation

$$\frac{\partial}{\partial \beta} \langle A \rangle = -\frac{1}{Z} \frac{\partial Z}{\partial \beta} \langle A \rangle + \frac{1}{Z} \sum_n \langle n|A|n \rangle (-E_n) \exp(-\beta E_n)$$
$$- \langle A \rangle \langle H \rangle - \langle AH \rangle. \tag{24.40}$$

This result is completely general.

Applying eq. (24.40) to the thermal average of the energy, we find

$$\frac{\partial}{\partial \beta} \langle H \rangle = \frac{\partial U}{\partial \beta} = -\left[\langle H^2 \rangle - \langle H \rangle^2 \right].$$

The derivative of the energy with respect to β is given by the negative of the variance of the energy.

If we apply this equation to the calculation of the specific heat, we find

$$c_V = \frac{1}{N} \frac{\partial U}{\partial T} = \frac{1}{N} \frac{\partial \beta}{\partial T} \frac{\partial U}{\partial \beta} = \frac{1}{N k_B T^2} \left[\langle H^2 \rangle - \langle H \rangle^2 \right]. \tag{24.41}$$

The proportionality between the specific heat and the variance of the energy is exactly the same in quantum and classical mechanics. (We derived the classical version in Section 19.9.)

The relationship between thermal fluctuations and thermodynamic derivatives is both deep and powerful. Using it can make seemingly difficult calculations become very easy. It is particularly useful in computer simulations.

24.7   Factorization of the Partition Function

The best trick in quantum statistical mechanics corresponds directly to the best trick in classical statistical mechanics, which was discussed in Section 19.12. In both cases, the form of the Hamiltonian allows us to factor the partition function, transforming a many-dimensional problem into many low-dimensional problems.

It is quite common to encounter problems in which the Hamiltonian can—at least approximately—be written as a sum of N terms

$$H = \sum_{j=1}^{N} H_j, \tag{24.42}$$

where the terms H_j all commute with one another. If this is true, the eigenvalue equation for H_j is

$$H_j |n_j\rangle = E_{n_j} |n_j\rangle, \tag{24.43}$$

and the eigenvector of the full system can be written as,

$$|n\rangle = \prod_{k=1}^{N} |n_k\rangle, \tag{24.44}$$

where we have used the index n to denote the full set of quantum numbers,

$$n \equiv \left\{ n_j | j = 1, \ldots, N \right\}. \tag{24.45}$$

Since the term H_j only acts on $|n_j\rangle$,

$$H_j |n\rangle = E_{n_j} \prod_{k=1}^{N} |n_k\rangle = E_{n_j} |n\rangle. \tag{24.46}$$

The full eigenvalue equation becomes,

$$H|n\rangle = \sum_{j=1}^{N} H_j \prod_{k=1}^{N} |n_k\rangle = \sum_{j=1}^{N} E_{n_j} \prod_{k=1}^{N} |n_k\rangle = \sum_{j=1}^{N} E_{n_j} |n\rangle = E_n |n\rangle, \tag{24.47}$$

where

$$E_n = \sum_{j=1}^{N} E_{n_j}.$$ (24.48)

We can use eq. (24.48) to cast the partition function in a very convenient form:

$$Z = \sum_{\{n_j\}} \exp\left(-\beta \sum_{j=1}^{N} E_{n_j}\right)$$

$$= \sum_{\{n_j\}} \prod_{j=1}^{N} \exp\left(-\beta E_{n_j}\right)$$

$$= \sum_{n_N} \exp\left(-\beta E_{n_N}\right) \cdots \sum_{n_2} \exp\left(-\beta E_{n_2}\right) \sum_{n_1} \exp(-\beta E_{n_1})$$

$$= \prod_{j=1}^{N}\left(\sum_{n_j} \exp(-\beta E_{n_j})\right).$$ (24.49)

Because this equation turns out to be extremely useful, we will repeat it without the intermediate steps,

$$Z = \sum_{\{n_j\}} \prod_{j=1}^{N} \exp(-\beta E_{n_j}) = \prod_{j=1}^{N}\left(\sum_{n_j} \exp(-\beta E_{n_j})\right).$$ (24.50)

This form of the partition function allows us to factor it into a product of terms, each of which is much easier to evaluate than the original expression. Just as for the corresponding equation in classical mechanics (see Section 19.12), difficult problems can become very easy by using eq. (24.50).

Because of the importance of eq. (24.50) and the frequency with which it is used, we should call attention to a mental hazard associated with it that can catch the unwary. In eq. (24.50) we are exchanging a sum and a product,

$$\sum_{\{n_j\}} \prod_{j=1}^{N} \longleftrightarrow \prod_{j=1}^{N} \sum_{n_j}.$$

However, the sums on the right and left sides of the equation are *not* over the same quantum numbers. The sum on the right, $\sum_{n_j}$, is only over the quantum numbers associated with the term H_j in the Hamiltonian. The sum on the left, $\sum_{\{n_j\}}$, is over the set of all quantum numbers for the entire system. In the heat of battle

continued

(while taking a test), it is not unusual to forget to write the indices of sums and products explicitly. Omitting the indices is always a bad way to save time, but it can be especially dangerous when using eq. (24.50).

We are now in a position to see why eq. (24.50) is so valuable. From eq. (24.22), the free energy is given by the logarithm of the partition function. When the Hamiltonian has the form given in eq. (24.42), the free energy becomes particularly simple,

$$
\begin{aligned}
F = -k_B T \ln Z &= -k_B T \ln \sum_{\{n_j\}} \prod_{j=1}^{N} \exp(-\beta E_{n_j}) \\
&= -k_B T \ln \prod_{j=1}^{N} \left(\sum_{n_j} \exp(-\beta E_{n_j}) \right) \\
&= -k_B T \sum_{j=1}^{N} \ln \left(\sum_{n_j} \exp(-\beta E_{n_j}) \right).
\end{aligned}
\tag{24.51}
$$

Since the sums in the last line of this equation are only over a single quantum number, rather than all combinations of N quantum numbers, they are much easier to carry out.

If the spectrum of eigenvalues is the same for every term H_j, eq. (24.50) simplifies further,

$$
F = -k_B T N \ln \sum_{n_1} \exp(-\beta E_{n_1}).
\tag{24.52}
$$

The calculation of Helmholtz free energy of the N-body system is reduced to the calculation of the free energy of a one-body system, which is much easier.

24.8 Special Systems

In the following sections we will discuss the simplest examples of component systems that we might find *after* factorization. These examples are important because they occur repeatedly in quantum statistical mechanics. We will see applications of the same mathematical forms in the analysis of magnets, vibrations in crystals, black-body radiation, fermions, and bosons. Consequently, these examples are to be studied very carefully; if you know them well, most of the mathematics in the rest of the book will be easy.

There are a number of many-body systems for which the partition function can be easily found because they have a small number of energy levels. For such systems, it is an easy matter to sum up a small number of terms to obtain the partition function.

There are two classes of systems that are so important in the development of statistical mechanics that we will devote the remaining sections in this chapter to their analysis. These systems are simple harmonic oscillators and any system with only two energy eigenstates.

24.9 Two-Level Systems

The simplest imaginable quantum system would have only one state—but it would not be very interesting.

The next simplest system has two quantum eigenstates for each element making up the system. There are two forms in which we will encounter a two-level system. The mathematics is essentially the same in both cases, but since the usual ways of expressing the results differ, they are both worth studying.

Because two-level systems can support negative temperatures, it is appropriate to use Massieu functions in their analysis.

24.9.1 Energies 0 and ϵ

The first form for a two-level system assigns the energies 0 and ϵ to the levels. Letting the quantum number n take on the values of 0 and 1, the Hamiltonian is simply

$$H = \epsilon n. \tag{24.53}$$

The partition function contains just two terms,

$$Z = \sum_{n=0}^{1} \exp(-\beta \epsilon n) = 1 + \exp(-\beta \epsilon). \tag{24.54}$$

The Massieu function, $\tilde{S}[\beta]$, and the energy are easily found,

$$\tilde{S}[\beta] = \ln Z = \ln(1 + \exp(-\beta \epsilon)) \tag{24.55}$$

$$U = -\frac{\partial \tilde{S}[\beta]}{\partial \beta} = \frac{\epsilon \exp(-\beta \epsilon)}{1 + \exp(-\beta \epsilon)} = \frac{\epsilon}{\exp(\beta \epsilon) + 1}. \tag{24.56}$$

Note that the average value of the quantum number n (average number of excitations in the system) is given by

$$\langle n \rangle = \sum_{n=0}^{1} n \frac{1}{Z} \exp(-\beta \epsilon n) = \frac{0 + \exp(-\beta \epsilon)}{1 + \exp(-\beta \epsilon)} = \frac{1}{\exp(\beta \epsilon) + 1}, \tag{24.57}$$

which is consistent with the average value of the energy

$$U = \langle H \rangle = \sum_{n=0}^{1} H \frac{1}{Z} \exp(-\beta \epsilon n) = \sum_{n=0}^{1} \epsilon n \frac{1}{Z} \exp(-\beta \epsilon n) = \epsilon \langle n \rangle. \tag{24.58}$$

It will be very important to remember the expression for $\langle n \rangle$ given in eq. (24.57), and *especially the positive sign of the exponent in the denominator*. Because the usual Boltzmann factor $\exp(-\beta H)$ contains a negative sign, confusing the sign is a well-known mental hazard.

To express the Planck entropy for the full N-body system, first note that the energy of the ℓ-th level is $\epsilon \ell$. The Planck entropy of the two-level of the N two-level objects is then

$$S_{P,\text{2-level}} = k_B \ln \left(\frac{N!}{\ell!(N-\ell)!} \right). \tag{24.59}$$

This expression is not exactly extensive, even though it should be since the system is composed of N independent objects. It is also defined only for energies that are integer multiples of ϵ. In the limit of infinite N, it agrees with the canonical entropy, which is given next.

The canonical entropy for the full N-body is

$$S_{C,\text{2-level}} = -Nk_B \left[y \ln y + (1-y) \ln(1-y) \right], \tag{24.60}$$

where,

$$y = \frac{U_{\text{2-level}}}{N\epsilon}. \tag{24.61}$$

It is obviously exactly extensive and a continuous function of energy. The derivation of this expression is the subject of Problem 24.4.

24.9.2   Spin-One-Half

The second form in which we encounter two-level systems is related to magnetism. If the operator σ takes on the values $+1$ and -1 (ignoring factors of Planck's constant), the Hamiltonian,

$$H = -b\sigma, \tag{24.62}$$

corresponds to a spin-one-half magnetic moment in a magnetic field b. In this equation b has units of energy, which makes life easier.

The partition function again contains just two terms:

$$Z = \sum_{\sigma = \pm 1} \exp(-\beta(-b\sigma)) = \exp(\beta b) + \exp(-\beta b). \qquad (24.63)$$

The free energy and the energy are again easily found,

$$\tilde{S}[\beta] = \ln Z = \ln\left[\exp(\beta b) + \exp(-\beta b)\right] \qquad (24.64)$$

$$U = -\frac{\partial(\tilde{S}[\beta])}{\partial \beta} = -b\left(\frac{\exp(\beta b) - \exp(-\beta b)}{\exp(\beta b) + \exp(-\beta b)}\right) = -b\tanh(\beta b). \qquad (24.65)$$

Note that the average value of σ (average magnetization) is given by

$$\langle \sigma \rangle = \sum_{\sigma = \pm 1} \sigma \frac{1}{Z} \exp(\beta b \sigma) = \frac{\exp(\beta b) - \exp(-\beta b)}{\exp(\beta b) + \exp(-\beta b)} = \tanh(\beta b), \qquad (24.66)$$

which is consistent with the average value of the energy.

It is traditional to use the hyperbolic tangent in magnetic problems and the exponential sums for other two-level systems. The two forms are, of course, equivalent. The mathematical expressions

$$\tanh x = \frac{e^x - e^{-x}}{e^x + e^{-x}} = \frac{1 - e^{-2x}}{1 + e^{-2x}}, \qquad (24.67)$$

are good choices for memorization.

The Planck entropy can be easily obtained from eq. (24.59), and the canonical entropy from eq. (24.60).

24.10 Simple Harmonic Oscillator

The Hamiltonian of a simple harmonic oscillator (SHO) in one dimension is given by

$$H = \frac{1}{2}Kx^2 + \frac{p^2}{2m} = \frac{1}{2}Kx^2 - \frac{\hbar^2}{2m}\frac{d^2}{dx^2}, \qquad (24.68)$$

where m is the mass of the particle and K is the spring constant.

The energy spectrum for a simple harmonic oscillator (SHO) has an infinite number of states, labeled by a quantum number that takes on non-negative integer values, $n = 0, 1, \ldots, \infty$,

$$E_n = \hbar\omega\left(n + \frac{1}{2}\right). \qquad (24.69)$$

The angular frequency, ω, in eq. (24.69) is identical to the classical value,

$$\omega = \sqrt{\frac{K}{m}}. \tag{24.70}$$

Even though there is an infinite number of states, the uniform spacing makes the evaluation of the partition function only slightly more difficult than for a two-level system,

$$Z = \sum_{n=0}^{\infty} \exp\left(-\beta\hbar\omega(n+1/2)\right)$$

$$= \frac{\exp(-\beta\hbar\omega/2)}{1 - \exp(-\beta\hbar\omega)}. \tag{24.71}$$

Sums of the form

$$\sum_{n=0}^{\infty} x^n = \frac{1}{1-x}$$

for $|x| < 1$, occur frequently in quantum statistical mechanics. The savvy student should expect it to appear on tests—and not only know it, but know how to derive it. [Hint: $y = \sum_{n=0}^{\infty} x^n$, $xy = y - 1$]

For high temperatures (small β), we expect the partition function to go to the classical value, $Z_{class} = 1/(\hbar\beta\omega)$, as found in Section 19.13,

$$Z = \frac{\exp(-\beta\hbar\omega/2)}{1 - \exp(-\beta\hbar\omega)}$$

$$\rightarrow (1 - \beta\hbar\omega/2 + \cdots)\left[1 - \left(1 - \beta\hbar\omega + \frac{1}{2}(\beta\hbar\omega)^2 + \cdots\right)\right]^{-1}$$

$$\rightarrow \frac{1}{\beta\hbar\omega}\left(\frac{1 - \beta\hbar\omega/2 + \cdots}{1 - \beta\hbar\omega/2 + \cdots}\right)$$

$$\rightarrow \frac{1}{\beta\hbar\omega}. \tag{24.72}$$

This agreement between the classical and quantum results for an SHO is the basic justification for the inclusion of the factors involving $1/h$ (where h is Planck's constant) in the classical definitions of the entropy in eqs. (7.3) and (7.30), and the partition function in eq. (19.27).

Returning to the full quantum partition function given in eq. (24.71), we can easily obtain the free energy and the energy,

$$
\begin{aligned}
F &= -k_B T \ln Z \\
&= -k_B T \ln \exp(-\beta\hbar\omega/2) + k_B T \ln(1 - \exp(-\beta\hbar\omega)) \\
&= \frac{1}{2}\hbar\omega + k_B T \ln(1 - \exp(-\beta\hbar\omega)) \quad\quad\quad (24.73)
\end{aligned}
$$

$$
\begin{aligned}
U &= \frac{\partial(\beta F)}{\partial\beta} \\
&= \frac{1}{2}\hbar\omega + \frac{\hbar\omega\exp(-\beta\hbar\omega)}{1 - \exp(-\beta\hbar\omega)} \\
&= \frac{1}{2}\hbar\omega + \frac{\hbar\omega}{\exp(\beta\hbar\omega) - 1}. \quad\quad\quad (24.74)
\end{aligned}
$$

The form of the energy in eq. (24.74) can be compared with the formal expression for the average of the energy

$$
\langle E_n \rangle = \frac{1}{2}\hbar\omega + \hbar\omega\langle n \rangle. \quad\quad\quad (24.75)
$$

Either from this comparison or from a direct calculation, we can see that the average number of excitations of the SHO is given by the expression

$$
\langle n \rangle = \frac{1}{\exp(\beta\hbar\omega) - 1}. \quad\quad\quad (24.76)
$$

As it was for the average number of excitations in a two-level system in eq. (24.57), it will be very important to remember the expression for $\langle n \rangle$ in eq. (24.76) for a quantum SHO. Here again, the positive sign of the exponent in the denominator can be easily forgotten.

The canonical entropy of a collection of N simple harmonic oscillators can be found from the equations for F and U.

24.11 Einstein Model of a Crystal

Our first application of the quantum simple harmonic oscillator is Einstein's model of the vibrations of a crystal. Einstein made the approximation that all atoms except one were fixed at their average locations. He then approximated the potential seen by the chosen atom as being parabolic, so that he could treat the motion of the remaining atom as a three-dimensional SHO.

Assuming for simplicity that the quadratic terms are isotropic, the Hamiltonian for the one moving particle is

$$H_1 = \frac{1}{2}K|\vec{r_1}|^2 + \frac{|\vec{p_1}|^2}{2m},$$
(24.77)

where the subscript 1 indicates that this Hamiltonian only applies to the one special particle.

The partition function for this system factorizes into three identical terms, each given by the partition function of the one-dimensional SHO from eq. (24.72),

$$Z_1 = \left(\frac{\exp(-\beta\hbar\omega/2)}{1 - \exp(-\beta\hbar\omega)}\right)^3.$$
(24.78)

The energy is just three times the energy of a single one-dimensional SHO,

$$U_1 = \frac{3}{2}\hbar\omega + \frac{3\hbar\omega}{\exp(\beta\hbar\omega) - 1}.$$
(24.79)

For N atoms in the crystal, the total energy is just N times the energy of a single atom,

$$U_N = \frac{3}{2}N\hbar\omega + \frac{3N\hbar\omega}{\exp(\beta\hbar\omega) - 1}.$$
(24.80)

The specific heat at constant volume is given by the usual derivative with respect to temperature,

$$
\begin{aligned}
c_V(T) &= \frac{1}{N}\left(\frac{\partial U_N}{\partial T}\right)_{V,N} \\
&= \frac{1}{N}\frac{\partial\beta}{\partial T}\frac{\partial}{\partial\beta}\left(\frac{3}{2}N\hbar\omega + \frac{3N\hbar\omega}{\exp(\beta\hbar\omega) - 1}\right) \\
&= -\frac{3\hbar\omega}{k_B T^2}\frac{\partial}{\partial\beta}\left(\frac{1}{\exp(\beta\hbar\omega) - 1}\right) \\
&= -\frac{3\hbar\omega}{k_B T^2}\frac{-\hbar\omega\exp(\beta\hbar\omega)}{(\exp(\beta\hbar\omega) - 1)^2} \\
&= 3k_B\left(\frac{\hbar\omega}{k_B T}\right)^2\frac{\exp(\beta\hbar\omega)}{(\exp(\beta\hbar\omega) - 1)^2}.
\end{aligned}
$$
(24.81)

The final expression for $c_V(T)$ in eq. (24.81) has interesting properties at high and low temperatures.

At high temperatures (small β), the factor of $\exp(\beta\hbar\omega)$ in the numerator goes to one, while the expression in the denominator can be expanded,

$$(\exp(\beta\hbar\omega) - 1)^2 = (1 + \beta\hbar\omega + \cdots - 1)^2 \to (\beta\hbar\omega)^2. \tag{24.82}$$

Using this to find the high-temperature limit of $c_V(T)$ gives

$$c_V(T) \to 3k_B \left(\frac{\hbar\omega}{k_BT}\right)^2 \frac{1}{(\beta\hbar\omega)^2} = 3k_B. \tag{24.83}$$

This constant value of the specific heat at high temperatures is just the well-known law of Dulong and Petit (Pierre Louis Dulong, French physicist and chemist, 1785–1837, and Alexis Thérèse Petit, French physicist, 1791–1820) It is identical to the specific heat of the corresponding classical model.

At low temperatures (large β), the factor of $\exp(\beta\hbar\omega)$ becomes extremely large, and the '-1' in the denominator can be neglected,

$$c_V(T) \to 3k_B (\beta\hbar\omega)^2 \exp(-\beta\hbar\omega). \tag{24.84}$$

Even though the factor of β^2 diverges at low temperature, the factor of $\exp(-\beta\hbar\omega)$ goes to zero much faster. The result is that the specific heat in the Einstein model goes to zero rapidly as the temperature goes to zero, as it must according to the Third Law of Thermodynamics.

The Einstein model was a great success in explaining the qualitative features of the specific heat of crystals at low temperatures and the quantitative behavior at high temperatures. Nevertheless, the specific heat of real crystals does not go to zero as rapidly as predicted by eq. (24.84). Instead, a T^3 behavior is seen in insulators, while the specific heat of metals is linear in T at low temperatures. The explanations of these observations will be given in Chapters 26 and 29.

The next chapter discusses black-body radiation, which might seem to be a bit of a detour. However, it turns out that the mathematical form of the equations we will find is very similar to, but simpler than, the equations needed to explain the low-temperature behavior of the specific heat of crystals, which will follow in Chapter 26.

24.12 Problems

PROBLEM 24.1

Quantum statistical mechanics: A spin in a magnetic field

Consider a magnetic moment with spin-one-half in a magnetic field. Being a lazy theorist, I prefer not to write $\hbar/2$ repeatedly, so I will use units in which the spin $\sigma = \pm 1$. I will also choose the units of the magnetic field such that the energy of the system is just

$$E = -b\sigma.$$

The system is in contact with a heat reservoir at temperature T.

1. Calculate the probability of the spin having the value $+1$.
2. Calculate the average magnetization $m = \langle \sigma \rangle$. Express your answer in terms of hyperbolic functions.
3. Calculate the two leading terms in a high-temperature expansion of the magnetization in powers of $\beta = 1/k_B T$.
4. Calculate the leading (non-constant) term in a low-temperature power series expansion of the magnetization in some variable at low temperatures. (Hint: it will be in the form of an exponential.)
5. Calculate the average energy. Plot it as a function of temperature.
6. Calculate the specific heat. Plot it as a function of temperature
7. Calculate the magnetic susceptibility,

$$\chi = \frac{dm}{db}.$$

PROBLEM 24.2
A computer simulation of a two-level system Quantum statistical mechanics: A spin in a magnetic field

Consider the same magnetic moment with spin-one-half in a magnetic field that we looked at in an earlier problem. The units of the magnetic field are again chosen such that the energy of the system is just

$$E = -b\sigma.$$

The system is in contact with a heat reservoir at temperature T.

   For this problem, instead of doing an analytic calculation, we will carry out a computer simulation using the Monte Carlo method. Although this might seem to be the hard way to do it, we will see later that the method can be generalized to solve problems that do not have analytic solutions.

1. Write a computer program to simulate this two-level system. The program should calculate the thermal probability of being in each state, as well as the average magnetization and the magnetic susceptibility. Have the program print out of the theoretical values for comparison.
2. Use your program to calculate the magnetization and magnetic susceptibility for a set of 'interesting' values of the magnetic field and temperature.

PROBLEM 24.3
Quantum statistical mechanics: Another two-level system

Although a two-level system might seem very simple, it is very important and occurs frequently in various guises. Here is another form that we will see often.

A system only has two states, which are both non-degenerate. The energies of these two states are $E = 0$ and $E = \epsilon > 0$.

The system is in contact with a heat reservoir at temperature T.

1. Calculate the probability of being in the excited state.
2. Calculate the average energy. Sketch it as a function of temperature.
3. Calculate the average specific heat. Sketch it as a function of temperature.
4. Calculate the two leading terms in a high-temperature expansion of the energy in powers of $\beta = 1/k_B T$.
5. Calculate the leading (non-constant) term in a low-temperature power series expansion of the energy in some variable at low temperatures. (Hint: the variable will be in the form of an exponential.)

PROBLEM 24.4
Derivation of S_C for N two-level systems

Consider a collection of N two-level quantum objects. The Hamiltonian is

$$H_{2\text{-level}} = \epsilon \sum_{k=1}^{N} n_k,$$

$n_k = 0$ or 1, and ϵ is the energy difference between the two levels in each object.

The entropy is denoted as $S = S(U, V, N)$, and the dimensionless entropy is $\tilde{S} = S/k_B$.

1. Find the partition function for this Hamiltonian in closed form.
2. Find the Massieu function $\tilde{S}[\beta]$
3. Find the energy $U = U(\beta)$ from $\tilde{S}_{2\text{-level}}[\beta]$.
4. Simplify the notation by defining a dimensionless energy

$$y = \frac{U_{2\text{-level}}}{N\epsilon}.$$

Find β as a function of y.
5. Plot β as a function of $y = U/N\epsilon$.
6. Find $\tilde{S}_{2\text{-level}}$ as a function of $y = U/N\epsilon$.
7. Plot $S_{2\text{-level}}$ as a function of $y = U/N\epsilon$.

PROBLEM 24.5
Quantum statistical mechanics: Return of the simple harmonic oscillator

The energy levels of a quantum SHO are given by

$$E_n = \hbar\omega\left(n + \frac{1}{2}\right)$$

where $n = 0, 1, 2, \ldots,$

$$\hbar = \frac{h}{2\pi}$$

and

$$\omega = \sqrt{\frac{K}{m}}.$$

1. Calculate the QM partition function.
2. Calculate the probability that a measurement of the energy will find the QM SHO in the n-th eigenstate.
3. Calculate the average energy.
4. Calculate the specific heat.
5. Calculate the high-temperature limit of the specific heat.
6. Calculate the leading term in a low-temperature expansion of the specific heat. (You should be able to figure out what a good quantity to expand in would be.)

PROBLEM 24.6
A Monte Carlo computer simulation of a quantum SHO

As we've seen before, the energy levels of a quantum SHO are given by

$$E_n = \hbar\omega \left(n + \frac{1}{2} \right)$$

where $n = 0, 1, 2, \ldots,$

$$\hbar = \frac{h}{2\pi}$$

and

$$\omega = \sqrt{\frac{K}{m}}.$$

This time we shall carry out a Monte Carlo simulation of this system. There are, of course, an infinite number of possible states. However, the probability of high-energy states is very small, making the simulation feasible.

1. Write a program to do a Monte Carlo simulation of a single quantum SHO. Label the state of the system with the quantum number n, as in the equations above. Let one MC step consist of an attempted move to increase or decrease n by ± 1. Remember to reject any attempted move that would make n negative.
2. Compute the energy and specific heat (the latter from fluctuations) for a range of interesting temperature. For convenience, you may take $\hbar\omega = 1.0$. Have the program print out the exact values for comparison.

PROBLEM 24.7
Quantum statistical mechanics: A many-spin system

Consider a macroscopic crystal with a spin-one quantum mechanical magnetic moment located on each of N atoms. Assume that we can represent the energy eigenvalues of the system with a Hamiltonian of the form

$$H = D \sum_{n=1}^{N} \sigma_n^2$$

where each σ_n takes on the values $-1, 0,$ or $+1,$ and D is a constant representing a 'crystal field'. The entire system is in contact with a thermal reservoir at temperature T.

1. Calculate the partition function for this system.
2. Calculate the free energy of this system.
3. Calculate the quantity

$$Q = \frac{1}{N} \left\langle \sum_{n=1}^{N} \sigma_n^2 \right\rangle.$$

4. Calculate the entropy per spin of this system.
5. Determine whether this system satisfies the Nernst Postulate for all values of the parameter D.

PROBLEM 24.8
Quantum statistical mechanics: A many-spin system (This is a slightly more difficult version of the previous problem).

Consider a macroscopic crystal with a spin-one quantum mechanical magnetic moment located on each of N atoms. Assume that we can represent the energy eigenvalues of the system with a Hamiltonian of the form

$$H = B \sum_{n=1}^{N} \sigma_n + D \sum_{n=1}^{N} \sigma_n^2$$

where each σ_n takes on the values $-1, 0,$ or $+1,$ and B and D are constants representing an external magnetic field and a 'crystal field', respectively. The entire system is in contact with a thermal reservoir at temperature T.

1. Calculate the partition function for this system.
2. Calculate the free energy of this system.
3. Calculate the magnetization per spin

$$m = \frac{1}{N} \left\langle \sum_{n=1}^{N} \sigma_n \right\rangle.$$

4. Calculate the entropy per spin of this system.
5. Determine whether this system satisfies the Nernst Postulate for all values of the parameters.

PROBLEM 24.9
A diatomic ideal gas Quantum statistical mechanics—but only when needed

Consider a dilute gas with N diatomic molecules that you can treat as ideal in the sense of neglecting interactions between the molecules. We shall assume that the molecules consist of two-point masses with a fixed distance between them. They can rotate freely, but they cannot vibrate.

The center of mass motion (translational degrees of freedom) can be treated classically. However, the rotational degrees of freedom must be treated quantum mechanically.

The quantum mechanical energy levels take the form $\epsilon(j) = j(j+1)\epsilon_o$, where $j = 0, 1, 2, \ldots$ and the degeneracy of the j-th level is given by $g(j) = 2j + 1$. (You can peek at a QM text to find the value of ϵ_o, but you do not need it for this assignment.)

Although you do not need it for this assignment, the parameter ϵ_o is given by

$$\epsilon_0 = \frac{\hbar^2}{2I}$$

where I is the moment of inertia of a molecule.

The whole system is in equilibrium with a thermal reservoir at temperature T.

1. Write down the canonical partition function, treating the translational degrees of freedom classically and the rotational degrees of freedom quantum mechanically.
2. Evaluate the energy and the specific heat for both high and low temperatures.
3. Sketch the energy and the specific heat as functions of temperature, indicating both the high- and low-temperature behavior.

PROBLEM 24.10
Another diatomic ideal gas—entirely classical this time, but in two dimensions

Consider a classical, dilute, diatomic gas with N molecules in two dimensions. The gas is again ideal in the sense of neglecting interactions between the molecules. Each molecule consist of two-point masses. Although there are no interactions between different molecules, there is an interaction between the atoms in the same molecule of the form

$$V(r) = \begin{cases} \mathcal{J}\ln(r) & r > a \\ \infty & r \leq a. \end{cases}$$

The gas is contained in a two-dimensional 'box' of area $A = L^2$. The whole system is in equilibrium with a thermal reservoir at temperature T.

1. What is the classical Hamiltonian for this system?
2. Write the canonical partition function as an integral over phase space.
3. Calculate the partition function in closed form, under the assumption that the molecule is much smaller than the box it is in. That is, let the limits on the integral over the separation between the atoms in a molecule extend to ∞.
4. This model is only valid for low temperatures. At what temperature do you expect it to break down?
5. Now calculate the average square separation $\langle r^2 \rangle$ between the atoms in a molecule.

PROBLEM 24.11
Two-level quantum systems

1. Consider a simple set of N two-level subsystems. The subsystems are all independent, and each has two allowed states with energies 0 (ground state) and ϵ (excited state) so that the full Hamiltonian can be written as

$$H = \sum_{n=1}^{N} E(n)$$

where $E(n) = 0$ or $E(n) = \epsilon$ The entire system is in thermal equilibrium at temperature T.

1. At what temperature is the average total energy equal to $\frac{1}{3}N\epsilon$?

2. At what temperature is the average total energy equal to $\frac{2}{3}N\epsilon$?

2. Suppose the subsystems in the part 1 of the problem had the same ground state with energy 0, but different values of the energy in the excited states. Assume that the energy of the excited state of the n-th subsystem is $n\epsilon$.

   1. What is the average energy of the total system?
   2. Suppose we are in the low-temperature regime, for which $k_B T \ll E_{max}$. Calculate the average energy. You may leave your answer in terms of a dimensionless integral, but you should obtain the temperature dependence.
   3. What is the heat capacity of the total system?

25

Black-Body Radiation

The spectral density of black body radiation ... represents something absolute, and since the search for the absolutes has always appeared to me to be the highest form of research, I applied myself vigorously to its solution.

Max Planck, German physicist (1858–1947), Nobel Prize 1918

25.1 Black Bodies

In physics, the expression 'black body' refers to an object that absorbs all radiation incident on it and reflects nothing. It is, of course, an idealization, but one that can be approximated very well in the laboratory.

A black body is *not* really black. Although it does not reflect light, it can and does radiate light arising from its thermal energy. This is, of course, necessary if the black body is ever to be in thermal equilibrium with another object.

The purpose of the current chapter is to calculate the spectrum of radiation emanating from a black body. The calculation was originally carried out by Max Planck in 1900 and published the following year. This was before quantum mechanics had been invented—or perhaps it could be regarded the first step in its invention. In any case, Planck investigated the consequences of the assumption that light could only appear in discrete amounts given by the quantity

$$\Delta \epsilon_\omega = h\nu = \hbar\omega, \tag{25.1}$$

where ν is the frequency, $\omega = 2\pi\nu$ is the angular frequency, h is Planck's constant, and $\hbar = h/2\pi$. This assumption is well accepted today, but it was pretty daring in 1900 when Max Planck introduced it.

25.2 Universal Frequency Spectrum

If two black bodies at the same temperature are in equilibrium with each other, the frequency spectrum must be the same for each object. To see why, suppose that two black bodies, A and B, are in equilibrium with each other, but that A emits more power than B in a particular frequency range. Place a baffle between the objects that transmits radiation

An Introduction to Statistical Mechanics and Thermodynamics. Robert H. Swendsen, Oxford University Press (2020).
© Robert H. Swendsen. DOI: 10.1093/oso/9780198853237.001.0001

well in that frequency range, but is opaque to other frequencies. This would have the effect of heating B and cooling A, in defiance of the Second Law of Thermodynamics. Since that cannot happen, the frequency spectrum must be the same for all black bodies.

Since the radiation spectrum does not depend on the object, we might as well take advantage of the fact and carry out our calculations for the simplest object we can think of.

25.3 A Simple Model

We will consider a cubic cavity with dimensions $L \times L \times L$. The sides are made of metal and it contains electromagnetic radiation, but no matter. Radiation can only come in and out of the cavity through a very small hole in one side. Since the radiation must be reflected off the walls many times before returning to the hole, we can assume that it has been absorbed along the way, making this object—or at least the hole—a black body.

The only thing inside the cavity is electromagnetic radiation at temperature T. We wish to find the frequency spectrum of the energy stored in that radiation, which will also give us the frequency spectrum of light emitted from the hole.

25.4 Two Types of Quantization

In analyzing the simple model described in the previous section, we must be aware of the two kinds of quantization that enter the problem.

As a result of the boundary conditions due to the metal walls of the container, the frequencies of allowed standing waves are quantized. This is an entirely classical effect, similar to the quantization of frequency in the vibrations of a guitar string.

The second form of quantization is due to quantum mechanics, which specifies that the energy stored in an electromagnetic wave with angular frequency ω comes in multiples of $\hbar\omega$ (usually called photons).

The theory of electrodynamics gives us the wave equation for the electric field $\vec{E}(\vec{r})$ in a vacuum,

$$\nabla^2 \vec{E}(\vec{r}, t) = \frac{1}{c^2} \frac{\partial^2 \vec{E}(\vec{r}, t)}{\partial t^2}. \tag{25.2}$$

In eq. (25.2),

$$\vec{\nabla} \equiv \left(\frac{\partial}{\partial x}, \frac{\partial}{\partial y}, \frac{\partial}{\partial z} \right) \tag{25.3}$$

and

$$\nabla^2 = \frac{\partial^2}{\partial x^2} + \frac{\partial^2}{\partial y^2} + \frac{\partial^2}{\partial z^2}. \tag{25.4}$$

The solutions to eq. (25.2) must satisfy the boundary conditions that the component of the electric field parallel to a wall must vanish at that wall.

Using the symmetry of the model, we find that there are solutions of the following form:

$$E_x(\vec{r}, t) = E_{x,o} \sin(\omega t) \cos(k_x x) \sin(k_y y) \sin(k_z z) \tag{25.5}$$

$$E_y(\vec{r}, t) = E_{y,o} \sin(\omega t) \sin(k_x x) \cos(k_y y) \sin(k_z z) \tag{25.6}$$

$$E_z(\vec{r}, t) = E_{z,o} \sin(\omega t) \sin(k_x x) \sin(k_y y) \cos(k_z z). \tag{25.7}$$

The values of $E_{x,o}$, $E_{y,o}$, and $E_{z,o}$ are the amplitudes of the corresponding components of the electric field in the cavity.

Eq. (25.5) through (25.7) were written to impose the boundary condition that the parallel components of the electric field vanish at the walls of the cube where x, y, or z is equal to zero. To impose the same boundary condition at the remaining walls, where x, y, or z is equal to L, we have the conditions

$$k_x L = n_x \pi \tag{25.8}$$

$$k_y L = n_y \pi \tag{25.9}$$

$$k_z L = n_z \pi \tag{25.10}$$

where n_x, n_y, n_z are integers. Only positive integers are counted, because negative integers give exactly the same solutions.

Substituting eqs. (25.5), (25.6), and (25.7) into eq. (25.2), we find a relationship between the frequency and the wave numbers,

$$k_x^2 + k_y^2 + k_z^2 = \frac{\omega^2}{c^2}. \tag{25.11}$$

This equation can also be written in terms of the integers n_x, n_y, n_z,

$$\left(\frac{n_x \pi}{L}\right)^2 + \left(\frac{n_y \pi}{L}\right)^2 + \left(\frac{n_z \pi}{L}\right)^2 = \frac{\omega^2}{c^2}. \tag{25.12}$$

Clearly the value of ω must depend on the vector $\vec{n} = (n_x, n_y, n_z)$, We will indicate the $\vec{n}$-dependence by a subscript and solve eq. (25.12) for $\omega_{\vec{n}}$

$$\omega_{\vec{n}}^2 = \left(n_x^2 + n_y^2 + n_z^2\right) \left(\frac{\pi c}{L}\right)^2. \tag{25.13}$$

Taking the square root to find $\omega_{\vec{n}}$, we obtain

$$\omega_{\vec{n}} = \frac{\pi c}{L} \sqrt{n_x^2 + n_y^2 + n_z^2} = \frac{n \pi c}{L} = \omega_n \tag{25.14}$$

where $n = |\vec{n}|$.

Because the wavelengths that contribute significantly to black-body radiation are very small in comparison with the size of the cavity, energy differences between neighboring points in $\vec{n}$-space are very small. This makes the frequency spectrum quasi-continuous and allows us to change sums over the discrete wavelengths into integrals. Furthermore, the dependence of the frequency on $\vec{n}$ shown in eq. (25.14) is rotationally symmetric, which simplifies the integrals further, as shown in the next section.

25.5 Black-Body Energy Spectrum

The first step in calculating the black-body energy spectrum is to find the density of states. From the solutions to the wave equation in Section 25.4, we expressed the individual modes in terms of the vectors $\vec{n}$. Note that the density of points in $\vec{n}$-space is one, since the components of $\vec{n}$ are all integers,

$$P_{\vec{n}}(\vec{n}) = 1. \tag{25.15}$$

To find the density of states as a function of frequency $P_\omega(\omega)$, integrate eq. (25.15) over $\vec{n}$-space,

$$P_\omega(\omega) = 2\frac{1}{8}\int_0^\infty 4\pi n^2\, dn\, \delta(\omega - nc\pi/L) = \pi \left(\frac{L}{c\pi}\right)^3 \omega^2. \tag{25.16}$$

The factor of 2 is for the two polarizations of electromagnetic radiation, and the factor of 1/8 corrects for counting both positive and negative values of the components of $\vec{n}$.

Since each photon with frequency ω has energy $\hbar\omega$, the average energy can be found by summing over all numbers of photons weighted by the Boltzmann factor $\exp(-\beta\hbar\omega)$. Since this sum is formally identical to that for the simple harmonic oscillator, we can just write down the answer,

$$\langle \epsilon_\omega \rangle = \frac{\hbar\omega}{\exp(\beta\hbar\omega) - 1}. \tag{25.17}$$

Note that eq. (25.17) does not include the ground-state energy that might be expected for a simple harmonic oscillator. The reason is a bit embarrassing. Since there is an infinite number of modes, the sum of the ground-state energies is infinite. The simplest way to deal with the problem is to ignore it on the grounds that a constant ground-state energy cannot affect the results for the radiation spectrum. That is what other textbooks do, and that is what I will do for the rest of the book. I suggest you do the same.

The energy density spectrum for black-body radiation as a function of the angular frequency ω is found by multiplying the density of states in eq. (25.16) by the average energy per state in eq. (25.17) and dividing by the volume $V = L^3$,

$$u_\omega = \left(\frac{1}{V}\right) \pi \left(\frac{L}{c\pi}\right)^3 \omega^2 \frac{\hbar\omega}{\exp(\beta\hbar\omega) - 1} = \frac{\hbar}{\pi^2 c^3} \omega^3 (\exp(\beta\hbar\omega) - 1)^{-1}. \tag{25.18}$$

Knowing the energy per unit volume contained in the black-body cavity from eq. (25.30), and the fact that light travels with the speed of light (if you will pardon the tautology), the energy per unit area radiated from the hole in the cavity, $\mathcal{J}_U$, is clearly proportional to cU/V. The actual equation includes a geometric factor of $1/4$, the calculation of which will be left to the reader.

Multiplying u_ω by the factor of $c/4$ to derive the radiated power gives us the Planck law for black-body radiation,

$$j_\omega = \frac{1}{4} c u_\omega = \left(\frac{\hbar}{4\pi^2 c^2}\right) \frac{\omega^3}{\exp(\beta\hbar\omega) - 1}. \tag{25.19}$$

Fig. 25.1 shows a plot of eq. (25.19) in dimensionless units; that is, $x^3/(\exp(x) - 1)$, where $x = \beta\hbar\omega$. The function has a maximum at $x_{max} \approx 2.82144$.

It is very important to understand how the spectrum of black-body radiation scales as a function of T, which will be discussed in the following subsections.

25.5.1   Frequency of Maximum Intensity

Since $\omega = x(k_B T/\hbar)$, the location of the maximum is

$$\omega_{max} = x_{max} k_B T/\hbar \approx 2.82144 \left(\frac{k_B T}{\hbar}\right), \tag{25.20}$$

which is proportional to the temperature, T.

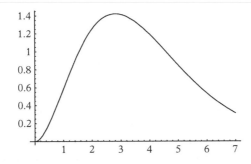

Fig. 25.1 *Plot of black-body radiation spectrum, $x^3/(\exp(x) - 1)$, in dimensionless units. For comparison with the black-body spectrum in eq. (25.19), note that $x = \beta\hbar\omega$. The integral under this dimensionless function was given as $\pi^4/15$ in eq. (25.29).*

The value of j_ω at its maximum is then

$$j_{\omega max} = \left(\frac{\hbar}{4\pi^2 c^2}\right) \frac{(x_{max} k_B T/\hbar)^3}{\exp(x_{max}) - 1} = \left(\frac{x_{max}^3}{4\pi^2 c^2 \hbar^2}\right) \frac{(k_B T)^3}{\exp(x_{max}) - 1}, \tag{25.21}$$

which is proportional to T^3. Clearly, the integral over the black-body spectrum is proportional to $T^3 \times T = T^4$, as expected from the Stefan–Boltzmann Law in eq. (25.31).

25.5.2 Low-Frequency Spectrum

The low-frequency energy spectrum is found for $x = \beta\hbar\omega << x_{max} \approx 2.82$. For small values of ω, we find

$$j_\omega = \frac{1}{4} c u_\omega \tag{25.22}$$

$$= \left(\frac{\hbar}{4\pi^2 c^2}\right) \frac{\omega^3}{\exp(\beta\hbar\omega) - 1}$$

$$\approx \left(\frac{\hbar}{4\pi^2 c^2}\right) \frac{\omega^3}{1 + \beta\hbar\omega - 1}$$

$$= \left(\frac{\hbar}{4\pi^2 c^2}\right) \frac{\omega^3}{\beta\hbar\omega}$$

$$= \left(\frac{1}{4\pi^2 c^2}\right) \omega^2 k_B T.$$

This expression can be understood because the small ω region corresponds to $\beta\hbar\omega \ll 1$, or $k_B T \gg \hbar\omega$, which is the condition that the classical theory is valid. In this limit,

$$\langle \epsilon_n \rangle \to k_B T, \tag{25.23}$$

independent of the value of ω. The factor of ω^2 in eq. (25.22) comes from the factor of ω^2 in eq. (25.16), which, in turn, came from the n^2 dependence of the surface of a sphere in $\vec{n}$-space.

25.5.3 High-Frequency Spectrum

At high frequencies, $\beta\hbar\omega \gg x_{max} > 1$, so that we can make the approximation that $\exp(\beta\hbar\omega) \gg 1$ in the expression for the spectrum of black-body radiation in eq. (25.22),

$$j_\omega = \left(\frac{\hbar}{4\pi^2 c^2}\right) \frac{\omega^3}{\exp(\beta\hbar\omega) - 1} \approx \left(\frac{\hbar}{4\pi^2 c^2}\right) \omega^3 \exp(-\beta\hbar\omega). \tag{25.24}$$

To understand the high-frequency behavior of the energy spectrum, first note that two factors of ω come from the n^2 dependence of the surface of a sphere in $\vec{n}$-space, as noted in the previous subsection. The third factor of ω comes from $\hbar\omega$, which is the energy of a single photon, while the factor $\exp(-\beta\hbar\omega)$ gives the relatively low probability of a high-frequency photon being excited.

Note that although the factor ω^3 diverges as ω increases, the factor of $\exp(-\beta\hbar\omega)$ goes to zero much more rapidly, giving the shape of the curve shown in Fig. 25.1.

25.6 Total Energy

The total quantum mechanical energy in the cavity radiation at temperature T is given by summing up the average energy in each mode. Frequency spectrum of a one-dimensional harmonic solid, as given in eq. (26.33).

$$U = 2\sum_{\vec{n}}\langle\epsilon_{\vec{n}}\rangle = 2\sum_{\vec{n}}\hbar\omega_{\vec{n}}(\exp(\beta\hbar\omega_{\vec{n}}) - 1)^{-1}. \tag{25.25}$$

The sum in eq. (25.25) is restricted to the positive octant in n-space in order to avoid double counting, and factor of two accounts for the two polarizations of light associated with every spatial mode. (Note that we have again omitted the ground-state energy for the electromagnetic modes for the same dubious reason given in Section 25.5.)

We again use the fact that the frequency spectrum is a quasi-continuum to write the sum in eq. (25.25) as an integral, which we can evaluate explicitly using the density of states, $P_\omega(\omega)$, found in eq. (25.16),

$$U = \int_0^\infty \langle\epsilon_\omega\rangle P_\omega(\omega)d\omega \tag{25.26}$$

$$= \pi\left(\frac{L}{c\pi}\right)^3 \int_0^\infty \langle\epsilon_\omega\rangle\omega^2 d\omega$$

$$= \pi\left(\frac{L}{c\pi}\right)^3 \int_0^\infty \frac{\hbar\omega}{\exp(\beta\hbar\omega) - 1}\omega^2 d\omega.$$

At this point we can simplify the equation for U by introducing a dimensionless integration variable,

$$x = \beta\hbar\omega. \tag{25.27}$$

The expression for U becomes

$$U = \pi\beta^{-1}\left(\frac{L}{\beta\hbar\pi c}\right)^3 \int_0^\infty dx\frac{x^3}{e^x - 1}. \tag{25.28}$$

By a stroke of good fortune, the dimensionless integral in this equation is known exactly

$$\int_0^\infty dx \frac{x^3}{e^x - 1} = \frac{\pi^4}{15},\tag{25.29}$$

Noting that the volume of the cavity is $V = L^3$, the average energy per volume in the cavity can be given exactly

$$\frac{U}{V} = u = \left(\frac{\pi^2}{15\hbar^3 c^3}\right)(k_B T)^4.\tag{25.30}$$

Since the energy of a black body is proportional to T^4, the specific heat per unit volume must be proportional to T^3. This might not seem terribly significant now, but keep it in mind for later in Chapter 26, when we discuss the specific heat of an insulator, which is also proportional to T^3 at low temperatures.

25.7 Total Black-Body Radiation

Multiplying the total energy by a factor of $c/4$, as explained in Section 25.5, we find the total black-body radiation density

$$\mathcal{J}_U = \frac{cU}{4V} = \frac{c}{4}\left(\frac{\pi^2}{15\hbar^3 c^3}\right)(k_B T)^4 = \sigma_B T^4\tag{25.31}$$

where

$$\sigma_B = \frac{\pi^2 k_B^4}{60\hbar^3 c^2}.\tag{25.32}$$

The constant σ_B is known as the Stefan–Boltzmann constant. It is named after the Austrian physicist Joseph Stefan (1835–1893), who first suggested that the energy radiated by a hot object was proportional to T^4, and his student Boltzmann, who found a theoretical argument for the fourth power of the temperature. The value of the Stefan–Boltzmann constant had been known experimentally long before Planck calculated it theoretically in 1900. Planck got the value right!

25.8 Significance of Black-Body Radiation

Perhaps the most famous occurrence of black-body radiation is in the background radiation of the universe, which was discovered in 1964 by Arno Allan Penzias (German physicist who became an American citizen, 1933–, Nobel Prize 1978) and Robert Woodrow Wilson (American astronomer, 1936–, Nobel Prize 1978).

Shortly after the Big Bang, the universe contained electromagnetic radiation at a very high temperature. With the expansion of the universe, the gas of photons cooled— much as a gas of particles would cool as the size of the container increased. Current measurements show that the background radiation of the universe is described extremely well by eq. (25.19) at a temperature of $2.725\,K$.

25.9 Problems

PROBLEM 25.1
Generalized energy spectra

For black-body radiation the frequency of the low-lying modes was proportional to the magnitude of the wave vector $\vec{k}$. Now consider a system in d dimensions for which the relationship between the frequencies of the modes and the wave vector is given by

$$\omega = Ak^s \text{ where } A \text{ and } s \text{ are constants.}$$

What is the temperature dependence of the specific heat at low temperatures?

PROBLEM 25.2
Radiation from the sun

1. The sun's radiation can be approximated by a black body. The surface temperature of the sun is about $5800\,K$ and its radius is $0.7 \times 10^9\,m$. The distance from the earth to the sun is $1.5 \times 10^{11}\,m$. The radius of the earth is $6.4 \times 10^6\,m$. Estimate the average temperature of the earth from this information. Be careful to state your assumptions and approximations explicitly.
2. From the nature of your assumptions in the previous question (rather than your knowledge of the actual temperature), would you expect a more accurate calculation with this data to give a higher or lower answer? Explain your reasons clearly to obtain one bonus point per reason.

PROBLEM 25.3
Black-body radiation

On the *same* graph, sketch the energy density of black-body radiation $u(\omega)$ vs. the frequency ω for the two temperatures T and $2T$. (Do *not* change the axes so that the two curves become identical.)

The graph should be *large*—filling the page, so that details can be seen. Small graphs are not acceptable.

26

The Harmonic Solid

It is not knowledge, but the act of learning, not possession but the act of getting there, which grants the greatest enjoyment.

Johann Carl Friedrich Gauss, German mathematician (1777–1855)

In this chapter we return to calculating the contributions to the specific heat of a crystal from the vibrations of the atoms. We have previously discussed a cruder approximation, the Einstein model, in Section 24.11 of Chapter 24. The harmonic solid is a model of lattice vibrations that goes beyond the Einstein model in that it allows all the atoms in the crystal to move simultaneously.

To simplify the mathematics, we will consider only a one-dimensional model in this book. The general extension to three dimensions does bring in some new phenomena, but it complicates the notation unnecessarily at this point. We will go into three dimensions only for particularly simple cases, in which the extension of the theory does not present any difficulties.

After going through the discussion in this chapter, it should be easy to follow the mathematics of the general three-dimensional case in any good textbook on solid-state physics.

26.1 Model of a Harmonic Solid

A one-dimensional model of a crystal lattice is described by uniformly spaced points along a line,

$$R_j = ja. \tag{26.1}$$

The spacing a is called the lattice constant and the index j is an integer. Atoms are located at points

$$r_j = R_j + x_j = ja + x_j, \tag{26.2}$$

where x_j is the deviation of the position of an atom relative to its associated lattice point.

An Introduction to Statistical Mechanics and Thermodynamics. Robert H. Swendsen, Oxford University Press (2020).
© Robert H. Swendsen. DOI: 10.1093/oso/9780198853237.001.0001

To write an expression for the kinetic energy we will need the time derivative of x_j which we will indicate by a dot over the variable,

$$\dot{x}_j \equiv \frac{\partial x_j}{\partial t} = \frac{\partial r_j}{\partial t} = \dot{r}_j. \tag{26.3}$$

The energy of a microscopic state is given by the following expression:

$$E = \frac{1}{2}m\sum_j \dot{r}_j^2 + \frac{1}{2}K\sum_j (a + r_j - r_{j+1})^2 \tag{26.4}$$

$$= \frac{1}{2}m\sum_j \dot{x}_j^2 + \frac{1}{2}K\sum_j (x_j - x_{j+1})^2.$$

The potential energy is a minimum for nearest-neighbor separations equal to the lattice constant, a, and is quadratic in the deviations from this optimum separation. As we will see, it is the assumption of a quadratic potential energy that makes this model tractable.

The general strategy to solve for the properties of the harmonic crystal is to transform the problem from N interacting particles to N independent simple harmonic oscillators, which enables us to factorize the partition function as discussed in Section 24.7. Once we have made this transformation, we need only copy the results for quantum SHOs from Section 24.10, and we have solved for the properties of the harmonic solid.

26.2 Normal Modes

The key to reducing the problem to independent oscillators is to Fourier transform[1] the position variables to find the normal modes. The Fourier transform of the positions of the particles is given by

$$\tilde{x}_k = N^{-1/2}\sum_j x_j \exp(-ikR_j), \tag{26.5}$$

where k is the (one-dimensional) wave vector or wave number. We can anticipate that the wave number is related to the wavelength λ by

$$k = \frac{2\pi}{\lambda}. \tag{26.6}$$

An obvious property of the Fourier transformed variables $\tilde{x}_k$ is that

$$\tilde{x}_{-k} = \tilde{x}_k^*, \tag{26.7}$$

where the superscript * indicates the complex conjugate.

---

[1] Jean Baptiste Joseph Fourier, French mathematician and physicist (1768–1830).

26.2.1   Inverse Fourier Transformation

The inverse of the Fourier transformation in eq. (26.5) is

$$x_j = N^{-1/2} \sum_k \tilde{x}_k \exp(ikR_j), \tag{26.8}$$

which we can confirm by direct substitution,

$$\tilde{x}_k = N^{-1/2} \sum_j x_j \exp(-ikR_j) \tag{26.9}$$

$$= N^{-1/2} \sum_j N^{-1/2} \sum_{k'} \tilde{x}_{k'} \exp(ik'R_j) \exp(-ikR_j)$$

$$= \sum_{k'} \tilde{x}_{k'} N^{-1} \sum_j \exp(i(k'-k)R_j)$$

$$= \sum_{k'} \tilde{x}_{k'} \delta_{k,k'} = \tilde{x}_k.$$

The identity

$$\sum_j \exp(i(k'-k)R_j) = N\delta_{k,k'}, \tag{26.10}$$

which was used in deriving eq. (26.9), will often prove useful in statistical mechanics. The validity of eq. (26.10) can be easily shown,

$$\sum_j \exp\left(i(k'-k)R_j\right) = \sum_j \exp\left(i(n'-n)\frac{2\pi}{L}ja\right) \tag{26.11}$$

$$= \sum_j \exp\left(2\pi i(n'-n)\frac{j}{N}\right).$$

If $n' = n$, so that $k' = k$, then the sum is clearly equal to N. If $n' - n = 1$, the sum is simply adding up the N complex roots of -1. Since they are uniformly distributed around the unit circle in the complex plane, they sum to zero. If $n' - n$ is any other integer (excepting a multiple of N), the angle between each root is multiplied by the same amount, they are still distributed uniformly, and they still sum to zero.

To complete the Fourier transform, we must specify the boundary conditions. There are two kinds of boundary condition in general use. Since they each have their own advantages, we will discuss them both in detail.

26.2.2   Pinned Boundary Conditions

Pinned boundary conditions for the harmonic solid are similar to the boundary conditions we used for black-body radiation. In that problem, the transverse electric field vanished at the boundaries of the metal box. For pinned boundary conditions, the vibrations vanish at the boundaries of the system.

To implement pinned boundary conditions, first extend the number of atoms in the model from N to $N+2$ by adding atoms with indices 0 and $N+1$. These two atoms enforce the boundary conditions by being fixed ('pinned') at the positions $R_0 = 0$ and $R_{N+1} = (N+1)a$.

The two solutions $\exp(ikR_j)$ and $\exp(-ikR_j)$ can be combined to give $\tilde{x}_k \propto \sin(kR_j)$ to satisfy the boundary condition that $x_0 = 0$ at $R_0 = 0$. From the other boundary condition that $x_{N+1} = 0$ at $R_{N+1} = (N+1)a = L$, we need

$$kL = k(N+1)a = n\pi, \tag{26.12}$$

or,

$$k = k(n) = \frac{n\pi}{L}, \tag{26.13}$$

for some integer n.

Since we have N atoms in the system, we expect to find N independent solutions. If we define

$$K \equiv \frac{2\pi}{a}, \tag{26.14}$$

we can see that k and $k+K$ correspond to the same solution:

$$\sin((k+K)R_j) = \sin(kR_j + jaK) \tag{26.15}$$
$$= \sin\left(kR_j + ja\frac{2\pi}{a}\right)$$
$$= \sin\left(kR_j + 2\pi j\right)$$
$$= \sin(kR_j).$$

Adding any integer multiple of K to k also produces the same solution.

For calculations, the standard choice for pinned boundary conditions is to use only those solutions corresponding to values of n from 1 to N.

26.2.3   Periodic Boundary Conditions

The other possible boundary conditions are periodic. In one dimension, this might be thought of as bending the line of atoms into a circle and tying the ends together, so that

$$R_{j+N} = R_j. \tag{26.16}$$

Note that the length of a system with periodic boundary conditions is

$$L = Na, \tag{26.17}$$

in contrast to the length of $(N+1)a$ for pinned boundary conditions.

In two dimensions, a rectangular lattice with periodic boundary conditions can be viewed as a torus. It is rather difficult to visualize periodic boundary conditions in three dimensions (it is called a Klein bottle), but the mathematics is no more difficult.

The great advantage of periodic boundary conditions is that the system becomes invariant under translations of any multiple of the lattice constant a. The system has no ends in one dimension (and no borders or surfaces in two or three dimensions).

For periodic boundary conditions, we will take solutions of the form $\exp(ikR_j)$. The condition that the system is periodic means that

$$\exp\left(ik(R_j + L)\right) = \exp(ikR_j), \tag{26.18}$$

which implies that

$$kL = 2n\pi, \tag{26.19}$$

or

$$k = k(n) = \frac{2n\pi}{L}, \tag{26.20}$$

for some integer value of n. Note that the values of k given in eq. (26.20) for periodic boundary conditions are spaced twice as far apart as the values of k for pinned boundary conditions given in eq. (26.13).

As was the case for pinned boundary conditions, the solution corresponding to k is identical to the solution corresponding to $k + K$, where K is given by eq. (26.14). We can confirm that k and $k + K$ correspond to the same solution,

$$\exp\left(i(k+K)R_j\right) = \exp\left(ikR_j + i\frac{2\pi}{a}ja\right) \tag{26.21}$$
$$= \exp\left(ikR_j + i2\pi j\right)$$
$$= \exp\left(ikR_j\right).$$

In eq. (26.21), do not forget that j is an integer index, while $i = \sqrt{-1}$.

Note that $n = 0$ is a solution for periodic boundary conditions, while it is not for pinned boundary conditions because $\sin(0) = 0$. The solution for $n = 0$ corresponds to moving all atoms together in the same direction.

It is customary to take both positive and negative values of $k(n)$, so that the central values of n we will use are

$$n = 0, \pm 1, \pm 2, \ldots, \pm(N-1)/2 \text{ for } N \text{ odd,} \tag{26.22}$$

and

$$n = 0, \pm 1, \pm 2, \ldots, \pm(N/2 - 1), N/2 \text{ for } N \text{ even.} \tag{26.23}$$

For even values of N, the state corresponding to $-N/2$ is identical to the state corresponding to $N/2$.

For periodic boundary conditions, the values of k lie at symmetric points in k-space, and the only independent states lie between $k = -K/2$ and $k = K/2$. This region is called the Brillouin Zone in honor of Léon Nicolas Brillouin (French physicist, 1889–1969, American citizen after 1949). All information in the Brillouin Zone is repeated throughout k-space with a periodicity of $K = 2\pi/a$.

When you take a course in solid state physics, you will find that three-dimensional Brillouin Zones can be rather complicated—and much more interesting. They are essential in understanding important properties of real materials, but they go beyond the scope of this book.

26.3   Transformation of the Energy

If we apply the Fourier transform in eq. (26.8) to the energy of the harmonic crystal given in eq. (26.4), we find that each Fourier mode is independent. This is the great simplification that justifies the bother of introducing Fourier transforms, because it enables us to factorize the partition function.

We will treat the kinetic and potential energy terms separately, beginning with the kinetic terms. The kinetic terms are, of course, already diagonal before the Fourier transform, but we have to demonstrate that they are also diagonal after the Fourier transform.

26.3.1   Kinetic Energy

Using the time derivative of eq. (26.8),

$$\dot{x}_j = N^{-1/2} \sum_k \dot{\tilde{x}}_k \exp(ikR_j), \tag{26.24}$$

we can carry out a Fourier transform of the kinetic energy term in eq. (26.4),

$$\text{K.E.} = \frac{1}{2}m\sum_j \dot{x}_j^2 = \frac{1}{2}m\sum_j \dot{x}_j\dot{x}_j$$

$$= \frac{1}{2}m\sum_j \left(N^{-1/2}\sum_k \dot{\tilde{x}}_k \exp(ikR_j)\right)\left(N^{-1/2}\sum_{k'} \dot{\tilde{x}}_{k'} \exp(ik'R_j)\right)$$

$$= N^{-1}\frac{1}{2}m\sum_k\sum_{k'} \dot{\tilde{x}}_k\dot{\tilde{x}}_{k'} \sum_j \exp(i(k+k')R_j)$$

$$= N^{-1}\frac{1}{2}m\sum_k\sum_{k'} \dot{\tilde{x}}_k\dot{\tilde{x}}_{k'} N\delta_{k+k',0}$$

$$= \frac{1}{2}m\sum_k \dot{\tilde{x}}_k\dot{\tilde{x}}_{-k} = \frac{1}{2}m\sum_k |\dot{\tilde{x}}_k|^2. \tag{26.25}$$

For the last equality, we have used eq. (26.7).

26.3.2 Potential Energy

For the potential energy term in eq. (26.4), separate the factors of $(x_j - x_{j+1})$ in the sum and use eq. (26.24) to introduce $\tilde{x}_k$,

$$\text{P.E.} = \frac{1}{2}K\sum_j (x_j - x_{j+1})^2 = \frac{1}{2}K\sum_j (x_j - x_{j+1})(x_j - x_{j+1})$$

$$= \frac{1}{2}K\sum_j \left(N^{-1/2}\sum_k \tilde{x}_k \left(\exp(ikR_j) - \exp(ikR_{j+1})\right)\right)$$

$$\times \left(N^{-1/2}\sum_{k'} \tilde{x}_{k'} \left(\exp(ik'R_j) - \exp(ik'R_{j+1})\right)\right)$$

$$= \frac{1}{2}K\sum_j \left(N^{-1/2}\sum_k \tilde{x}_k \exp(ikja)\,(1 - \exp(ika))\right)$$

$$\times \left(N^{-1/2}\sum_{k'} \tilde{x}_{k'} \exp(ik'ja)\,(1 - \exp(ik'a))\right) \tag{26.26}$$

Next, collect terms and simplify using the identity eq. (26.10):

$$P.E. = \frac{1}{2}KN^{-1}\sum_k\sum_{k'}\tilde{x}_k\tilde{x}_{k'}\left(1 - \exp(ika)\right)\left(1 - \exp(ik'a)\right)$$

$$\times \sum_j \exp(i(k-k')ja)$$

$$= \frac{1}{2}K\sum_k \tilde{x}_k\tilde{x}_{-k}(1 - \exp(ika))(1 - \exp(-ika))$$

$$= \frac{1}{2}\sum_k K(k)|\tilde{x}_k|^2. \tag{26.27}$$

The function

$$K(k) = K(1 - \exp(ika))(1 - \exp(-ika)) \tag{26.28}$$

gives the effective spring constant for the mode with wave number k.

Adding eqs. (26.25) and (26.27) together, we find the energy of the harmonic crystal from eq. (26.4) in terms of the Fourier transformed variables

$$E = \frac{1}{2}m\sum_k |\dot{\tilde{x}}_k|^2 + \frac{1}{2}\sum_k K(k)|\tilde{x}_k|^2 = \sum_k \left(\frac{1}{2}m|\dot{\tilde{x}}_k|^2 + \frac{1}{2}K(k)|\tilde{x}_k|^2\right). \tag{26.29}$$

The problem has been transformed from one with N variables (the positions of the atoms) to N problems, each with a single variable (the amplitude of the mode $\tilde{x}_k$ with wave number k). Each mode represents a simple harmonic oscillator with mass m (the same as the original atomic mass) and spring constant $K(k)$. For each mode, the square of the frequency is given by the usual ratio of the spring constant to the mass,

$$\omega^2(k) = \frac{K(k)}{m}. \tag{26.30}$$

26.4 The Frequency Spectrum

To analyze the spectrum of frequencies, $\omega(k)$, we can rewrite eq. (26.28) for $K(k)$ in a more convenient form:

$$\begin{aligned}
K(k) &= K(1 - \exp(ika))(1 - \exp(-ika)) \\
&= K(1 - \exp(ika) - \exp(-ika) + 1) \\
&= 2K(1 - \cos(ka)) \\
&= 4K\sin^2(ka/2).
\end{aligned} \tag{26.31}$$

The angular frequency is then given by

$$\omega^2(k) = \frac{4K\sin^2(ka/2)}{m},$$ (26.32)

or,

$$\omega(k) = 2\tilde{\omega}\left|\sin\left(\frac{ka}{2}\right)\right|,$$ (26.33)

where $\tilde{\omega} = \sqrt{K/m}$. This spectrum is plotted in Fig. 26.1.

For small wave number $(k \ll \pi/a)$, the frequency of a mode in eq. (26.33) becomes linear in the wave number k,

$$\omega(k) = 2\tilde{\omega}|\sin(ka/2)| \approx \tilde{\omega}ka.$$ (26.34)

The speed of a sound wave $v(k)$ is given by the product of the frequency $\nu(k)$ times the wavelength for long wavelengths (small k). From eq. (26.34), we see that the speed of a sound waves is a constant,

$$\frac{\omega(k)}{k} = \frac{2\pi\nu(k)}{2\pi/\lambda} = \nu(k)\lambda = v(k) = a\sqrt{\frac{K}{m}}.$$ (26.35)

The linearity of the function $\omega(k)$ can also be seen near the origin in Fig. 26.1.

The generalization of these calculations to a three-dimensional harmonic crystal is not particularly difficult, although the mathematics becomes rather more messy because of the necessary introduction of vector notation. The three-dimensional form of the Brillouin Zone is also somewhat more complicated. However, the generalization of the frequency spectrum to three dimensions has a particularly simple form,

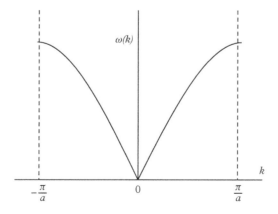

Fig. 26.1 *Frequency spectrum of a one-dimensional harmonic solid, as given in eq. (26.33).*

$$\omega^2(\vec{k}) = 2K \sum_{\vec{\delta}} \sin^2\left(\frac{\vec{k}\cdot\vec{\delta}}{2}\right).$$

(26.36)

The sum in this equation goes over the vectors $\vec{\delta}$ connecting a site to its nearest neighbors. Eq. (26.36) is valid for all lattices with nearest-neighbor harmonic interactions.

Interactions with more distant neighbors requires only a small change in eq. (26.36),

$$\omega^2(\vec{k}) = 2K \sum_{\alpha} \sum_{\vec{\delta}_\alpha} \sin^2\left(\frac{\vec{k}\cdot\vec{\delta}_\alpha}{2}\right).$$

(26.37)

The index α goes over more distant types of neighbors.

26.5   Alternate Derivation: Equations of Motion

An alternative derivation of the frequency spectrum of the harmonic solid is given by the equations of motion. This approach is somewhat more flexible in dealing with problems involving multiple sublattices or nonuniform interactions, such as Problems 26.2 and 26.3 at the end of this chapter.

Consider the equation of motion for the coordinate x_j,

$$m\ddot{x}_j = -K\left(x_j - x_{j-1} + x_j - x_{j+1}\right).$$

(26.38)

Look for solutions of the form

$$x_j = X\exp(ik2aj)\exp(-i\omega t).$$

(26.39)

Substituting this into the equation of motion, we find

$$-m\omega^2 X = -K\left(X - X\exp(-ika) + X - X\exp(+ika)\right),$$

(26.40)

or,

$$m\omega^2 = K\left(2 - 2\cos(ka)\right).$$

(26.41)

This can be simplified further to

$$\omega^2 = \frac{4K\sin^2(ka/2)}{m},$$

(26.42)

which agrees with eq. (26.32).

26.6 The Energy in the Classical Model

Since the SHOs in eq. (26.29) are not particles that can move from one system to another, the classical partition function given here does not contain a factor of $1/N!$

$$Z_{class} = \frac{1}{h^N} \int dp \int dq \exp(-\beta H). \qquad (26.43)$$

The limits for the integrals in eq. (26.43) have been omitted (which is not usually a good idea) because they all go from $-\infty$ to ∞.

Because of the diagonal form of the Fourier transformed energy, the partition function factors. Since all the integrals are gaussian, we can evaluate them immediately,

$$Z_{class} = \prod_k \frac{1}{h} \left[(2\pi m k_B T)^{1/2} (2\pi k_B T / K(k))^{1/2} \right]. \qquad (26.44)$$

Simplifying this expression, we find

$$Z_{class} = \prod_k \left[\frac{2\pi}{h} k_B T \left(\frac{m}{K(k)} \right)^{1/2} \right] = \prod_k (\beta \hbar \omega(k))^{-1}, \qquad (26.45)$$

where

$$\omega(k) = \sqrt{\frac{K(k)}{m}}, \qquad (26.46)$$

is the frequency of the k-th mode.

As usual, the average energy is found by taking a derivative with respect to β,

$$U = \frac{\partial(\beta F)}{\partial \beta} \qquad (26.47)$$

$$= -\frac{\partial(\ln Z_{class})}{\partial \beta}$$

$$= -\frac{\partial}{\partial \beta} \left[-N \ln \beta - \sum_k \ln(\hbar \omega(k)) \right]$$

$$= N \frac{1}{\beta} = N k_B T.$$

As expected, the specific heat has the constant value of k_B, which is the Law of Dulong and Petit.

For a classical harmonic crystal, the spectrum of frequencies as a function of the wave number k is unimportant for equilibrium properties. In the next section we will see that this is not the case for a quantum mechanical harmonic crystal.

26.7 The Quantum Harmonic Crystal

Formal expressions for the quantum mechanical properties of a harmonic lattice are obtained just as easily as the classical properties—though their evaluation is rather more challenging. Because the system factorizes into N SHOs, the partition function is found by inserting the results from Section 24.10,

$$Z_{QM} = \prod_k \left[\exp(-\beta\hbar\omega(k)/2)(1 - \exp(-\beta\hbar\omega(k)))^{-1} \right]. \tag{26.48}$$

The energy of the system is just the sum of the energies of the individual modes, which can be confirmed by taking the negative logarithmic derivative of eq. (26.48) with respect to β, as in eq. (24.12),

$$U_{QM} = \sum_k \left[\frac{1}{2}\hbar\omega(k) + \frac{\hbar\omega(k)}{\exp(\beta\hbar\omega(k)) - 1} \right]. \tag{26.49}$$

Because the wave numbers are closely spaced, eq. (26.49) can be turned into an integral without introducing a significant error,

$$U_{QM} = \frac{L}{\pi} \int_{-\pi/a}^{\pi/a} dk \left[\frac{1}{2}\hbar\omega(k) + \frac{\hbar\omega(k)}{\exp(\beta\hbar\omega(k)) - 1} \right]. \tag{26.50}$$

For high temperatures ($k_BT \gg \hbar\omega_k$ for all modes), the average energy of every mode goes to k_BT, and we recover the classical answer and the Law of Dulong and Petit.

For low temperatures the situation is more complicated, because different modes have different frequencies. The contribution of each mode to the specific heat depends on the ratio $\hbar\omega_k/k_BT$; when the ratio is small, the contribution to the specific heat is k_B, but when the ratio is large, the contribution becomes very small. As the temperature is lowered, the contributions to the specific heat from a larger fraction of the modes becomes negligible, and the total specific heat goes to zero.

To calculate how the specific heat of the harmonic crystal goes to zero, we need the frequency spectrum found in Section 26.4. The formal solution for the specific heat is found by substituting eq. (26.33) into eq. (26.49) and differentiating with respect to temperature. This procedure is easily carried out numerically on a modern computer.

On the other hand, it turns out to be both useful and enlightening to develop an approximation that allows us to investigate the low-temperature behavior analytically, as we will do in the following section on the Debye approximation.

26.8 Debye Approximation

We mentioned previously that the Brillouin Zone is very simple in one dimension; it is just the region from $k = -\pi/a$ to π/a. In three dimensions it takes on more complicated forms that reflect the symmetry of the particular lattice: face-centered cubic, body-centered cubic, hexagonal, and so on. This makes it necessary to perform a three-dimensional integral over the Brillouin Zone.

When the theory of the quantum mechanical harmonic lattice was first developed in the early part of the twentieth century, computers were non-existent; all calculations had to be accomplished by hand. This made it very important to simplify calculations as much as possible, and led to the approximation that we will discuss in this section.

However, even today, when we can easily and rapidly evaluate eq. (26.50) and its generalization to three dimensions on a computer, there is still considerable value in simplifying the calculation as an aid to understanding its significance. Such simplifying approximations will also show the relationship of the harmonic crystal to black-body radiation, which might otherwise be much less obvious.

The Debye approximation—named after its inventor, the Dutch physicist Peter Joseph William Debye (born Petrus Josephus Wilhelmus Debije, 1884–1966)—can be viewed as a kind of interpolation between the known high- and low-temperature regions.

For low temperatures, only low-energy modes will be excited significantly because the high-energy modes will be cut off by the exponential term $\exp(\beta\hbar\omega(k))$ in the denominator of the integrand in eq. (26.50). Therefore, we only need to look at the low-frequency region of the energy spectrum. As shown in Section 26.4, the low-frequency spectrum is given by a simple linear relation

$$\epsilon(\vec{k}) = \hbar\omega(\vec{k}) \approx \hbar v \left|\vec{k}\right|. \tag{26.51}$$

In this equation, v is the speed of sound, and we have generalized the result to an arbitrary number of dimensions.

Eq. (26.51) is a very nice result because it is spherically symmetric. If the Brillouin Zone were also spherically symmetric, we could use that symmetry to reduce the three-dimensional integral for the energy to a one-dimensional integral. For this reason, Debye introduced an approximation that does exactly that. He approximated the energy spectrum by $\epsilon(\vec{k}) = \hbar v|k|$ and the true shape of the Brillouin Zone by a sphere!

It's easy to see how Debye got away with this approximation at low temperatures. Only the low-energy modes made a significant contribution to the average energy, and eq. (26.51) was an excellent approximation for those modes. The high-energy modes that would be affected by true functional form of $\omega(\vec{k})$ and the shape of the Brillouin Zone did not contribute anyway, so that distortion did not matter.

However, at high temperatures neither of these two arguments is valid. The high-energy modes are important, and they do not have the form given in eq. (26.51). Fortunately, a different argument comes into play to save the approximation.

Recall that for high temperatures, when $k_B T \gg \hbar \omega(\vec{k})$, the energy of each mode is just given by the classical value of $k_B T$. Since each mode contributes $k_B T$, the energy spectrum is irrelevant; the only thing that matters is *how many modes there are*. For this reason, Debye fixed the size of his (approximate) spherical Brillouin Zone so that it had exactly $3N$ modes—the same as the true Brillouin Zone.

The radius of the approximate Brillouin Zone is most easily found by going back into n-space, in which the points are distributed uniformly with a density equal to one. Each point in the positive octant (n_x, n_y, n_z are all positive) represents three modes (two for the transverse modes, and one for the longitudinal mode). The total number of modes must equal $3N$,

$$3N = \frac{3}{8} \int_0^{n_D} 4\pi n^2 \, dn = \frac{3}{8} \frac{4}{3} \pi n_D^3 = \frac{1}{2} \pi n_D^3. \tag{26.52}$$

The Debye radius in $\vec{n}$-space is

$$n_D = \left(\frac{6N}{\pi} \right)^{1/3}. \tag{26.53}$$

This also gives us results for the corresponding values of the Debye wave number and the Debye energy,

$$k_D = \frac{\pi}{L} \left(\frac{6N}{\pi} \right)^{1/3} \tag{26.54}$$

$$\epsilon_D = \frac{\pi}{L} \left(\frac{6N}{\pi} \right)^{1/3} \hbar v. \tag{26.55}$$

Note that v in eq. (26.55) is the speed of sound, as in eq. (26.51).

Since the ground-state energy plays no role in the thermodynamics of the harmonic solid, we will ignore it and write the energy in the Debye approximation for a three-dimensional harmonic crystal as a one-dimensional integral,

$$U_{Debye} = \frac{3\pi}{2} \int_0^{n_D} n^2 \, dn \frac{\hbar \omega(n)}{\exp(\beta \hbar \omega(n)) - 1}. \tag{26.56}$$

At this point we will follow the same procedure for simplifying the integral that we used in Section 25.6 for black-body radiation. We define a dimensionless integration variable to transform the integral in eq. (26.56) to a more convenient form,

$$x = \beta \hbar \omega(n) = \left(\frac{\pi \hbar v}{L k_B T} \right) n. \tag{26.57}$$

The upper limit of the integral in eq. (26.56) will become

$$x_d = \frac{\Theta_D}{T},$$ (26.58)

where the Debye temperature, Θ_D, is given by

$$\Theta_D = \frac{\hbar v}{k_B} \left(\frac{6\pi^2 N}{L^3} \right)^{1/3}.$$ (26.59)

Using these substitutions, eq. (26.56) takes on the following form:

$$U_{Debye} = \frac{3\pi}{2} \left(\frac{Lk_B T}{\pi \hbar v} \right)^3 k_B T \int_0^{\Theta_D/T} dx\, x^3 (\exp(x) - 1)^{-1}$$ (26.60)

$$= \frac{3L^3 (k_B T)^4}{2\pi^2 \hbar^3 v^3} \int_0^{\Theta_D/T} dx\, x^3 (\exp(x) - 1)^{-1}.$$

The great advantage of eq. (26.60) when Debye first derived it was that it could be evaluated numerically with pencil and paper (and a bit of effort). However, even today Debye's equation remains very useful for gaining insights into the behavior of a crystal at both high and low temperatures, as we will show in the following subsections.

26.8.1 Debye Approximation for $T \gg \Theta_D$

At high temperatures we expect the energy to go to $U \approx 3Nk_B T$, because that was how we selected the size of the spherical Brillouin Zone. But it is always good to check to make sure that we did not make an error in the calculation.

For high temperatures, $\Theta_D/T \ll 1$, so that the upper limit of the integral in eq. (26.60) is small. For small values of x in the integrand, we can make the approximation

$$x^3 (\exp(x) - 1)^{-1} \approx x^3 (1 + x - 1)^{-1} = x^2,$$ (26.61)

which is easy to integrate. Inserting this approximation into eq. (26.60), we find

$$U_{Debye} \approx \frac{3L^3 (k_B T)^4}{2\pi^2 \hbar^3 v^3} \int_0^{\Theta_D/T} dx\, x^2$$ (26.62)

$$= \frac{3L^3 (k_B T)^4}{2\pi^2 \hbar^3 v^3} \frac{1}{3} \left(\frac{\Theta_D}{T} \right)^3$$

$$= \frac{3L^3(k_BT)^4}{2\pi^2\hbar^3 v^3}\frac{1}{3}\left(\frac{\hbar v}{k_B}\left(\frac{6\pi^2 N}{L^3}\right)^{1/3}\frac{1}{T}\right)^3$$

$$= \frac{L^3(k_BT)^4}{2\pi^2\hbar^3 v^3}\frac{\hbar^3 v^3}{k_B^3}\left(\frac{6\pi^2 N}{L^3 T^3}\right)$$

$$= 3Nk_BT.$$

Since the Debye temperature was chosen to include $3N$ modes in the spherical approximate Brillouin Zone, the high-temperature expression for the energy is as expected, and the specific heat is $3k_B$, consistent with the Law of Dulong and Petit.

26.8.2   Debye Approximation for $T \ll \Theta_D$

For low temperatures, $\Theta_D/T \gg 1$, and the upper limit of eq. (26.60) is large. Since the integrand goes to zero as $x^3 \exp(-x)$, taking the upper limit to be infinite is a good approximation,

$$U_{Debye} \approx \frac{3L^3(k_BT)^4}{2\pi^2\hbar^3 v^3}\int_0^\infty dx\, x^3 (\exp(x) - 1)^{-1}. \tag{26.63}$$

The integral in eq. (26.63) should be familiar. It is exactly the same as the integral we found in eq. (25.29) during the analysis of black-body radiation. This is not a coincidence, as discussed next,

$$\int_0^\infty dx \frac{x^3}{e^x - 1} = \frac{\pi^4}{15}. \tag{26.64}$$

Inserting the value of the integral into eq. (26.63), we find the low-temperature behavior of the energy of the harmonic lattice in closed form,

$$U_{Debye} \approx \frac{3L^3(k_BT)^4}{2\pi^2\hbar^3 v^3}\frac{\pi^4}{15} \tag{26.65}$$

$$= L^3 \frac{\pi^2}{10\hbar^3 v^3}(k_BT)^4$$

$$= V \frac{\pi^2}{10\hbar^3 v^3}(k_BT)^4.$$

The energy per unit volume is then given by

$$\frac{U}{V} = u = \left(\frac{\pi^2}{10\hbar^3 v^3}\right)(k_BT)^4. \tag{26.66}$$

The energy of the harmonic crystal at low temperatures, for which the Debye approximation is accurate, is very similar to the energy of black-body radiation. Recall eq. (25.30) from Chapter 25 for the energy per unit volume of black-body radiation

$$\frac{U_{BB}}{V} = u = \left(\frac{\pi^2}{15\hbar^3 c^3}\right)(k_B T)^4. \tag{26.67}$$

The form of eqs. (26.66) and (26.67) is the same, because for both black-body radiation and the low-temperature harmonic crystal, the energy spectrum is linear in the wave number: $\omega = vk$ for the harmonic solid, and $\omega = ck$ for black-body radiation (light waves).

There are only two differences between eq. (26.66) for the low-temperature behavior of the harmonic solid and eq. (25.30) or (26.67) for black-body radiation. The expression for the harmonic crystal contains the speed of sound v, instead of the speed of light c, and the factor in the denominator is 10 instead of 15 because there are three polarizations for sound, but only two for light.

From eq. (26.66) we see that the energy of a harmonic crystal is proportional to T^4, which is analogous to the T^4 factor in the Stefan–Boltzmann Law, eq. (25.31). Taking a derivative of the energy with respect to temperature, we see that the specific heat is proportional to T^3, which is observed to be the correct behavior of insulating crystals at low temperatures.

The specific heats of metallic crystals, on the other hand, are linear in the temperature. The origin of this difference between conductors and insulators will be discussed in Chapter 29.

26.8.3   Summary of the Debye Approximation

There are three steps in deriving the Debye approximation.

1. Replace the true energy spectrum with an approximate spectrum that is linear in the wave number, k, and spherically symmetric: $\epsilon(\vec{k}) = \hbar v |\vec{k}|$.

2. Replace the true Brillouin Zone by a spherically symmetric Brillouin Zone.

3. Choose the size of the spherically symmetric Brillouin Zone so that it contains exactly N k-values and $3N$ modes.

The Debye approximation is excellent at both high and low temperatures. Do not be fooled, however: in between, it is only mediocre. The qualitative behavior of the contributions to the specific heat from lattice vibrations in real crystals is monotonically increasing with temperature, starting from T^3-behavior at low temperatures and going to the constant value $3k_B$ (Law of Dulong and Petit) at high temperatures. The detailed form of the function depends on details of the true energy spectrum and the shape of the true Brillouin Zone. Fortunately, the full, three-dimensional integrals are easy to do on a computer, so that we need not rely on the Debye approximation as much as researchers did in the middle of the last century.

26.9 Problems

PROBLEM 26.1
Fourier transform of the particle positions

In Section 26.2 an argument was given for the validity of eq. (26.10),

$$\sum_j \exp(i(k' - k)R_j) = N\delta_{k,k'}.$$

This argument can be made into a proof by carrying out the sum exactly.

1. Extend the derivation of the infinite sum $\sum_{n=0}^{\infty} x^n$ to calculate the finite sum $\sum_{n=0}^{N} x^n$.
2. Carry out the proof for periodic boundary conditions with odd N.

PROBLEM 26.2
The harmonic crystal

1. Show that for a harmonic crystal the integral

$$\int_0^{\infty} [C_V(\infty) - C_V(T)] \, dT$$

   is exactly equal to the zero-point energy of the solid.
2. Interpret the result graphically. It might help to relate this result to an earlier assignment on the properties of quantum SHOs.

PROBLEM 26.3
The harmonic crystal with alternating spring constants

Most crystals contain more than one kind of atom. This leads to both quantitative and qualitative differences in the vibration spectrum from the results derived in class. These differences occur whenever the periodicity is altered, whether by changing the masses of the atoms or changing the spring constants.

As an example of the kinds of differences that can arise when the periodicity changes, consider the problem of a one-dimensional lattice with alternating spring constants K_1 and K_2. For simplicity, you can assume that all the masses are equal to m.

1. If the distance between atoms is a, the periodicity is $2a$. What is the size of the Brillouin Zone?

2. Calculate the vibration spectrum of the lattice. You might find it more convenient to write down equations of motion for the two sublattices.
3. Sketch the vibration spectrum in the Brillouin Zone based on your calculations.
4. Sometimes the vibration spectrum of a crystal with more than one kind of atom is approximated by a combination of an Einstein model and a Debye model. Explain why this might make sense and, give appropriate parameters for such a description in the present case. In answering this question, consider the high- and low-temperature behavior of the specific heat of this model.

PROBLEM 26.4
The harmonic crystal with alternating masses

Most crystals contain more than one kind of atom. This leads to both quantitative and qualitative differences in the vibration spectrum from the results derived in class. These differences occur whenever the periodicity is altered, whether by changing the masses of the atoms or changing the spring constants.

As an example of the kinds of differences that can arise when the periodicity changes, consider the problem of a one-dimensional lattice with alternating masses m_1 and m_2.

1. If the distance between atoms is a, the periodicity is $2a$. What is the size of the Brillouin Zone?
2. Calculate the vibration spectrum of the lattice. You might find it more convenient to write down equations of motion for the two sublattices.
3. Sketch the vibration spectrum in the Brillouin Zone based on your calculations.

27

Ideal Quantum Gases

Out of perfection nothing can be made. Every process involves breaking something up.

Joseph Campbell

This chapter discusses the theory of ideal quantum gases. It develops general equations that will be valid for gases that consist of either bosons or fermions. Since the consequences of these equations are both non-trivial and very different for the two cases of bosons and fermions, the details of each of these cases will be dealt with in subsequent chapters.

This chapter will also discuss the interesting case of a quantum gas of distinguishable particles. Although all atoms are either bosons or fermions, and therefore indistinguishable, there are nevertheless real systems that are composed of distinguishable particles. In particular, colloidal particles can each be composed of about 10^9 molecules. The number of molecules in each particle will vary, as will the arrangement and number of each type of molecule. In short, while the particles in a colloid might be similar in size and composition, they do not have identical properties and they are not indistinguishable. While the properties of colloids can usually be obtained accurately using classical statistical mechanics, it is also of interest to see how quantum mechanics might affect the results.

We begin in the next section by discussing many-particle quantum states, which are needed to describe a macroscopic quantum system.

27.1 Single-Particle Quantum States

As is the case for most of statistical mechanics, we would like to put the theory of ideal quantum gases into a form that allows us to factorize the partition function. For ideal gases, that points to building up the macroscopic state in terms of single-particle states, which will then lead to such a factorization.

Consider a quantum ideal gas in a box of volume V. For simplicity, we will assume that the box is cubic, with length L, so that $V = L^3$. The sides of the box will be assumed to be impenetrable, also for simplicity.

An Introduction to Statistical Mechanics and Thermodynamics. Robert H. Swendsen, Oxford University Press (2020).
© Robert H. Swendsen. DOI: 10.1093/oso/9780198853237.001.0001

For quantum ideal gases, we can write the Hamiltonian of a single particle of mass m as

$$H = \frac{|\vec{p}|^2}{2m} = -\frac{\hbar^2}{2m}\vec{\nabla}^2,$$
(27.1)

where

$$\vec{\nabla} = \left(\frac{\partial}{\partial x}, \frac{\partial}{\partial y}, \frac{\partial}{\partial z}\right).$$
(27.2)

The eigenfunctions of eq. (27.1) can be written in a form that is very similar to that of the electric field for the black-body problem in eq. (25.5), but is even simpler because the wave function ψ is a scalar, while the electric field is a vector

$$\psi_{\vec{k}}(\vec{r}) = \sqrt{\frac{8}{L^3}} \sin(k_x x) \sin(k_y y) \sin(k_z z).$$
(27.3)

(The derivation of the normalization constant is left to the reader.)

The wave numbers in eq. (27.3) are determined by the condition that the wave function vanishes at the boundaries

$$k_x = \frac{n_x \pi}{L}$$
(27.4)

$$k_y = \frac{n_y \pi}{L}$$
(27.5)

$$k_z = \frac{n_z \pi}{L}.$$
(27.6)

If the values of n_x, n_y, and n_z are integers, the boundary conditions are fulfilled. Since the overall sign of the wave function is unimportant, negative values of these integers do not represent distinct states; only non-zero, positive integers represent physically distinct states. The wave equation can also be written in terms of the values of $\vec{n} = \{n_x, n_y, n_z\}$,

$$\psi_{\vec{n}}(\vec{r}) = \sqrt{\frac{8}{L^3}} \sin\left(\frac{n_x \pi}{L}x\right) \sin\left(\frac{n_y \pi}{L}y\right) \sin\left(\frac{n_z \pi}{L}z\right).$$
(27.7)

The energy eigenvalues for a single-particle state are easily found,

$$H\psi_{\vec{k}}(\vec{r}) = -\frac{\hbar^2}{2m}\vec{\nabla}^2\psi_{\vec{k}}(\vec{r}) = \frac{\hbar^2}{2m}\left(k_x^2 + k_y^2 + k_z^2\right)\psi_{\vec{k}}(\vec{r}) = \epsilon_{\vec{k}}\psi_{\vec{k}}(\vec{r})$$
(27.8)

$$\epsilon_{\vec{k}} = \frac{\hbar^2}{2m}\left(k_x^2 + k_y^2 + k_z^2\right) = \frac{\hbar^2\pi^2}{2mL^2}\left(n_x^2 + n_y^2 + n_z^2\right) = \epsilon_{\vec{n}}.$$
(27.9)

(The symbol ϵ will be reserved for the energies of single-particle states, or single-particle energy levels, while E will be used for the energy of an N-particle state.)

If we define

$$k^2 = k_x^2 + k_y^2 + k_z^2, \tag{27.10}$$

and

$$n^2 = n_x^2 + n_y^2 + n_z^2, \tag{27.11}$$

we can express eq. (27.9) more compactly,

$$\epsilon_{\vec{k}} = \frac{\hbar^2}{2m}k^2 = \frac{\hbar^2\pi^2}{2mL^2}n^2 = \epsilon_{\vec{n}}. \tag{27.12}$$

Eq. (27.9) gives the single-particle energy eigenvalues for a particle in a three-dimensional, cubic box with sides of length L. This is the case on which we will concentrate. However, since we can preserve generality with little cost, I will replace the quantum number vector $\vec{n}$ with a general quantum number α for most of the rest of this chapter. This means that the equations in this chapter will be valid for any system of non-interacting particles, even if those particles are moving in an external potential.

The energy eigenstates in eq. (27.9) are often called 'orbitals' for historical reasons; the early work on fermions concentrated on electrons orbiting around a nucleus, which led to the use of the term. It might seem strange to see the term 'orbitals' when nothing is orbiting, but tradition is tradition. I will refer to them as single-particle states, which is less compact, but also (I believe) less confusing.

27.2 Density of Single-Particle States

Just as we needed the photon density of states to calculate the properties of black bodies, we will need the single-particle density of states to calculate the properties of a system of particles.

The density of states in $\vec{n}$-space is easy to find. Since there is one state at every point with integer components, the density of states is 1. To calculate the density of states as a function of energy, we make a transformation using a Dirac delta function, following the methods developed in Chapter 5:

$$D(\epsilon) = \int_0^\infty dn_x \int_0^\infty dn_y \int_0^\infty dn_z \delta(\epsilon - \epsilon_{\vec{n}}) \tag{27.13}$$

$$= \frac{1}{8} \int_0^\infty 4\pi n^2 \, dn \, \delta\left(\epsilon - \frac{\hbar^2 \pi^2}{2mL^2} n^2\right)$$

$$= \frac{\pi}{4} \left(\frac{2mL^2}{\hbar^2 \pi^2}\right)^{3/2} \int_0^\infty x^{1/2} \, dx \, \delta(\epsilon - x)$$

$$= \frac{V}{4\pi^2} \left(\frac{2m}{\hbar^2}\right)^{3/2} \epsilon^{1/2}.$$

In the third line of eq. (27.13) we introduced a new variable of integration:

$$x = \left(\frac{\hbar^2 \pi^2}{2mL^2}\right) n^2. \tag{27.14}$$

In the fourth line of eq. (27.13) we replaced L^3 by the volume of the system, $V = L^3$.

Eq. (27.13) is valid for non-interacting fermions, bosons, or distinguishable particles (without spin). It will play a central role in Chapters 28, 29, and 30.

27.3 Many-Particle Quantum States

An N-particle wave function can be constructed from eq. (27.3) by multiplying single-particle wave functions,

$$\psi_N = \prod_{j=1}^N \psi_{\alpha_j}(\vec{r}_j). \tag{27.15}$$

In this equation, α_j represents the quantum number(s) that describe the wave function for the j th particle.

Eq. (27.15) is a correct wave function for distinguishable particles, and we will return to it in Section 27.9. However, for indistinguishable particles, the wave function must be symmetrized for Bose–Einstein statistics[1] (bosons) or anti-symmetrized for Fermi–Dirac statistics[2] (fermions).

The wave function for bosons can be written as a sum over all permutations, P, of the assignments of particles to wave functions in the expression in eq. (27.15),

---

[1] Satyendranath Bose (1894–1974), Indian physicist, and Albert Einstein (1879–1956), German, Swiss, Austrian, and American physicist, Nobel Prize 1921.
[2] Enrico Fermi (1901–1954), Italian physicist, Nobel Prize 1938, and Paul Adrien Maurice Dirac (1902–1984), British physicist, Nobel Prize 1933.

$$\psi_N^{BE} = X_{BE} \sum_P P \left[\prod_{j=1}^N \psi_{\alpha_j}(\vec{r}_j) \right]. \tag{27.16}$$

The normalization constant, X_{BE}, can be calculated, but does not play a significant role and will be left to the interested reader.

The wave function for fermions can be represented by a similar expression if we include a factor of $(-1)^P$, in which the exponent denotes the number of two-particle exchanges needed to construct the permutation P,

$$\psi_N^{FD} = X_{FD} \sum_P (-1)^P P \left[\prod_{j=1}^N \psi_{\alpha_j}(\vec{r}_j) \right]. \tag{27.17}$$

X_{FD} is the normalization constant for Fermi–Dirac statistics, which will also be left to the reader.

Note that the many-particle wave function for fermions vanishes if two or more particles are in the same single-particle state ($\alpha_i = \alpha_j$ for $i \neq j$). Each single-particle state can either be empty or contain one particle. There is no such restriction for bosons, and any number of bosons can occupy a single-particle state.

The Hamiltonian of an N-particle system of independent particles is just

$$H_N = \sum_{j=1}^N H_j, \tag{27.18}$$

where H_j is the single-particle Hamiltonian for the j-th particle.

The N-particle eigenvalue equation is then

$$H_N \psi_N = E \psi_N = \sum_{j=1}^N \epsilon_{\alpha_j} \psi_N \tag{27.19}$$

for bosons, fermions, or distinguishable particles.

To construct the Boltzmann factor we need to apply the operator $\exp(-\beta H_N)$ to the wave function. Fortunately, this is easy. Simply expand the exponential in a series in powers of H_N, apply the eigenvalue equation, eq. (27.19), and sum the power series in E,

$$\exp(-\beta H_N)\psi_N = \exp(-\beta E)\psi_N = \exp\left(-\beta \sum_{j=1}^N \epsilon_{\alpha_j} \right) \psi_N. \tag{27.20}$$

27.4 Quantum Canonical Ensemble

In principle, we could go to the canonical ensemble and sum the Boltzmann factor in eq. (27.20) over all N-particle states, just as we did for black bodies in Chapter 25 and harmonic solids in Chapter 26. Unfortunately, if we try to do this the canonical partition function does not factor. The difficulty has to do with enumerating the states. If the total number of particles is fixed, we can only put a particle in one state if we simultaneously take it out of some other state. We did not run into this problem for either black bodies or harmonic solids because neither photons nor phonons are conserved.

To be able to sum freely over the number of particles in each single-particle state—and thereby factorize the partition function—we need to introduce a reservoir that can exchange particles with the system of interest. This brings us to the grand canonical ensemble, which is the subject of the next section.

27.5 Grand Canonical Ensemble

The grand canonical partition function was introduced for classical statistical mechanics in Sections 20.1 and 20.2. The equations describing the quantum grand canonical ensemble are the same as those for the classical case, as long as we now interpret $\Omega(E,V,N)$ to mean the degeneracy of an N-particle quantum state with energy E.

Eq. (20.9) gives the quantum mechanical grand canonical probability distribution for E and N,

$$P(E,N) = \frac{1}{\mathcal{Z}}\Omega(E,V,N)\exp\left[-\beta E + \beta\mu N\right], \tag{27.21}$$

and eq. (20.10) gives us the quantum grand canonical partition function,

$$\mathcal{Z} = \sum_{N=0}^{\infty}\sum_{E}\Omega(E,V,N)\exp\left[-\beta E + \beta\mu N\right]. \tag{27.22}$$

The sum over E in eq. (27.22) denotes the sum over all energy levels of the N-particle system, with $\Omega(E,V,N)$ representing the degeneracy. It replaces the integral over the continuum of energies in the classical case that was used in eq. (20.10).

For a quantum system of non-interacting particles, the N-particle energies are given by

$$E = \sum_{j=1}^{N}\epsilon_{\alpha_j}, \tag{27.23}$$

so that

$$\exp[-\beta E + \mu N] = \exp[-\beta \sum_{j=1}^{N}(\epsilon_{\alpha_j} - \mu)] = \prod_{j=1}^{N} \exp[-\beta(\epsilon_{\alpha_j} - \mu)]. \tag{27.24}$$

We can also rewrite eq. (27.22) in terms of sums over the eigenstates, thereby eliminating the degeneracy factor $\Omega(E, V, N)$

$$\mathcal{Z} = \sum_{N=0}^{\infty} \sum_{\{\alpha_j\}} \prod_{j=1}^{N} \exp[-\beta(\epsilon_{\alpha_j} - \mu)]. \tag{27.25}$$

27.6   A New Notation for Energy Levels

To simplify the expression for the grand canonical partition function, it is useful to change the notation. So far, we have been describing an N-particle state by listing the set of single-particle quantum numbers that specify that state. If more than one particle were in a particular single-particle state, the corresponding quantum number would occur more than once in the list.

An alternative representation would be to specify for each state the number of times it appears in the list, which we will call the occupation number.

Perhaps an example would help illustrate the two representations. In the following table I have indicated each state by a line and indicated the each time it occurs by an 'x'. To represent the data in Table 27.1, we could have a list of quantum numbers,

$$\{\alpha_j\} = \{\alpha_1, \alpha_2, \alpha_2, \alpha_4, \alpha_4, \alpha_4, \alpha_5\}, \tag{27.26}$$

Table 27.1 *Representations of energy levels.*

quantum number α	Energy levels	Occupation number, n_α
6	–	0
5	×	1
4	× × ×	3
3	–	0
2	× ×	2
1	×	1

or, equivalently, a list of occupation numbers for the energy levels,

$$\{n_\alpha\} = \{1,2,0,3,1,0\}. \tag{27.27}$$

Both representations contain exactly the same information. However, the representation in terms of occupation numbers turns out to be more useful for the evaluation of the grand canonical partition function.

With this change in notation, the expression for the grand canonical partition function becomes

$$\mathcal{Z} = \sum_{N=0}^{\infty} \sum_{\{n_\alpha\}}^{\sum n_\alpha = N} \prod_\epsilon \exp[-\beta(\epsilon_\alpha - \mu)n_\alpha], \tag{27.28}$$

where the sum over n_α is understood to included only allowed values of n_α.

Now note that the sum over $\{n_\alpha\}$ is limited to those cases in which $\sum n_\alpha = N$, but the result is then summed over all values of N. We can simplify this double sum by removing the sum over N and removing the limit on the sum over $\{n_\alpha\}$,

$$\mathcal{Z} = \sum_{\{n_\alpha\}} \prod_\alpha \exp[-\beta(\epsilon_\alpha - \mu)n_\alpha]. \tag{27.29}$$

The limitation to allowed values of n_α is, of course, still in effect.

Eq. (27.29) can be greatly simplified by exchanging sums and products in much the same way as we exchanged integrals and products for classical statistical mechanics in Section 19.12. Exchanging the sum and product is the key step that leads to the factorization of the grand canonical partition function. This trick is just as important in quantum statistical mechanics as it is in classical statistical mechanics, and so deserves its own section.

27.7 Exchanging Sums and Products

In eq. (27.29) we found a sum of products of the form $\sum_{\{n_\alpha\}} \prod_\alpha \exp[-\beta(\epsilon_\alpha - \mu)n_\alpha]$, which can be transformed into a product of sums. This is another application of what we have called the best trick in statistical mechanics,

$$\sum_{\{n_\alpha\}} \prod_\alpha \exp[-\beta(\epsilon_\alpha - \mu)n_\alpha], = \sum_{n_{\alpha_1}} \sum_{n_{\alpha_2}} \cdots e^{-\beta(\alpha_1-\mu)n_{\alpha_1}} e^{-\beta(\alpha_2-\mu)n_{\alpha_2}} \cdots$$

$$= \sum_{n_{\alpha_1}} e^{-\beta(\alpha_1-\mu)n_{\alpha_1}} \sum_{n_{\alpha_2}} e^{-\beta(\alpha_2-\mu)n_{\alpha_2}} \cdots$$

$$= \prod_\alpha \sum_{n_\alpha} \exp[-\beta(\epsilon_\alpha - \mu)n_\alpha]. \tag{27.30}$$

As in Section 19.12, where we discussed the classical version of this trick, care must be taken with the indices of the sums on each side of the equality. The sum on the left side of eq. (27.30) is over the entire set $\{n_\alpha\}$ for the N-particle system, while the sum in the last line on the right is only over the values taken on by the single-particle quantum number n_α.

27.8   Grand Canonical Partition Function for Independent Particles

Inserting the equality in eq. (27.30) into eq. (27.29), we find the following expression for the grand canonical partition function

$$\mathcal{Z} = \prod_\alpha \sum_{n_\alpha} \exp[-\beta(\epsilon_\alpha - \mu)n_\alpha]. \tag{27.31}$$

The evaluation of the grand canonical partition function has been reduced to the product of sums over the number of particles occupying each of the states. Each of these sums will turn out to be quite easy to evaluate for ideal gases, making the transition to the grand canonical ensemble definitely worth the effort.

The logarithm of the grand canonical partition function gives us the Legendre transform of the energy with respect to temperature and chemical potential, just as it did in the case of classical statistical mechanics

$$\ln \mathcal{Z} = -\beta U[T, \mu] = \tilde{S}[\beta, (\beta\mu)] \tag{27.32}$$

$$U[T, \mu] = U - TS - \mu N. \tag{27.33}$$

Once the grand canonical potential has been calculated, all thermodynamic information can be obtained by the methods discussed in Part 2.

In particular, if the system is extensive, which is the case in most problems of interest, Euler's equation tells us that

$$\ln \mathcal{Z} = -\beta U[T, \mu] = \tilde{S}[\beta, (\beta\mu)] = \beta PV. \tag{27.34}$$

This provides a good method of calculating the pressure in a quantum gas.

Before we continue with the derivation of the basic equations for bosons and fermions, we will make a slight detour in the next two sections to discuss the properties of distinguishable quantum particles. We will return to identical particles in Section 27.11.

27.9 Distinguishable Quantum Particles

The most common systems of distinguishable particles are colloids, in which the particles are usually composed of many atoms. Because of the large mass of colloidal particles, it is rarely necessary to treat them quantum mechanically. On the other hand, it is instructive to see how the quantum grand canonical partition function for distinguishable particles differs from that of fermions or bosons.

The distinguishing feature of a many-particle wave function for distinguishable particles is that it is not generally symmetric or antisymmetric. This means that any wave function of the form given in eq. (27.15) is valid. This equation is repeated here for convenience, with the notation modified to denote the product over single-particle states as a product over values of α_j,

$$\psi_N = \prod_{j=1}^{N} \psi_{\alpha_j}(\vec{r}_j).$$ (27.35)

Since the particles are distinguishable and can move between the system and the reservoir, we must include a factor giving the number of ways we can have N particles in the system and the remaining N_R particles in the reservoir. Letting the total number of particles be $N_T = N + N_R$, the factor is

$$\frac{N_T!}{N!N_R!}.$$ (27.36)

This is exactly the same factor that we introduced in Chapter 4 for a classical ideal gas of distinguishable particles; the combinatorics are identical in the two cases.

We must also avoid double counting when two or more particles are in the same single-particle state. This leads to a factor of

$$\frac{N!}{\prod_\alpha n_\alpha!}.$$ (27.37)

If we include both of these factors, and note that the factor of $N_T!$ and $N_R!$ are not included in the entropy—or the grand canonical partition function—we find an expression for $\mathcal{Z}$ of the following form:

$$\mathcal{Z} = \prod_\alpha \sum_{n_\alpha=0}^{\infty} \frac{1}{n_\alpha!} \exp[-\beta(\epsilon_\alpha - \mu)n_\alpha] = \prod_\alpha \exp\left(\exp[-\beta(\epsilon_\alpha - \mu)]\right).$$ (27.38)

If we take the logarithm of the partition function, this expression simplifies,

$$\ln \mathcal{Z} = \sum_\alpha \exp[-\beta(\epsilon_\alpha - \mu)] = \beta PV. \tag{27.39}$$

The last equality is, of course, valid only in the case that the system is extensive.

27.10 Sneaky Derivation of $PV = Nk_BT$

The form of eq. (27.39) has a curious consequence. If we calculate the predicted value of N from the grand canonical partition function (see eq. (27.42)), we find

$$\langle N \rangle = k_B T \frac{\partial}{\partial \mu} \ln \mathcal{Z}$$

$$= k_B T \frac{\partial}{\partial \mu} \sum_\alpha \exp[-\beta(\epsilon_\alpha - \mu)]$$

$$= \sum_\alpha \exp[-\beta(\epsilon_\alpha - \mu)] = \ln \mathcal{Z} = \beta PV. \tag{27.40}$$

where the last equality comes from eq. (27.39), and is valid for extensive systems. Eq. (27.40) is, of course, equivalent to

$$PV = Nk_BT, \tag{27.41}$$

if the fluctuations of N are neglected.

This derivation of the ideal gas law depends only on distinguishability, extensivity, and the total wave function's being a product of the single-particle eigenstates. Extensivity is necessary to eliminate, for example, a system in a gravitational field.

The most interesting feature about this derivation is that it is fully quantum mechanical. The ideal gas law does *not* depend on taking the classical limit!

27.11 Equations for $U = \langle E \rangle$ and $\langle N \rangle$

In this section we return to calculations of the properties of bosons and fermions. Eq. (27.31) for the grand canonical partition function is, in principle, sufficient to perform such calculations. However, it turns out not to be the most efficient method for obtaining the properties of fermions and bosons. In this section we will derive the equations for the energy and number of particles, which will be the basis for the methods discussed in the rest of the chapter, as well as Chapters 28 and 29.

Beginning with the formal expression for the grand canonical partition function in eq. (27.31),

$$Z = \prod_\alpha \sum_{n_\alpha} \exp[-\beta(\epsilon_\alpha - \mu)n_\alpha], \tag{27.42}$$

we can easily find an expression for the average number of particles by taking a logarithmic derivative with respect to μ,

$$\frac{\partial}{\partial \mu} \ln Z = \sum_\alpha \frac{\partial}{\partial \mu} \ln \sum_{n_\alpha} \exp[-\beta(\epsilon_\alpha - \mu)n_\alpha] \tag{27.43}$$

$$= \sum_\alpha \left[\frac{\sum_{n_\alpha} \beta n_\alpha \exp[-\beta(\epsilon_\alpha - \mu)n_\alpha]}{\sum_{n_\alpha} \exp[-\beta(\epsilon_\alpha - \mu)n_\alpha]} \right]$$

$$= \beta \sum_\alpha \langle n_\alpha \rangle$$

$$= \beta \langle N \rangle.$$

The quantity $\langle n_\alpha \rangle$ clearly represents the average number of particles in the state α,

$$\langle n_\alpha \rangle = \frac{\sum_{n_\alpha} n_\alpha \exp[-\beta(\epsilon_\alpha - \mu)n_\alpha]}{\sum_{n_\alpha} \exp[-\beta(\epsilon_\alpha - \mu)n_\alpha]}. \tag{27.44}$$

The equation for the average number of particles as a derivative of the grand canonical partition function is, of course, related to the corresponding thermodynamic identity in terms of the Legendre transform of the energy with respect to T and μ,

$$\langle N \rangle = \frac{1}{\beta} \frac{\partial}{\partial \mu} \ln Z = \frac{1}{\beta} \frac{\partial}{\partial \mu}(-\beta U[T,\mu]) = \frac{\partial(PV)}{\partial \mu}. \tag{27.45}$$

The average number of particles is then

$$\langle N \rangle = \sum_\alpha \langle n_\alpha \rangle. \tag{27.46}$$

We will leave the proof of the corresponding equation for the energy to the reader.

$$U = \langle E \rangle = \sum_\alpha \epsilon_\alpha \langle n_\alpha \rangle. \tag{27.47}$$

Eqs. (27.46) and (27.47) turn out to be key to a relatively easy calculation of the properties of fermions and bosons. Since we are going to use them frequently in the next two chapters, we will need to evaluate $\langle n_\alpha \rangle$ for both fermions and bosons, which will be done in the next two sections.

27.12 $\langle n_\alpha \rangle$ for Bosons

For bosons, there are an infinite number of possibilities for the occupation of a single-particle state. This makes the evaluation of eq. (27.44) only slightly more difficult than the corresponding sum for fermions, since we can carry out the sum over all positive integers,

$$\sum_{n_\alpha=0}^{\infty} \exp[-\beta(\epsilon_\alpha - \mu)n_\alpha] = (1 - \exp[-\beta(\epsilon_\alpha - \mu)])^{-1} \tag{27.48}$$

The numerator of eq. (27.44) for bosons can be obtained by differentiating eq. (27.48) with respect to β and cancelling a factor of $(\epsilon_\alpha - \mu)$ on both sides,

$$\sum_{n_\alpha=0}^{\infty} n_\alpha \exp[-\beta(\epsilon_\alpha - \mu)n_\alpha] = \exp[-\beta(\epsilon_\alpha - \mu)](1 - \exp[-\beta(\epsilon_\alpha - \mu)])^{-2} \tag{27.49}$$

The average occupation number is then given by the ratio of these quantities, according to eq. (27.44),

$$\langle n_\alpha \rangle = \frac{\exp[-\beta(\epsilon_\alpha - \mu)](1 - \exp[-\beta(\epsilon_\alpha - \mu)])^{-2}}{(1 - \exp[-\beta(\epsilon_\alpha - \mu)])^{-1}} \tag{27.50}$$
$$= (\exp[\beta(\epsilon_\alpha - \mu)] - 1)^{-1}.$$

Note the strong similarity between the expression for the occupation number for bosons and the average number of excited particles in a simple harmonic oscillator, which we derived in eq. (24.76) in Section 24.10. They are both manifestations of the same mathematical structure, which we first encountered in the solution to the quantum simple harmonic oscillator in Section 24.10.

There is also an important property of eq. (27.50), which will play en essential role in determining the properties of a Bose gas in Chapter 28. The average number of particles, $\langle n_\alpha \rangle$, must be a positive number, but the last line of eq. (27.50) is negative unless $\epsilon_\alpha - \mu > 0$ for all energy levels ϵ_α. Therefore, the chemical potential of a Bose gas must have a value below the lowest energy level.

27.13 $\langle n_\alpha \rangle$ for Fermions

Since there are only two possibilities for the occupation of a single-particle state by fermions, the evaluation of eq. (27.44) is quite easy. The denominator and the numerator are calculated first,

$$\sum_{n_\alpha=0}^{1} \exp[-\beta(\epsilon_\alpha - \mu)n_\alpha] = 1 + \exp[-\beta(\epsilon_\alpha - \mu)] \tag{27.51}$$

$$\sum_{n_\epsilon=0}^{1} n_\alpha \exp[-\beta(\epsilon_\alpha - \mu)n_\alpha] = 0 + \exp[-\beta(\epsilon_\alpha - \mu)] \tag{27.52}$$

The average occupation number is then given by the ratio of these quantities, according to eq. (27.44),

$$\langle n_\alpha \rangle = \frac{\exp[-\beta(\epsilon_\alpha - \mu)]}{1 + \exp[-\beta(\epsilon_\alpha - \mu)]} = (\exp[\beta(\epsilon_\alpha - \mu)] + 1)^{-1}. \tag{27.53}$$

Note the strong similarity between the expression for the occupation number for fermions and the average number of excited particles in a two-level system, which we derived in eq. (24.57) in Section 24.9.

27.14   Summary of Equations for Fermions and Bosons

We can summarize the equations for $\langle n_\alpha \rangle$ by writing

$$\langle n_\alpha \rangle = (\exp[\beta(\epsilon_\alpha - \mu)] \pm 1)^{-1}, \tag{27.54}$$

where the upper (plus) sign refers to fermions and the lower (minus) sign refers to bosons.

Eqs. (27.46) and (27.47) can then be written compactly:

$$\langle N \rangle = \sum_\alpha (\exp[\beta(\epsilon_\alpha - \mu)] \pm 1)^{-1} \tag{27.55}$$

$$U = \sum_\alpha \epsilon_\alpha (\exp[\beta(\epsilon_\alpha - \mu)] \pm 1)^{-1} \tag{27.56}$$

For the rest of the discussion in this chapter, as well as the following chapters on fermions and bosons, we will write N for $\langle N \rangle$. While this is not really correct, and we should keep the distinction between N and $\langle N \rangle$ in mind, it is tiresome to constantly write the brackets.

We can express the grand canonical partition function for bosons by inserting eq. (27.48) in eq. (27.42),

$$\mathcal{Z} = \prod_\alpha (1 - \exp[-\beta(\epsilon_\alpha - \mu)])^{-1}. \tag{27.57}$$

We can express the grand canonical partition function for fermions by inserting eq. (27.51) in eq. (27.42),

$$\mathcal{Z} = \prod_{\alpha} (1 + \exp[-\beta(\epsilon_\alpha - \mu)]) . \tag{27.58}$$

Finally, we can express the logarithm of the grand canonical partition function in a compact form for both cases,

$$\ln \mathcal{Z} = \pm \sum_{\alpha} \ln (1 \pm \exp[-\beta(\epsilon_\alpha - \mu)]) = \beta PV. \tag{27.59}$$

The upper signs refer to fermions, and the lower signs to bosons. The last equality is valid only when the system is extensive, but that will be true of most of the cases we will consider.

If we use the density of states that we calculated in Section 27.2, we can write the logarithm of the grand canonical partition function in terms of an integral.

$$\ln \mathcal{Z} = \pm \int_0^\infty D(\epsilon) \ln (1 \pm \exp[-\beta(\epsilon - \mu)]) \, d\epsilon = \beta PV. \tag{27.60}$$

27.15   Integral Form of Equations for *N* and *U*

In Section 27.2 we calculated the density of single-particle states, $D(\epsilon)$, for non-interacting particles in a box, with the result given in eq. (27.13),

$$N = \int_0^\infty D(\epsilon) (\exp[\beta(\epsilon - \mu)] \pm 1)^{-1} \, d\epsilon \tag{27.61}$$

$$U = \int_0^\infty D(\epsilon) (\exp[\beta(\epsilon - \mu)] \pm 1)^{-1} \epsilon \, d\epsilon. \tag{27.62}$$

For more general systems, which we will discuss later in this chapter, the density of states, $D(\epsilon)$, will have a different structure, which might be considerably more complex. However, we will still be able to express eqs. (27.55) and (27.56) in integral form.

The integral form of the equations can even be used if the energy spectrum is partially discrete, if we represent the discrete part of the spectrum with a delta function of the form $X\delta(\epsilon - \epsilon_1)$, where the constant X represents the degeneracy of the energy level at ϵ_1.

One of the most interesting aspects of eqs. (27.61) and (27.62) is that they completely separate the quantum statistics (which enter as the factor $(\exp[\beta(\epsilon - \mu)] \pm 1)^{-1}$) and the effects of the Hamiltonian (which are reflected in the density of states, $D(\epsilon)$). We will spend considerable effort in understanding the properties of systems of non-interacting particles, which might seem to be very special cases. Nevertheless, the methods we develop will apply directly to much more general cases, as long as we are able to calculate the density of states, $D(\epsilon)$. We will even be able to obtain very general results in the following chapters for the consequences of the various kinds of structure in $D(\epsilon)$ that can arise.

27.16   Basic Strategy for Fermions and Bosons

The basic strategy for dealing with problems of fermions and bosons differs from that usually employed for other problems in quantum statistical mechanics, where the first task is almost always to evaluate the canonical partition function. Although it is possible to solve problems with fermions and bosons by evaluating the grand canonical partition function, it is almost always better to use eqs. (27.55) and (27.56) (or eqs. (27.61) and (27.62)) for the average number of particles and the average energy.

The first step in solving problems involving fermions and bosons is to use eq. (27.55) (or eq. (27.61) to obtain N as a function of T and μ, that is $N = N(T, \mu)$.

Next, note that we rarely know the value of the chemical potential, while we almost always do know the number of particles in the system. Furthermore, the number of particles is generally fixed during experiments, while the chemical potential is not. This suggests inverting the function $N = N(T, \mu)$ that we found from eq. (27.55) or eq. (27.61) to give us $\mu = \mu(T, N)$.

Finally, we use eq. (27.56) or eq. (27.62) to find the energy as a function of T and N (the latter through the function $\mu = \mu(T, N)$). This gives us two equations of state that will probably answer any questions in which we are interested. If we need a complete fundamental relation, we can—at least in principle—find it by integrating the equations of state, as discussed in Chapter 13. A fundamental relation can be found from eq. (27.59), which also provides a quick way of calculating the pressure for extensive systems.

The details of how these calculations can be carried out in practice will be discussed in Chapters 28 and 29.

27.17 $P = 2U/3V$

The equation in the title of this section is easy to derive for a three-dimensional, monatomic, classical, ideal gas or a quantum gas of distinguishable particles. We need only combine the ideal gas law,

$$PV = Nk_BT,$$ (27.63)

with the equation for the energy,

$$U = \frac{3}{2}Nk_BT,$$ (27.64)

to get

$$P = \frac{2}{3}\frac{U}{V}.$$

(27.65)

The surprising thing about eq. (27.65) is that it is also true for ideal Fermi and Bose gases, even though eqs. (27.63) and (27.64) are not.

The proof begins with the integral form of the equations for the grand canonical partition function from eq. (27.60),

$$\ln Z = \pm \int_0^\infty D(\epsilon)\ln(1 \pm \exp[-\beta(\epsilon - \mu)])\, d\epsilon = \beta PV,$$

(27.66)

and the average energy from eq. (27.62)

$$U = \int_0^\infty D(\epsilon)(\exp[\beta(\epsilon - \mu)] \pm 1)^{-1}\epsilon\, d\epsilon.$$

(27.67)

In both of these equations, the density of states is given by eq. (27.13),

$$D(\epsilon) = \frac{V}{4\pi^2}\left(\frac{2m}{\hbar^2}\right)^{3/2}\epsilon^{1/2} = X\epsilon^{1/2},$$

(27.68)

where we have introduced a constant X to simplify the algebra.

To derive eq. (27.65) we integrate the expression for $\ln Z$ by parts, so that

$$\ln Z = \pm X \int_0^\infty \epsilon^{1/2}\ln(1 \pm \exp[-\beta(\epsilon - \mu)])\, d\epsilon,$$

(27.69)

becomes

$$\ln Z = \pm X\left[\frac{2}{3}\epsilon^{3/2}\ln(1 \pm \exp[-\beta(\epsilon - \mu)])\right]_0^\infty$$

(27.70)

$$\mp X\int_0^\infty \frac{2}{3}\epsilon^{3/2}(1 \pm \exp[-\beta(\epsilon - \mu)])^{-1}(\mp\beta\exp[-\beta(\epsilon - \mu)])\, d\epsilon.$$

The first term on the right in eq. (27.70) vanishes exactly. Comparing eq. (27.70) with eq. (27.67) leaves us with

$$\ln Z = \frac{2X\beta}{3}\int_0^\infty \epsilon^{3/2}(\exp[\beta(\epsilon - \mu)] \pm 1)^{-1}\, d\epsilon = \frac{2\beta}{3}U.$$

(27.71)

Since $\ln Z = \beta PV$ for an extensive system, this gives us eq. (27.65).

It is rather remarkable that eq. (27.65) holds for any form of statistics in either classical or quantum mechanics. The factor 2/3 does depend on the system being three-dimensional, but that proof will be left to the reader.

27.18 Problems

....

PROBLEM 27.1
Identities in the grand canonical ensemble

We have seen that derivatives of partition functions can be used to express various quantities of interest. Derive identities for the following in terms of derivatives of the quantum-mechanical, grand canonical partition function.

1. The average number of particles in the system, $\langle N \rangle$, in terms of a derivative with respect to μ.
2. The average number of particles in the system, $\langle N \rangle$, in terms of a derivative with respect to the fugacity, $\lambda = \exp(\beta\mu)$.
3. The average energy as a function of β and μ in terms of derivatives.

PROBLEM 27.2
More on the grand canonical ensemble

We have previously derived a general expression for the grand canonical partition function for a system of independent identical particles

$$\mathscr{Z} = \prod_\epsilon \sum_{n_\epsilon} \exp\left[-\beta(\epsilon - \mu)n_\epsilon\right]$$

where the product over ϵ is actually the product over the single-particle eigenstates. Each eigenstate must be counted separately, even when several eigenstates have the same energy (also denoted by ϵ).

The sum over n_ϵ is only over *allowed* values of n_ϵ.

For bosons, an arbitrary number of particles can be in the same state, so n_ϵ can take on any non-negative integer value, $n_\epsilon = \{0, 1, 2, \ldots\}$.

For fermions, no more than one particle can be in a given state, so n_ϵ can only take on the values 0 and 1, $n_\epsilon = \{0, 1\}$.

Do each of the following for both fermions and bosons:

1. Carry out the sum in the grand canonical partition function explicitly.
2. Using the result from the previous question, calculate the average number of particles $\langle N \rangle$ in terms of β and μ.
3. Calculate the average number of particles in an eigenstate with energy ϵ, denoted by $\langle n_\epsilon \rangle$.

368   *Ideal Quantum Gases*

4. Express the average number of particles $\langle N \rangle$ in terms of a sum over $\langle n_\epsilon \rangle$.
5. Calculate the average energy in terms of a sum over $\langle n_\epsilon \rangle$.

PROBLEM 27.3
Fermion and boson fluctuations

Prove that the fluctuations about the average number of particles in an eigenstate, $\langle n \rangle$, is given by

$$\delta^2 \langle n \rangle = \langle n \rangle \left(1 \pm \langle n \rangle \right)$$

where one of the signs is for fermions and the other is for bosons.

28

Bose–Einstein Statistics

> *The world is full of magical things patiently waiting for our wits to grow sharper.*
>
> Bertrand Russell (1872–1970), philosopher and mathematician

Perhaps the most startling property of systems governed by Bose–Einstein statistics is that they can exhibit a phase transition in the absence of interactions between the particles. In this chapter we will explore the behavior of an ideal gas of bosons and show how this unusual phase transition arises.

28.1 Basic Equations for Bosons

We begin with the Bose–Einstein equations for N and U, taken from eqs. (27.55) and (27.56),

$$N = \sum_{\alpha} \langle n_{\alpha} \rangle = \sum_{\alpha} (\exp[\beta(\epsilon_{\alpha} - \mu)] - 1)^{-1} \tag{28.1}$$

$$U = \sum_{\alpha} \epsilon_{\alpha} \langle n_{\alpha} \rangle = \sum_{\alpha} \epsilon_{\alpha} (\exp[\beta(\epsilon_{\alpha} - \mu)] - 1)^{-1}. \tag{28.2}$$

We are interested in understanding experiments in which the chemical potential μ is unknown and the number of particles N is fixed. On the other hand, as discussed in Chapter 27, eqs. (28.1) and (28.2) were derived under the assumption that μ is known. This leads us to begin calculations by finding $N = N(T, \mu)$ from eq. (28.1), and then inverting the equation to obtain the chemical potential, $\mu = \mu(T, N)$. After we have found μ, we can use eq. (28.2) to calculate the energy and the specific heat for a fixed number of particles.

28.2 $\langle n_{\alpha} \rangle$ for Bosons

The occupation number for bosons was derived in Chapter 27 and given in eq. (27.50),

$$f_{BE}(\epsilon_{\alpha}) = \langle n_{\alpha} \rangle = (\exp[\beta(\epsilon_{\alpha} - \mu)] - 1)^{-1}. \tag{28.3}$$

An Introduction to Statistical Mechanics and Thermodynamics. Robert H. Swendsen, Oxford University Press (2020).
© Robert H. Swendsen. DOI: 10.1093/oso/9780198853237.001.0001

Although the expressions for the occupation numbers for bosons and fermions only differ by a plus or minus sign in the denominator, the consequences of this small difference are enormous. The first consequence of the minus sign in the denominator of eq. (28.3) for bosons is a limitation on the allowed values of the chemical potential. Since $f_{BE}(\epsilon_\alpha)$ gives the average number of particles in a single-particle state with energy ϵ_α, it cannot be negative,

$$(\exp[\beta(\epsilon_\alpha - \mu)] - 1)^{-1} > 0. \tag{28.4}$$

This immediately requires that

$$\epsilon_\alpha > \mu, \tag{28.5}$$

for *all* values of ϵ_α. In particular, if $\epsilon_\alpha = 0$ is the lowest energy of a single-particle state, then

$$\mu < 0. \tag{28.6}$$

The chemical potential of a gas of bosons must be negative.

Eq. (28.6) is *almost* a general result. It assumes that the lowest energy level is zero, which might not always be true. Because this can be a trap for the unwary, it is very important to know *why* μ is algebraically less than the energy of the lowest single-particle state, instead of just remembering that it is less than zero.

28.3   The Ideal Bose Gas

For systems with a continuous energy spectrum, we expect to be able to change the sums in eqs. (28.1) and (28.2) into integrals and use eqs. (27.61) and (27.62) to calculate the properties of the system. For bosons, these equations take the following form:

$$N = \int_0^\infty D_{BE}(\epsilon)\,(\exp[\beta(\epsilon - \mu)] - 1)^{-1}\,d\epsilon \tag{28.7}$$

$$U = \int_0^\infty D_{BE}(\epsilon)\,\epsilon\,(\exp[\beta(\epsilon - \mu)] - 1)^{-1}\,d\epsilon. \tag{28.8}$$

A particularly important case, which we will discuss in detail, is an ideal gas of bosons. We have already derived the single-particle density of states for particles in a box in eq. (27.13),

$$D_{BE}(\epsilon) = \frac{V}{4\pi^2} \left(\frac{2m}{\hbar^2}\right)^{3/2} \epsilon^{1/2}. \tag{28.9}$$

Eq. (28.9) completes the equations we need to investigate the properties of free bosons. We will begin to calculate those properties by looking at the low-temperature behavior of the chemical potential in the next section.

28.4   Low-Temperature Behavior of μ

We can simplify the problem of finding the chemical potential from eq. (28.7) by making the integral dimensionless. To do this, we introduce a dimensionless variable $x = \beta\epsilon$,

$$N = \int_0^\infty \frac{V}{4\pi^2} \left(\frac{2m}{\hbar^2}\right)^{3/2} \epsilon^{1/2} \left(\exp[\beta(\epsilon - \mu)] - 1\right)^{-1} d\epsilon \tag{28.10}$$

$$= \frac{V}{4\pi^2} \left(\frac{2m}{\hbar^2}\right)^{3/2} (k_B T)^{3/2} \int_0^\infty x^{1/2} \left(\exp(-\beta\mu)\exp(x) - 1\right)^{-1} dx.$$

It is convenient at this point to introduce the fugacity

$$\lambda = \exp(\beta\mu). \tag{28.11}$$

Note that because $\mu < 0$, the fugacity must be less than 1, and its reciprocal, which enters into eq. (28.10), must be greater than 1,

$$\lambda^{-1} > 1 \tag{28.12}$$

In terms of the fugacity, the equation for N becomes

$$N = \frac{V}{4\pi^2} \left(\frac{2m}{\hbar^2}\right)^{3/2} (k_B T)^{3/2} \int_0^\infty x^{1/2} \left(\lambda^{-1}\exp(x) - 1\right)^{-1} dx. \tag{28.13}$$

The form of eq. (28.13) is extremely important in understanding the properties of bosons. As the temperature is decreased, the factor of $T^{3/2}$ in front of the integral also decreases. Since the total number of particles is fixed, the dimensionless integral in eq. (28.13) must increase. The only parameter in the integral is the inverse fugacity, so it must vary in such a way as to increase the value of the integral. Since the inverse fugacity is in the denominator of the integrand, the only way to increase the value of the integral is to decrease the value of the inverse fugacity. However, because of eq. (28.12) we cannot decrease the value of the inverse fugacity below $\lambda^{-1} = 1$. If the value of the dimensionless integral in eq. (28.13) diverged as $\lambda^{-1} \to 1$, there would be no problem.

However, although the *integrand* diverges at $x = 0$ when $\lambda = 1$, the value of the integral is finite:

$$\int_0^\infty x^{1/2} (\exp(x) - 1)^{-1} \, dx = \frac{\sqrt{\pi}}{2} \zeta\left(\frac{3}{2}\right) = 1.306\sqrt{\pi} = 2.315. \tag{28.14}$$

It is traditional to express the value of the integral in eq. (28.14) in terms of a ζ-function instead of simply 2.315. This is largely because the theory of bosons was developed before computers were available. At that time, instead of simply evaluating integrals numerically, great efforts were made to express them in terms of 'known' functions; that is, functions whose values had already been tabulated. Although retaining the ζ-function is a bit of an anachronism, I have included it and the factor of $\sqrt{\pi}$ in eq. (28.14) in the following equations for the sake of tradition (and comparison with other textbooks).

The consequence of the dimensionless integral in eq. (28.14) having an upper bound is that the equation cannot be correct below a temperature T_E that is determined by setting the integral equal to its maximum value,

$$N = \frac{V}{4\pi^2} \left(\frac{2m}{\hbar^2}\right)^{3/2} (k_B T_E)^{3/2} 1.306\sqrt{\pi}. \tag{28.15}$$

This equation can also be written as

$$N = 2.315 \frac{V}{4\pi^2} \left(\frac{2m}{\hbar^2}\right)^{3/2} (k_B T_E)^{3/2}. \tag{28.16}$$

The temperature T_E is known as the Einstein temperature,

$$k_B T_E = \left(\frac{2\pi\hbar^2}{m}\right) \left(\frac{N}{2.612\,V}\right)^{2/3}. \tag{28.17}$$

For fixed N and temperatures less than T_E, eq. (28.13) has no solution, even though you can certainly continue to cool a gas of bosons below T_E. This apparent contradiction has an unusual origin. Even though the energy levels are very closely spaced, so that changing from a sum in eq. (28.1) to the integral in eq. (28.7) would not normally be expected to cause difficulties, eq. (28.7) is not valid for $T < T_E$.

In the next section we will show how to remove the contradiction by modifying eq. (28.7) to treat the lowest energy state explicitly.

28.5 Bose–Einstein Condensation

We can trace the strange behavior of bosons at low temperatures to the form of the boson occupation number in eq. (28.3),

$$f_{BE}(\epsilon) = \langle n_\epsilon \rangle = (\exp[\beta(\epsilon - \mu)] - 1)^{-1}. \tag{28.18}$$

The difficulty arises from the fact that the occupation of the lowest energy level, $\epsilon = 0$, has an occupation number

$$f_{BE}(0) = \langle n_0 \rangle = N_0 = (\exp[-\beta\mu] - 1)^{-1}. \tag{28.19}$$

If we were to set $\mu = 0$, the occupation number would be infinite. This means that for very small but non-zero values of μ, the occupation number of the $\epsilon = 0$ state can be arbitrarily large. In fact, it can even contain all the bosons in the system!

As will be justified in Section 28.8, the only modification we need to make in eq. (28.7) is to include the number of particles in the $\epsilon = 0$ state, which we denote as N_0,

$$N = N_0 + \int_0^\infty D_{BE}(\epsilon) \, (\exp[\beta(\epsilon - \mu)] - 1)^{-1} \, d\epsilon. \tag{28.20}$$

For free bosons this becomes

$$N = N_0 + \frac{V}{4\pi^2} \left(\frac{2m}{\hbar^2}\right)^{3/2} (k_B T)^{3/2} \int_0^\infty x^{1/2} \left(\lambda^{-1} \exp(x) - 1\right)^{-1} dx. \tag{28.21}$$

Above the temperature T_E, the occupation N_0 of the $\epsilon = 0$ state is much smaller than the total number N of particles in the system, and may be neglected. Below T_E, the value of N_0 is comparable to N. That is, a significant fraction of the particles in the system are in the single-particle state with the lowest energy. The transition at T_E is called the Bose–Einstein condensation, because a significant fraction of the particles 'condense' into the $\epsilon = 0$ state at temperatures below T_E.

28.6 Below the Einstein Temperature

The behavior of the Bose gas below T_E turns out to be remarkably simple and easy to calculate. If we compare eq. (28.21) with eq. (28.15), which determines the Einstein temperature, we see that we can rewrite eq. (28.21) as

$$N = N_0 + N\left(\frac{T}{T_E}\right)^{3/2}. \tag{28.22}$$

We can then solve this equation for the occupation of the $\epsilon = 0$ state below T_E,

$$N_0 = N\left[1 - \left(\frac{T}{T_E}\right)^{3/2}\right]. \tag{28.23}$$

Eq. (28.23) shows that $N_0 \to N$ as $T \to 0$; for zero temperature all particles are in the lowest single-particle energy state.

Eq. (28.23) also shows that as $T \to T_E$, $N_0 \to 0$. This result is, of course, only an approximation. What it means is that above T_E, $N_0 \ll N$, so that it can be safely ignored.

Note that as $T \to T_E$ from eq. (28.24), N_0 goes linearly to zero:

$$\frac{N_0}{N} = \frac{3}{2}\left(\frac{T_E - T}{T_E}\right) + \cdots. \tag{28.24}$$

If we combine eq. (28.23) with eq. (28.19), we can find the temperature dependence of the chemical potential (taking the lowest single-particle energy state to be zero)

$$N_0 = [\exp(-\beta\mu) - 1]^{-1} = N\left[1 - \left(\frac{T}{T_E}\right)^{3/2}\right]. \tag{28.25}$$

Since we know that $\beta\mu$ is very small below T_E, we can expand $\exp(-\beta\mu)$ in eq. (28.25) to find a very good approximation for μ,

$$\mu \approx -\frac{k_B T}{N}\left[1 - \left(\frac{T}{T_E}\right)^{3/2}\right]^{-1}. \tag{28.26}$$

Because of the factor of $1/N$ in eq. (28.26), the chemical potential μ is extremely small below the Einstein temperature T_E; if N is of the order of Avogadro's number, $|\mu|$ is smaller than $10^{-23} k_B T$.

The approximation in eq. (28.26) also shows that as $T \to T_E$, the value of μ falls sharply. If T is close enough to the Einstein temperature, the approximation breaks down. However, this only happens when T is extremely close to T_E. For all practical purposes, eq. (28.26) can be used for all temperatures below T_E.

Now that we have found the chemical potential for an ideal gas of bosons, we are in a position to calculate the energy and the specific heat, which we will do in the next section.

28.7 Energy of an Ideal Gas of Bosons

Given the chemical potential, we can calculate the energy of a gas of bosons from eq. (28.8),

$$U = U_0 + \int_0^\infty D_{BE}(\epsilon) \, \epsilon \, (\exp[\beta(\epsilon - \mu)] - 1)^{-1} \epsilon \, d\epsilon. \tag{28.27}$$

In this equation,

$$U_0 = 0 \tag{28.28}$$

is the energy of a particle in the lowest energy level is $\epsilon = 0$.

Introducing the dimensionless variable $x = \beta \epsilon$, as we did in eq. (28.13), we can write an equation for U in terms of a dimensionless integral

$$U = \frac{V}{4\pi^2} \left(\frac{2m}{\hbar^2} \right)^{3/2} (k_B T)^{5/2} \int_0^\infty x^{3/2} \left(\lambda^{-1} \exp(x) - 1 \right)^{-1} dx. \tag{28.29}$$

Below T_E, eq. (28.29) simplifies because $\lambda^{-1} = 1$. The dimensionless integral can be evaluated numerically,

$$\int_0^\infty x^{3/2} (\exp(x) - 1)^{-1} dx = \zeta \left(\frac{5}{2} \right) \Gamma \left(\frac{5}{2} \right) = 1.341 \left(\frac{3}{4} \right) \pi^{1/2} = 1.7826. \tag{28.30}$$

In this equation, $\Gamma(\cdot)$ indicates the gamma function and $\zeta(\cdot)$ again indicates the zeta function.

Using eq. (28.30), eq. (28.29) becomes

$$U = 1.7826 \frac{V}{4\pi^2} \left(\frac{2m}{\hbar^2} \right)^{3/2} (k_B T)^{5/2}. \tag{28.31}$$

Since we know from eq. (28.15) that

$$N = 2.315 \frac{V}{4\pi^2} \left(\frac{2m}{\hbar^2} \right)^{3/2} (k_B T_E)^{3/2}, \tag{28.32}$$

we find that the energy per particle has a simple form,

$$\frac{U}{N} = \left(\frac{1.7826}{2.315} \right) \left(\frac{T}{T_E} \right)^{3/2} k_B T = .7700 \left(\frac{T}{T_E} \right)^{3/2} k_B T. \tag{28.33}$$

The specific heat at constant volume is found by differentiating with respect to temperature,

$$c_V = 1.925\, k_B \left(\frac{T}{T_E}\right)^{3/2}. \tag{28.34}$$

Notice that for $T = T_E$, the specific heat takes on the value of $1.925\, k_B$, which is greater than the classical value of $1.5\, k_B$ that it takes on in the limit of high temperatures.

For temperatures above T_E, the specific heat of an ideal boson gas decreases monotonically to its asymptotic value of $1.5\, k_B$. The derivation involves expanding the integral in eq. (28.8) as a function of $\beta\mu$. Since the expansion involves interesting mathematics, but little physics, I will refer the readers to the many textbooks that give the details. The fastest and easiest way to obtain the function today is to carry out the integral in eq. (28.8) numerically, which will be done in Section 28.11.

28.8   What About the Second-Lowest Energy State?

Since we have seen that the occupation of the lowest energy level has such a dramatic effect on the properties of a gas of bosons, it might be expected that the second-lowest energy state would also play a significant role. Oddly enough, it does not, even though the second-lowest energy state lies only slightly higher than the lowest energy state.

The energy levels are given by eq. (27.7). Since the wave function in eq. (27.3) vanishes if n_x, n_y, or n_z is zero, the lowest energy level, ϵ_o, corresponds to $\vec{n} = (1,1,1)$. This gives us

$$\epsilon_0 = \epsilon_{(1,1,1)} = \frac{\hbar^2 \pi^2}{2mL^2}\left(1^2 + 1^2 + 1^2\right) = 3\frac{\hbar^2 \pi^2}{2mL^2}. \tag{28.35}$$

The next lowest energy level, ϵ_1, corresponds to $\vec{n} = (1,1,2)$, or $(1,2,1)$, or $(2,1,1)$ These three states each have the energy

$$\epsilon_1 = \epsilon_{(1,1,2)} = \frac{\hbar^2 \pi^2}{2mL^2}\left(1^2 + 1^2 + 2^2\right) = 6\frac{\hbar^2 \pi^2}{2mL^2}. \tag{28.36}$$

The difference in energy between the two states is

$$\Delta\epsilon = \epsilon_1 - \epsilon_0 = 3\frac{\hbar^2 \pi^2}{2mL^2}. \tag{28.37}$$

The energy difference found in eq. (28.37) is extremely small. Suppose we consider a system consisting of ^4He in a cubic container with $L = 1\, cm$. The energy difference is only about $2.5 \times 10^{-26}\, erg$. Expressing this as a temperature, we have

$$\frac{\Delta \epsilon}{k_B} \approx 1.8 \times 10^{-14} \, K. \tag{28.38}$$

Despite the fact that this energy difference is so small, it has a large effect on the occupation number of the state because it is substantially larger than the difference between the chemical potential and the lowest single-particle energy level.

Define $\tilde{\mu} = \mu - \epsilon_0$ as the difference between the chemical potential and the lowest single-particle energy level. This will allow us to use eq. (28.26) with only the substitution $\mu \to \tilde{\mu}$.

From eq. (28.26) we see that $\beta\tilde{\mu} \approx -1/N \approx -10^{-23}$, so that the occupation number, $\langle n_1 \rangle$, of the second-lowest energy levels is

$$\langle n_1 \rangle = (\exp(\beta(\Delta\epsilon - \tilde{\mu})) - 1)^{-1} \tag{28.39}$$

$$\approx (\exp(\beta\Delta\epsilon) - 1)^{-1}$$

$$\approx \frac{1}{\beta\Delta\epsilon}$$

$$\approx \frac{T}{1.8 \times 10^{-14} \, K},$$

$$\approx (5.5 \times 10^{13} \, K^{-1})T,$$

where the last two lines come from eq. (28.38). The occupation of the second lowest states is of the order of 10^{12} or more, but this is still very small in comparison to the total number of particles, $N \approx 10^{23}$, or the number of particles in the lowest energy state below T_E. Because the numbers of particles in each of the states above the lowest-energy state are so much smaller than the number in the lowest-energy state, it is completely sufficient to use the integral in eq. (28.20) to calculate their total contribution.

28.9 The Pressure below $T < T_E$

Eq. (27.65) has an interesting consequence for the pressure of a boson gas below the Bose–Einstein condensation. If we insert eq. (27.65) into eq. (28.31), we obtain

$$P = \frac{2}{3}\frac{U}{V} = 0.2971 \frac{1}{\pi^2} \left(\frac{2m}{\hbar^2}\right)^{3/2} (k_B T)^{5/2}. \tag{28.40}$$

This means that the pressure along any isotherm for a gas of bosons below T_E is constant. It depends on the mass of the particles and the temperature, but nothing else. Another way of expressing this is to say that the isothermal compressibility of an ideal Bose gas is infinite below the Bose–Einstein transition.

28.10   Transition Line in *P-V* Plot

We can combine eq. (28.17) with eq. (28.40) to find the Bose–Einstein transition line in a *P-V* plot,

$$P = 0.9606\,\hbar^2 m\pi^{1/2} \left(\frac{N}{V}\right)^{5/3}. \tag{28.41}$$

As eq. (28.41) shows, the pressure is proportional to the (5/3)-power of the number density along the transition line.

28.11   A Numerical Approach to Bose–Einstein Statistics

This concludes the derivation of the integral equations describing the Bose–Einstein transition. The treatment is correct, but not entirely satisfying. The derivation jumps from the original equations involving discrete sums, eqs. (28.1) and (28.2), to new equations with the sums replaced by integrals, eqs. (28.7) and (28.8), then back to the sums to extract to lowest energy state, then back to the integrals for all higher energy states.

As an alternative, a completely numerical approach will be presented in the first five problems at the end of the chapter. The approach only involves the sums in eqs. (28.1) and (28.2). It was developed by a student, Tyson Price, and has appeared in the literature.[1] The alternative approach does not treat the lowest energy state differently than the rest. The large number of particles in the lowest energy state is simply a result of the numerical solution. Temperatures above and below the Einstein temperature are handled in the same way.

The inversion of the equation for the number of particles, $N = N(\mu, T)$, must also be performed numerically, because a solution for $\mu = \mu(N, T)$ is not available. This is the subject of Problem 28.2. A plot of μ as a function of temperature is shown in Fig. 28.1.

Problem 28.4 guides you through the calculation of the specific heat as a function of temperature. Fig. 28.2 shows a plot of the result for $N = 10000$, compared to a plot in the same figure of the analytic result of the low-temperature specific heat from eq. (28.34). The specific heat above T_E is considerably more difficult to obtain with an analytic approximation, but numerically it is easy. The plot for $N = 10000$ shows a smooth maximum, typical of a finite system. The height of the maximum is greater than

---

[1] T. Price, R. H. Swendsen, 'Numerical computation for teaching quantum statistics,' *Am. J. Phys.*, 81: 866–72 (2013). Earlier work by other authors treated bosons confined by harmonic traps: S. Grossmann and M. Holthaus, 'On Bose–Einstein condensation in harmonic traps,' *Phys. Lett.* A 208, 188–92 (1995). M. Ligare, 'Numerical analysis of Bose–Einstein condensation in a three-dimensional harmonic oscillator potential,' *Am. J. Phys.*, 66: 185–90 (1998). M. Ligare, 'Comment on "Numerical analysis of Bose–Einstein condensation in a three-dimensional harmonic oscillator potential"—An extension to anisotropic traps and lower dimensions,' *Am. J. Phys.*, 70: 76–8 (2002).

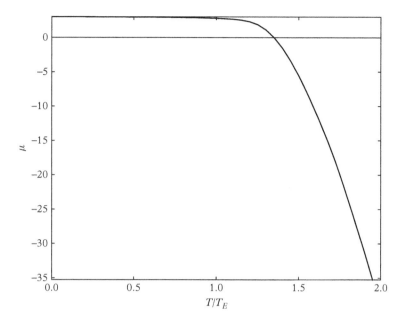

Fig. 28.1 *The chemical potential, μ, in units of $\hbar^2\pi^2/2mL^2 = \epsilon_0/3$ is plotted as a function of T/T_E for $N = 1000$ from the numerical calculation described in the text.*

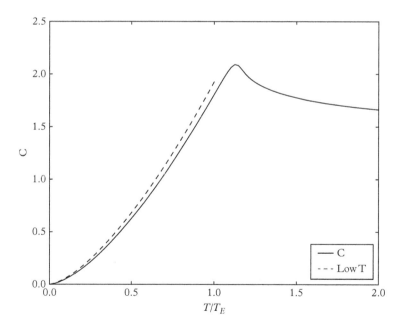

Fig. 28.2 *The specific heat in units of k_B as a function of T/T_E for $N = 10000$ from the numerical calculation described in the text. The analytic low-temperature result is shown as a dashed line for comparison.*

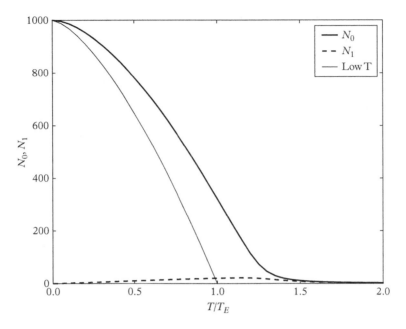

Fig. 28.3 *The occupation numbers, N_0 and N_1, are plotted as a function of T/T_E for $N = 1000$ from the numerical calculation described in the text. The analytic low-temperature result, Eq. (28.19) for N_0 is shown for comparison.*

the specific heat at $T = T_E$, which is a finite-size effect. The height of the maximum is greatest for about $N = 1000$, and decreases for larger systems.

Problem 29.5 has you calculate the occupation of the two lowest single-particle states, and compare them to the asymptotic curves. A plot is shown in Fig. 28.3.

28.12 Problems

PROBLEM 28.1
Boson Gas Problem: Numerical computation of $N = N(\mu, T)$

Write a program to compute the average number of particles as a function of the chemical potential μ and the temperature T by direct summation without using the integral approximation.

Consider an ideal quantum gas of N particles, each with mass m, in a cubic box with sides of length L. The N-particle Hamiltonian is

$$H = \sum_{j=1}^{N} \frac{|\vec{p}_j|^2}{2m}.$$

The single-particle energy is

$$\epsilon(\vec{n}) = An^2,$$

where $\vec{n} = (n_x, n_y, n_z)$ and $n^2 = n_x^2 + n_y^2 + n_z^2$. The value of the constant A is

$$A = \frac{\hbar^2}{2m}\frac{\pi^2}{L^2},$$

but we will use dimensionless variables, with $A = 1$ and $k_B = 1$. [Your program should still contain the constants A and k_B in case we want to use values with dimensions in the future.]

The occupation number of a single-particle state with energy ϵ

$$\langle n_\epsilon \rangle = (\exp(\beta(\epsilon - \mu)) + \sigma)^{-1},$$

where for

Bose − −Einstein	$\sigma = -1$
Fermi − −Dirac	$\sigma = +1$
Maxwell − −Boltzmann	$\sigma = 0$.

Your program should contain σ as a parameter, so that you can treat any kind of statistics.

The basic equation for the average number of particles $\langle N \rangle$ (which can be written as N for simplicity) as a function of μ and T is

$$N(\mu, T) = \sum_{\vec{n}} \langle n_{\epsilon_{\vec{n}}} \rangle = \sum_{\vec{n}} (\exp(\beta(\epsilon(\vec{n}) - \mu)) + \sigma)^{-1}.$$

You will not need the equation for U for this assignment.

1. Since the sum in the equation for $N(\mu, T)$ is over an infinite number of terms, it must be truncated to a finite number of terms for the program. Find a criterion for truncating the sum such that the neglected terms are each smaller than some δ. In practice, you will find that you can choose $\delta = 10^{-8}$, or even smaller, without making the run time program too long.
2. Write a function using Python (or another programming language of your choice) to carry out the truncated sum in the equation for $N(\mu, T)$.
3. Write a loop using the function you programed in answer to the previous question to compute $N(\mu, T)$ for uniformly spaced values of μ between arbitrary values μ_1 and μ_2.
4. Include a plot of N vs. μ in your program for the values you've computed. For this assignment, plot individual points.
5. For the special case of bosons ($\sigma = -1$) and $T = 1$, run your program for 37 values of μ between -1.0 and 7.0 (in units of $A = 1$).
6. Are there any strange features of the results? If there are, can you explain them?
7. Why did I specify 37 values of μ instead of 36 in the previous problem?

PROBLEM 28.2
Boson Gas Problem: Numerical computation of $\mu = \mu(T,N)$

In the previous problem, you wrote a function to compute the average number of bosons, $N = \langle N \rangle$, given the temperature T and the chemical potential μ. In this assignment, you will invert the function $N = N(\mu, T)$ to compute $\mu = \mu(T,N)$.

We will only be considering bosons for this assignment, but write your program using the more general occupation number of a single-particle state with energy ϵ

$$\langle n_\epsilon \rangle = (\exp(\beta(\epsilon - \mu)) + \sigma)^{-1}.$$

To invert the equation $N = N(\mu, T)$ to find $\mu = \mu(N, T)$, we need the fact that N is a monotonically increasing function of μ, as found in the previous assignment.

The basic algorithm begins with a choice of interval $[\mu_-, \mu_+]$ large enough to contain the desired value of $\mu(T,N)$. Fortunately, the choice is not very delicate. We then follow an iterative procedure until we find a value of μ that gives the correct number of particles. The suggested criterion is that $|\mu_+ - mu_-| < \delta\mu$, where $\delta\mu \approx 10^{-4}$.

1. Calculate $\mu_{trial} = (\mu_- + \mu_+)/2$.
2. Calculate $N_{trial} = N(\mu_{trial}, T)$.
3. If $N_{trial} > N$ set $\mu_+ = \mu_{trial}$
   If $N_{trial} < N$ set $\mu_- = \mu_{trial}$
4. Check to see if $\mu_+ - \mu_- < \delta\mu$. If not, return to step 1.

QUESTIONS:

1. Write a VPython function to compute $\mu = \mu(T,N)$.
2. Write a loop using the function you programed in answer to the previous question to compute $\mu(T,N)$ for uniformly spaced values of T between arbitrary values of T_- and T_+.
3. Calculate the Einstein temperature as a function of the parameter A.
4. For $N = 5$, 50, and 500, use your program to compute $\mu(T)$ for uniformly spaced values of T between $T = 0$ and $T = 2T_E$, where T_E is estimated from the expression you found in answer to the previous question.

PROBLEM 28.3
Quantum Gas Problem: A thermodynamic identity

A quantity that we wish to compute and plot is the heat capacity at constant N and V

$$C_{V,N} = T\left(\frac{\partial S}{\partial T}\right)_{V,N} = \left(\frac{\partial U}{\partial T}\right)_{V,N}.$$

Keeping V constant is not a problem, we will ignore the volume dependence to simplify the notation, writing C_N when we really mean $C_{V,N}$.

The difficulty is that the natural quantity to calculate is

$$\left(\frac{\partial U}{\partial T}\right)_\mu.$$

QUESTIONS:

1. Prove the following thermodynamic identity.

$$C_N = \left(\frac{\partial U}{\partial T}\right)_\mu - \left(\frac{\partial U}{\partial \mu}\right)_T \left(\frac{\partial N}{\partial T}\right)_\mu \bigg/ \left(\frac{\partial N}{\partial \mu}\right)_T$$

2. Find explicit expressions for all quantities on the right side of Eq. (1) in terms of sums over $\vec{n}$.

PROBLEM 28.4
Boson Gas Problem: The heat capacity of an ideal gas of bosons

Using the thermodynamic identity you derived in the previous problem, write a program to compute the specific heat at constant particle number as a function of temperature,

$$C_{V,N} = \left(\frac{\partial U}{\partial T}\right)_{V,N}.$$

Include the integral approximation to $C_N(T)$ in your plot for $T < T_E$. You can find the appropriate formula in the textbook.

As previously, you will find it helpful to write the computation of C_N as a function in Python. In fact, much of the code for the function can be copied from earlier programs.

Plot your results for T between zero and $4T_E$, for $N = 5$, $N = 50$, and $N = 500$.

PROBLEM 28.5
Boson Gas Problem: Comparisons with the large-N results

We have derived closed form equations for some properties of the Bose–Einstein ideal gas in the large-N limit. In particular, we found that below T_c, the occupation of the ground state is

$$N_0 = N\left(1 - \left(\frac{T}{T_E}\right)^{3/2}\right)$$

and the specific heat is

$$c_{V,N} = 1.925 k_B \left(\frac{T}{T_E}\right)^{3/2}.$$

1. Modify a copy of your program to compute $\mu(T,N)$ to make a plot of the ground state occupation N_0 as a function of T. In this new plot, include a plot of eq. (28.12) for comparison.

Make plots for at least two different numbers of particles to see how eq. (28.12) is approached as N is increased.
2. Modify a copy of your program to compute $c_N(T)$ to include a plot of eq. (28.12) for comparison.

Make plots for at least two different numbers of particles to see how eq. (28.12) is approached as N is increased.

The temperature range in the new plots should again be from $T = 0$ to $T = 2T_E$, but the functions in eq. (28.12) and (28.12) should each only extend from $T = 0$ to $T = T_E$.

PROBLEM 28.6
Bosons in two dimensions

Consider an ideal boson gas in two dimensions. The N particles in the gas each have mass m and are confined to a box of dimensions $L \times L$.

1. Calculate the density of states for the two-dimensional, ideal Bose gas.
2. Calculate the Einstein temperature in two dimensions in terms of the given parameters and a dimensionless integral. You *do* have to evaluate the dimensionless integral.

PROBLEM 28.7
Bosons in two dimensions with a bound state

Consider a simple model of bosons in a system with a single bound state. To simplify the math, we will consider the problem in two-dimensions.

If the boson is not in the bound state, it is a free particle in a two-dimensional box of length L. The density of states is a constant, $\mathscr{D}(\epsilon) = D$, for $\epsilon > 0$.

[You do not have to evaluate D.]

The boson binding potential is represented by a single, **non-degenerate**bound state with energy $-\epsilon_o < 0$, *below the band.*

Assume that the number of bosons is $N \gg 1$.

1. Write down the equation or equations that determine the chemical potential.
2. What is the limit on possible values of the chemical potential? Explain your answer.
      To simplify the algebra, assume that $N = \epsilon_o D$ for the following questions.
3. Calculate the Einstein transition temperature, T_E.
4. For temperatures below the Einstein temperature, calculate the occupation of the ground state as a function of temperature.
5. For $T < T_E$, calculate the chemical potential as a function of temperature.

PROBLEM 28.8
Bosons in four dimensions

Consider an ideal boson gas in four dimensions. The N particles in the gas each have mass m and are confined to a box of dimensions $L \times L \times L \times L$.

1. Calculate the density of states for the four-dimensional, ideal Bose gas.

2. Calculate the Einstein temperature in four dimensions in terms of the given parameters and a dimensionless integral. You do not have to evaluate the dimensionless integral.
3. Below the Einstein temperature, calculate the occupation number of the lowest-energy, single-particle state as a function of temperture, in terms of the Einstein temperature and the total number of particles.
4. Calculate the chemical potential below the Einstein temperature.

PROBLEM 28.9
Energy of an ideal Bose gas in four dimensions

Again consider an ideal boson gas in four dimensions. The N particles in the gas each have mass m and are confined to a box of dimensions $L \times L \times L \times L$. Calculate the energy and the specific heat as functions of temperature below the Einstein temperature.

29

Fermi–Dirac Statistics

> *There are two possible outcomes: If the result confirms the hypothesis, then you've*
> *made a measurement. If the result is contrary to the hypothesis, then you've made a*
> *discovery.*
>
> Enrico Fermi

In the previous chapter we investigated the properties of bosons, based on the general equations for quantum gases derived in Chapter 27. The same equations—with a positive sign in the denominator—govern fermions. Nevertheless, the properties of fermions are dramatically different from those of bosons. In particular, fermions do not exhibit any phase transition that might correspond to the Bose–Einstein condensation.

The most important fermions are electrons, and understanding the properties of electrons is central to understanding the properties of all materials. In this chapter we will study the ideal Fermi gas, which turns out to explain many of the properties of electrons in metals. In Chapter 30, we will see how Fermi–Dirac statistics also explains the basic features of insulators and semiconductors.

We will begin in the next section with the fundamental equations for fermions.

29.1 Basic Equations for Fermions

The Fermi–Dirac equations for N and U, taken from eqs. (27.55) and (27.56), are given below,

$$N = \sum_\alpha \langle n_{\epsilon_\alpha} \rangle = \sum_\alpha (\exp[\beta(\epsilon_\alpha - \mu)] + 1)^{-1} \tag{29.1}$$

$$U = \sum_\alpha \epsilon_\alpha \langle n_{\epsilon_\alpha} \rangle = \sum_\alpha \epsilon_\alpha (\exp[\beta(\epsilon_\alpha - \mu)] + 1)^{-1}. \tag{29.2}$$

To find solutions to these equations, we follow the same basic procedure as we did for bosons in the previous chapter by finding $N = N(T, \mu)$ from eq. (29.1), and then inverting the equation to obtain the chemical potential, $\mu = \mu(T, N)$. However, both the

An Introduction to Statistical Mechanics and Thermodynamics. Robert H. Swendsen, Oxford University Press (2020).
© Robert H. Swendsen. DOI: 10.1093/oso/9780198853237.001.0001

mathematical methods and the physical results for fermions are quite different than what they are for bosons.

29.2   The Fermi Function and the Fermi Energy

The average occupation number, $\langle n_\epsilon \rangle$, for fermions in an eigenstate with energy ϵ was calculated in eq. (27.53). It is called the Fermi function, and it is written as

$$f_{FD}(\epsilon) = \langle n_\epsilon \rangle = (\exp[\beta(\epsilon - \mu)] + 1)^{-1}. \tag{29.3}$$

To solve for the properties of a Fermi gas, it is extremely helpful to have a clear idea of the form of the Fermi function, which is shown for a moderately low temperature in Fig. 29.1.

For lower temperatures (higher $\beta = 1/k_B T$), the Fermi function becomes steeper at $\epsilon = \mu$. The derivative of the Fermi function is given by

$$\frac{\partial f_{FD}(\epsilon)}{\partial \epsilon} = f'_{FD}(\epsilon) = -\beta \exp[\beta(\epsilon - \mu)] \, (\exp[\beta(\epsilon - \mu)] + 1)^{-2}. \tag{29.4}$$

Evaluating the derivative at $\epsilon = \mu$ gives the maximum value of the slope,

$$f'_{FD}(\mu) = -\frac{\beta}{4} = -\frac{1}{4k_B T}. \tag{29.5}$$

It is clear from the form of the function that the width of the non-constant part of the function near $\epsilon = \mu$ is of the order of $2k_B T$. As the temperature is lowered, the Fermi

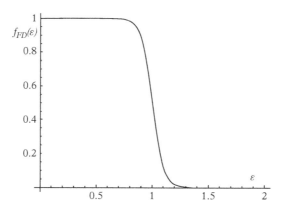

Fig. 29.1 *The form of the Fermi function. The units of energy have been chosen such that $\mu = 1$, and $\beta = 10$ in this figure.*

function approaches a step function. At $T = 0$, the Fermi function is equal to 1 for $\epsilon < \mu$, 0 for $\epsilon > \mu$, and $1/2$ for $\epsilon = \mu$.

For most problems of interest, $k_B T \ll \mu$, so that treating the Fermi function as a step function is a very good first approximation. Indeed, the Fermi function for most materials of common interest at room temperature is much closer to a step function than the curve shown in Fig. 29.1.

As mentioned above, the total number of particles, N, is fixed in any system we will discuss, so that the chemical potential is a function of N and the temperature T. However, as we will see later, the chemical potential is a rather weak function of temperature, so that its zero-temperature limit usually provides a very good approximation to its value at non-zero temperatures. This limit is so important that it has a name, the Fermi energy, and is denoted by ϵ_F,

$$\epsilon_F = \lim_{T \to 0} \mu(T, N). \tag{29.6}$$

Because the Fermi function becomes a step function at $T = 0$, the Fermi energy must always lie between the energy of the highest occupied state and that of the lowest unoccupied state. In fact, as the problems in this section will show, it always lies *exactly* half way between the energies of the highest occupied state and the lowest unoccupied state. This gives a simple rule for finding the Fermi energy that greatly simplifies fermion calculations.

The definition in eq. (29.6) agrees with that of Landau (Lev Davidovich Landau, Russian physicist, 1908–1968) and Lifshitz (Evgeny Mikhailovich Lifshitz, Russian physicist, 1915–1985) in their classic book on statistical mechanics. Unfortunately, many textbooks give a different definition (the location of the highest occupied state at $T = 0$), which only agrees with the Landau–Lifshitz definition when energy levels are quasi-continuous. The alternative definition loses the connection between the Fermi energy and the chemical potential for systems with a discrete energy spectrum, or for insulators and semiconductors. This is a serious handicap in solving problems. Using the definition of ϵ_F given in eq. (29.6) makes it much easier to understand the behavior of such systems. The rule of thumb that the Fermi energy lies exactly half way between energies of the highest occupied state and the lowest unoccupied state is easy to remember and use in calculations. Finding ϵ_F should always be the first step in solving a fermion problem.

29.3   A Useful Identity

The following identity is very useful in calculations with fermions:

$$(\exp[\beta(\epsilon - \mu)] + 1)^{-1} = 1 - (\exp[-\beta(\epsilon - \mu)] + 1)^{-1}. \tag{29.7}$$

The left side of eq. (29.7) is, of course, the Fermi function, f_{FD}. On the right side, the second term gives the deviation of the Fermi function from 1.

It is easy to prove the identity in eq. (29.7), and I strongly recommend that you do so yourself. It might be easier to prove it again during an examination than to remember it. [Hint: First prove $(x+1)^{-1} + (1/x+1)^{-1} = 1$.]

For energies more than a couple of factors of $k_B T$ *above* the chemical potential μ, the Fermi function is very small, and can be approximated well by neglecting the '+1' in the denominator,

$$f_{FD} = (\exp[\beta(\epsilon - \mu)] + 1)^{-1} \approx \exp[-\beta(\epsilon - \mu)]. \tag{29.8}$$

For energies more than a couple of factors of $k_B T$ *below* the chemical potential μ, the Fermi function is close to 1, and the second term on the right side of eq. (29.7) can be approximated well by neglecting the '+1' in its denominator,

$$f_{FD} = 1 - (\exp[-\beta(\epsilon - \mu)] + 1)^{-1} \approx 1 - \exp[\beta(\epsilon - \mu)]. \tag{29.9}$$

These simple approximations are the key to solving fermion problems with a discrete spectrum or a gap in a continuous spectrum.

Up to this point we have been discussing the effects of Fermi statistics on the occupation number as a function of the energy of a given state. Now we will investigate how Fermi statistics affect the properties of systems with different distributions of energy levels.

29.4 Systems with a Discrete Energy Spectrum

We will first consider systems with a discrete spectrum, like that of the hydrogen atom. Much of the early work on the properties of fermions dealt with electronic behavior in atoms, and built on the exact quantum solution for the hydrogen atom to construct approximations in which electrons were taken to be subject to the field of the nucleus, but otherwise non-interacting. Since the electrons were orbiting the nucleus, the single-particle eigenstates were called orbitals—a name that has stuck, even for systems in which the electrons are not orbiting anything.

The problems at the end of the chapter contain several examples of fermion systems with discrete energy levels. One very simple example includes just two energy levels and a single particle. The lower level (energy 0) is non-degenerate and the upper level (energy ϵ) is two-fold degenerate. Find the chemical potential as a function of temperature.

The only equation needed is eq. (29.1), which for this problem becomes

$$N = 1 = \sum_{\alpha} (\exp[\beta(\epsilon_\alpha - \mu)] + 1)^{-1} \tag{29.10}$$

$$= (\exp[\beta(0 - \mu)] + 1)^{-1} + 2(\exp[\beta(\epsilon - \mu)] + 1)^{-1}.$$

We expect the Fermi energy, ϵ_F, to be $\epsilon/2$, which we must check in the final solution. Assuming that $\epsilon_F = \epsilon/2$, at low temperatures we should have $\mu \approx \epsilon_F$, $-\beta\mu << -1$, and $\beta(\epsilon - \mu) >> 1$. This implies that

$$1 = (1 - \exp[\beta(-\mu)]) + 2\exp[-\beta(\epsilon - \mu)]. \tag{29.11}$$

This can be easily solved to find

$$\mu = \frac{1}{2}\epsilon - \frac{k_B T}{2} \ln 2. \tag{29.12}$$

As $T \to 0$, $\mu \to \epsilon_F = \epsilon/2$, as expected.

It is so much more valuable to work out the properties of systems with discrete spectra yourself than to see somebody else's solutions, that I am not going to present any more examples at all—just problems at the end of the chapter.

29.5 Systems with Continuous Energy Spectra

For systems with a continuous energy spectrum, we can use the integral forms in eqs. (27.61) and (27.62) to calculate the properties of the system. For fermions, these equations take the following form:

$$N = \int_0^\infty D_{FD}(\epsilon)(\exp[\beta(\epsilon - \mu)] + 1)^{-1} d\epsilon \tag{29.13}$$

$$U = \int_0^\infty D_{FD}(\epsilon)\epsilon(\exp[\beta(\epsilon - \mu)] + 1)^{-1} d\epsilon. \tag{29.14}$$

The Fermi energy is given by the zero-temperature limit of the chemical potential, as shown in eq. (29.6). In that limit, the Fermi function becomes a step function, and N is determined by the equation

$$N = \int_0^{\epsilon_F} D_{FD}(\epsilon)d\epsilon. \tag{29.15}$$

If we know the density of states, $D(\epsilon)$, we can solve eq. (29.15) for the Fermi energy ϵ_F. In the following sections we will carry this out explicitly for an ideal Fermi gas.

29.6　Ideal Fermi Gas

The case of an ideal gas of fermions in three dimensions is extremely important. This might seem surprising, since electrons certainly interact through the Coulomb force, and that interaction might be expected to dominate the behavior in a macroscopic object. Fortunately, for reasons that go well beyond the scope of this book, electrons in a solid behave as if they were very nearly independent. You will understand why this is true when you take a course in many-body quantum mechanics. In the meantime, just regard it as a stroke of good luck that simplifies your life.

We have already derived the single-particle density of states for particles in a box in eq. (27.13). To apply it to electrons we need only multiply it by 2 to take into account the spin of the electron, which takes on two eigenvalues,

$$D_{FD}(\epsilon) = \frac{V}{2\pi^2}\left(\frac{2m}{\hbar^2}\right)^{3/2}\epsilon^{1/2}.$$ (29.16)

29.7　Fermi Energy

Since the Fermi energy will prove to be a good first approximation to the chemical potential, we will begin by calculating it from the density of states.

In the limit of zero temperature, the Fermi function becomes a step function, and we use eq. (29.15) for the number of particles to find an equation for ϵ_F,

$$N = \int_0^{\epsilon_F} D_{FD}(\epsilon)d\epsilon = \int_0^{\epsilon_F} \frac{V}{2\pi^2}\left(\frac{2m}{\hbar^2}\right)^{3/2}\epsilon^{1/2}d\epsilon = \frac{V}{2\pi^2}\left(\frac{2m}{\hbar^2}\right)^{3/2}\frac{2}{3}\epsilon_F^{3/2}.$$ (29.17)

This equation can then be inverted to find ϵ_F as a function of N,

$$\epsilon_F = \frac{\hbar^2}{2m}\left(\frac{3\pi^2 N}{V}\right)^{2/3}.$$ (29.18)

Note that the Fermi energy depends on the two-thirds power of the number density N/V, and inversely on the particle mass.

Because of the frequency with which the product $\beta\epsilon_F$ appears in calculations, the Fermi energy is often expressed in terms of the Fermi temperature, T_F,

$$\epsilon_F = k_B T_F$$ (29.19)

$$\beta\epsilon_F = \frac{T_F}{T}.$$ (29.20)

Fermi temperatures turn out to be remarkably high. T_F for most metals ranges from about $2 \times 10^4 K$ to $15 \times 10^4 K$. For comparison, the temperature of the surface of the sun is only about $6 \times 10^3 K$. For experiments at room temperature ($300K$), this means that $T/T_F \approx 0.01$, justifying the statement made in Section 29.2 that the Fermi function is very close to a step function for real systems of interest. Again for comparison, Fig. 29.1 was drawn for the case of $T/T_F \approx 0.1$.

Once we have the Fermi energy, the total energy of the system at $T = 0$ can be obtained from eq. (29.14), using the step function that is the zero-temperature limit of the Fermi function,

$$U = \int_0^{\epsilon_F} D_{FD}(\epsilon)\epsilon \, d\epsilon = \frac{V}{2\pi^2}\left(\frac{2m}{\hbar^2}\right)^{3/2}\int_0^{\epsilon_F} \epsilon^{3/2} d\epsilon = \frac{V}{2\pi^2}\left(\frac{2m}{\hbar^2}\right)^{3/2}\frac{2}{5}\epsilon_F^{5/2}. \quad (29.21)$$

Combining eq. (29.17) with eq. (29.21), we find that the energy per particle takes on a particularly simple form

$$\frac{U}{N} = \frac{3}{5}\epsilon_F. \quad (29.22)$$

The proportionality of U/N to ϵ_F is quite general, but the factor of $3/5$ depends on the dimension of the system.

29.8 Compressibility of Metals

Metals are hard to compress. They also contain electrons that are fairly well described by the ideal gas of fermions we have been discussing. Perhaps surprisingly, these two facts are closely related.

If we approximate the energy of an ideal gas of fermions by the Fermi energy in eq. (29.22) and insert the expression for the Fermi energy from eq. (29.18), we find an expression for the energy as a function of the volume V,

$$\frac{U}{N} = \frac{3}{5}\frac{\hbar^2}{2m}\left(\frac{3\pi^2 N}{V}\right)^{2/3}. \quad (29.23)$$

Recalling the definition of the pressure as a derivative of the energy,

$$P = -\frac{\partial U}{\partial V}, \quad (29.24)$$

we find

$$P = -N\frac{3}{5}\frac{\hbar^2}{2m}\left(3\pi^2 N\right)^{2/3}\frac{\partial}{\partial V}V^{-2/3} = -N\frac{3}{5}\frac{\hbar^2}{2m}\left(\pi^2 N\right)^{2/3}\left(\frac{-2}{3}\right)V^{-5/3}, \quad (29.25)$$

or

$$P = \frac{2}{5}\frac{\hbar^2}{2m}\pi^{4/3}\left(\frac{N}{V}\right)^{5/3}.$$

(29.26)

Recalling the definition of the compressibility from eq. (14.14),

$$\kappa_T = -\frac{1}{V}\left(\frac{\partial V}{\partial P}\right)_{T,N} = -1 \Bigg/ \left[V\left(\frac{\partial P}{\partial V}\right)_{T,N}\right],$$

(29.27)

we find that

$$\kappa_T = \frac{3m}{\hbar^2}\pi^{-4/3}\left(\frac{V}{N}\right)^{5/3}.$$

(29.28)

This can also be written in terms of the bulk modulus,

$$B = \frac{1}{\kappa_T} = -V\left(\frac{\partial P}{\partial V}\right)_{T,N} = \frac{\hbar^2}{3m}\pi^{4/3}\left(\frac{N}{V}\right)^{5/3}.$$

(29.29)

Comparing eq. (29.29) with the expression for the Fermi energy in eq. (29.18)

$$\epsilon_F = \frac{\hbar^2}{2m}\left(\frac{3\pi^2 N}{V}\right)^{2/3}$$

(29.30)

we find a surprisingly simple result,

$$B = \frac{1}{\kappa_T} = \frac{5}{3}P = \frac{2}{3}\frac{N}{V}\epsilon_F.$$

(29.31)

The numerical predictions of eq. (29.31) turn out to be within roughly a factor of 2 of the experimental results for metals, even though the model ignores the lattice structure entirely. This is remarkably good agreement for a very simple model of a metal. It shows that the quantum effects of Fermi–Dirac statistics are responsible for a major part of the bulk modulus of metals.

29.9 Sommerfeld Expansion

While we have obtained important information about Fermi gases from the zero-temperature limit of the Fermi function in the previous section, the behavior at non-zero temperatures is even more interesting. The difficulty in doing calculations for

$T \neq 0$ is that the temperature dependence of the integrals in eqs. (29.13) and (29.14) is not trivial to extract.

The problem was solved by the German physicist Arnold Sommerfeld (1868–1951). His solution is known today as the Sommerfeld expansion, and it is the topic of the current section. The results obtained from the Sommerfeld expansion are essential for understanding the properties of metals.

First note that both eq. (29.13) and eq. (29.14) can be written in the form,

$$\mathcal{I} = \int_0^\infty \phi(\epsilon) f_F(\epsilon) d\epsilon = \int_0^\infty \phi(\epsilon) \left(\exp[\beta(\epsilon - \mu)] + 1 \right)^{-1} d\epsilon \qquad (29.32)$$

where $\phi(\epsilon) = D_{FD}(\epsilon)$ to calculate N, and $\phi(\epsilon) = \epsilon D_{FD}(\epsilon)$ to calculate U.

Break the integral in eq. (29.32) into two parts, and use the identity in eq. (29.7) to rewrite the integrand for $\epsilon < \mu$,

$$\mathcal{I} = \int_0^\mu \phi(\epsilon) \left(\exp[\beta(\epsilon - \mu)] + 1 \right)^{-1} d\epsilon$$

$$+ \int_\mu^\infty \phi(\epsilon) \left(\exp[\beta(\epsilon - \mu)] + 1 \right)^{-1} d\epsilon$$

$$= \int_0^\mu \phi(\epsilon) \left[1 - \left(\exp[-\beta(\epsilon - \mu)] + 1 \right)^{-1} \right] d\epsilon$$

$$+ \int_\mu^\infty \phi(\epsilon) \left(\exp[\beta(\epsilon - \mu)] + 1 \right)^{-1}$$

$$= \int_0^\mu \phi(\epsilon) d\epsilon - \int_0^\mu \phi(\epsilon) \left(\exp[-\beta(\epsilon - \mu)] + 1 \right)^{-1} d\epsilon$$

$$+ \int_\mu^\infty \phi(\epsilon) \left(\exp[\beta(\epsilon - \mu)] + 1 \right)^{-1}. \qquad (29.33)$$

In the last line of this equation, the first integral represents the contributions of a step function, while the second and third integrals represent the deviations of the Fermi function from the step function.

The next step is to substitute the dimensionless integration variable $z = -\beta(\epsilon - \mu)$ in the second integral, and $z = \beta(\epsilon - \mu)$ in the third integral,

$$\mathcal{I} = \int_0^\mu \phi(\epsilon) d\epsilon - \beta^{-1} \int_0^{\beta\mu} \phi(\mu - \beta^{-1}z) \left(e^z + 1 \right)^{-1} dz$$

$$+ \beta^{-1} \int_0^\infty \phi(\mu + \beta^{-1}z) \left(e^z + 1 \right)^{-1} dz. \qquad (29.34)$$

At low temperatures, $\beta\mu \approx \beta\epsilon_F = T_F/T \gg 1$. Since the upper limit in the second integral in eq. (29.34) is large and the integrand goes to zero exponentially rapidly as $z \to \infty$, replacing the upper limit by infinity should be a good approximation,

$$\mathcal{I} \approx \int_0^\mu \phi(\epsilon) d\epsilon \qquad (29.35)$$

$$+ \beta^{-1} \int_0^\infty \left[\phi(\mu + \beta^{-1} z) - \phi(\mu - \beta^{-1} z) \right] (e^z + 1)^{-1} dz.$$

Note that while the replacement of the upper limit of $\beta\mu$ in eq. (29.34) by infinity is almost always an excellent approximation, it is not exact.

For the next step we will assume that the integrand in eq. (29.35) is analytic in some region around $z = 0$ (or $\epsilon = \mu$), so that we can expand the function $\phi(\mu + \beta^{-1} z)$ in powers of z,

$$\phi(\mu + \beta^{-1} z) = \sum_{j=0}^\infty \frac{1}{j!} \phi^{(j)}(\mu) \beta^{-j} z^j. \qquad (29.36)$$

The assumption of analyticity is essential to the Sommerfeld expansion. While the density of states $D_{FD}(\epsilon)$ is generally analytic in some region around the Fermi energy, it is not analytic everywhere. This means that there are always small corrections to the Sommerfeld expansion, in addition to the one that comes from the extension of the upper limit of the integral in eq. (29.34) to infinity. These corrections are so small that they are often not even mentioned in textbooks. However, if the density of states is zero at $\epsilon = \mu$, all terms in the Sommerfeld expansion vanish, and only the non-analytic corrections are left.

Inserting eq. (29.36) in eq. (29.35), we find

$$\mathcal{I} = \int_0^\mu \phi(\epsilon) d\epsilon$$

$$+ \beta^{-1} \sum_{j=0}^\infty \frac{1}{j!} \phi^j(\mu) \beta^{-j} \int_0^\infty \left[z^j - (-z)^j \right] (e^z + 1)^{-1} dz$$

$$= \int_0^\mu \phi(\epsilon) d\epsilon$$

$$+ 2 \sum_{n=0}^\infty \frac{1}{(2n+1)!} \phi^{(2n+1)}(\mu) \beta^{-2n-2} \int_0^\infty z^{2n+1} (e^z + 1)^{-1} dz. \qquad (29.37)$$

The integrals on the first line of eq. (29.37) vanish for even values of j, which led us to define $n = (j-1)/2$ for odd values of j and rewrite the sum as shown in the second line.

The dimensionless integrals in eq. (29.37) can be evaluated exactly (with a little effort) to obtain the first few terms of the Sommerfeld expansion,

$$\mathcal{I} = \int_0^\mu \phi(\epsilon)d\epsilon + \frac{\pi^2}{6}\phi^{(1)}(\mu)(k_B T)^2 + \frac{7\pi^4}{360}\phi^{(3)}(\mu)(k_B T)^4 + \cdots . \tag{29.38}$$

In eq. (29.38), $\phi^{(1)}(\mu)$ denotes the first derivative of $\phi(\epsilon)$, evaluated at μ, and $\phi^{(3)}(\mu)$ denotes the corresponding third derivative.

As we will see in the next section, the Sommerfeld expansion converges rapidly for low temperatures.

29.10   General Fermi Gas at Low Temperatures

As discussed in Section 27.16 of the previous chapter, the first step in calculating the properties of a Fermi gas is to use eq. (29.13) to find N as a function of T and μ, and then invert the equation to obtain $\mu = \mu(T, N)$. Using the Sommerfeld expansion, eq. (29.38), derived in the previous section, we can expand eq. (29.13) to find N as a power series in T,

$$N = \int_0^\infty D_{FD}(\epsilon)(\exp[\beta(\epsilon - \mu)] + 1)^{-1} d\epsilon \tag{29.39}$$

$$= \int_0^\mu D_{FD}(\epsilon)d\epsilon + \frac{\pi^2}{6}D_{FD}^{(1)}(\mu)(k_B T)^2 + \frac{7\pi^4}{360}D_{FD}^{(3)}(\mu)(k_B T)^4 + \cdots .$$

Recall from eq. (29.15) that the Fermi energy is defined as the zero-temperature limit of the chemical potential,

$$N = \int_0^{\epsilon_F} D_{FD}(\epsilon)d\epsilon . \tag{29.40}$$

Since $\mu - \epsilon_F$ is small at low temperatures, we can approximate the integral in eq. (29.39) as

$$\int_0^\mu D_{FD}(\epsilon)d\epsilon \approx \int_0^{\epsilon_F} D_{FD}(\epsilon)d\epsilon + (\mu - \epsilon_F)D_{FD}(\epsilon_F). \tag{29.41}$$

Putting this approximation into eq. (29.39), we find an equation for the leading term in $\mu - \epsilon_F$,

$$\mu - \epsilon_F = -\frac{\pi^2}{6}(k_B T)^2 D_{FD}^{(1)}(\mu)/D_{FD}(\epsilon_F) + \cdots . \tag{29.42}$$

We see that the deviation of the chemical potential from the Fermi energy goes to zero as T^2 for low temperatures, justifying our earlier assertion that ϵ_F is a good approximation for μ at low temperatures.

The next step is to use eq. (29.42) to find a low-temperature expansion for the energy. Defining the zero-temperature energy of the system to be

$$U_0 = \int_0^{\epsilon_F} D_{FD}(\epsilon)\epsilon \, d\epsilon, \tag{29.43}$$

the Sommerfeld expansion of eq. (29.14) can be written as

$$U = U_0 + (\mu - \epsilon_F)\epsilon_F D_{FD}(\epsilon_F) \tag{29.44}$$

$$+ \frac{\pi^2}{6}\left[D_{FD}(\epsilon_F) + \epsilon_F D_{FD}^{(1)}(\epsilon_F)\right](k_B T)^2 + \cdots .$$

Inserting eq. (29.42) for $\mu - \epsilon_F$, this becomes

$$U = U_0 - \frac{\pi^2}{6}(k_B T)^2 \epsilon_F D_{FD}^{(1)}(\epsilon_F) \tag{29.45}$$

$$+ \frac{\pi^2}{6}\left[D_{FD}(\epsilon_F) + \epsilon_F D_{FD}^{(1)}(\epsilon_F)\right](k_B T)^2 + \cdots$$

$$= U_0 + \frac{\pi^2}{6}D_{FD}(\epsilon_F)(k_B T)^2 + \cdots .$$

The heat capacity at constant volume is found by differentiating of eq. (29.45) with respect to temperature,

$$C_V = \left(\frac{\partial U}{\partial T}\right)_{N,V} = \frac{\pi^2}{3}D_{FD}(\epsilon_F)k_B^2 T + \cdots . \tag{29.46}$$

Eq. (29.46) has very interesting properties. First, it shows that as long as $D_{FD}(\epsilon_F)$ does not vanish, the low-temperature heat capacity is proportional to the temperature T. Since the low-temperature contributions of the phonons to the heat capacity have a T^3 temperature-dependence, we can clearly separate the contributions of the phonons and electrons experimentally.

Next, we have answered the question of why the heat capacity of materials does not have a contribution of $3k_B/2$ from every particle, whether electrons or nuclei, which would be expected from classical theory. We will see this again below in eq. (29.54).

Finally, eq. (29.46) shows that the only thing we need to know to calculate the low-temperature heat capacity is the density of states at the Fermi energy. Or, turning the equation around, by measuring the low-temperature heat capacity we can determine the density of states at the Fermi energy from experiment.

29.11   Ideal Fermi Gas at Low Temperatures

While the equations for the specific heat derived so far in this chapter apply to a system of fermions with any density of states, it is particularly instructive to look at the special case of an ideal Fermi gas.

Recall that the density of states for an ideal gas of electrons was given in eq. (29.16),

$$D_{FD}(\epsilon) = \frac{V}{2\pi^2}\left(\frac{2m}{\hbar^2}\right)^{3/2}\epsilon^{1/2} = A\epsilon^{1/2}. \tag{29.47}$$

In eq. (29.47) we defined a constant

$$A = \frac{V}{2\pi^2}\left(\frac{2m}{\hbar^2}\right)^{3/2}, \tag{29.48}$$

to simplify further calculations.

Eq. (29.39) for N becomes

$$N = \int_0^\mu A\epsilon^{1/2}d\epsilon + \frac{\pi^2}{6}A\frac{1}{2}\mu^{-1/2}(k_BT)^2 + \frac{7\pi^4}{360}A\frac{3}{8}\mu^{-5/2}(k_BT)^4 + \cdots \tag{29.49}$$

$$= \frac{2}{3}A\mu^{3/2} + \frac{\pi^2}{12}A\mu^{-1/2}(k_BT)^2 + \frac{7\pi^4}{960}A\mu^{-5/2}(k_BT)^4 + \cdots$$

$$= \frac{2}{3}A\mu^{3/2}\left[1 + \frac{\pi^2}{8}\left(\frac{k_BT}{\mu}\right)^2 + \frac{7\pi^4}{640}\left(\frac{k_BT}{\mu}\right)^4 + \cdots\right].$$

Since

$$N = \frac{2}{3}A\epsilon_F^{3/2}, \tag{29.50}$$

Eq. (29.49) gives an equation for ϵ_F in terms of μ,

$$\epsilon_F^{3/2} \approx \mu^{3/2}\left[1 + \frac{\pi^2}{8}\left(\frac{k_BT}{\mu}\right)^2 + \frac{7\pi^4}{640}\left(\frac{k_BT}{\mu}\right)^4\right]. \tag{29.51}$$

Rewriting this for μ in terms of ϵ_F, we find

$$\mu \approx \epsilon_F\left[1 + \frac{\pi^2}{8}\left(\frac{k_BT}{\mu}\right)^2 + \frac{7\pi^4}{640}\left(\frac{k_BT}{\mu}\right)^4\right]^{-2/3}. \tag{29.52}$$

$$\approx \epsilon_F \left[1 - \frac{\pi^2}{12} \left(\frac{k_B T}{\epsilon_F} \right)^2 \right] = \epsilon_F \left[1 - \frac{\pi^2}{12} \left(\frac{T}{T_F} \right)^2 \right].$$

Since a typical value of T_F is of the order of $6 \times 10^4 K$, the T^2-term in eq. (29.52) is of the order of 10^{-4} at room temperature, confirming that the chemical potential is very close to the Fermi energy at low temperatures. This also justifies the omission of the next term in the expansion, which is of order $(T/T_F)^4 \approx 10^{-8}$.

The heat capacity can be obtained by inserting eq. (29.47) into eq. (29.46),

$$C_V = \frac{\pi^2}{2} N k_B \left(\frac{T}{T_F} \right). \tag{29.53}$$

This equation can be rewritten in a form that shows the effect of Fermi statistics on the heat capacity,

$$C_V = \frac{3}{2} N k_B \left(\frac{\pi^2}{3} \frac{T}{T_F} \right). \tag{29.54}$$

The factor in parentheses is the quantum correction, which shows explicitly that the Fermi heat capacity at low temperatures is much smaller than that of a classical ideal gas, as claimed in the previous section.

We have seen in Chapter 26 that the specific heat due to lattice vibrations has a T^3-behavior at low temperatures. Eq. (29.54) shows that when the density of states at the Fermi energy, $D(\epsilon_F)$, is non-zero, there is an additional T-dependence, which dominates at very low temperatures. This is indeed observed experimentally for metals. However, it is not observed for insulators or semiconductors, for reasons we will discuss in the next chapter.

29.12 Problems

PROBLEM 29.1
Numerical Fermi–Dirac Quantum Gas Problem: μ and n_0 and n_1 for a small number of particles

Using the programs that you've written for quantum gases, carry out the following calculations for Fermi–Dirac statistics.

1. For the extreme cases of $N = 1, 2, 3$, and 4, compute the chemical potential μ and the occupations of the two lowest energy states as functions of the temperature.

   Don't include temperatures below $T = 0.1$ in your computations. There are numerical difficulties at very low temperatures that are not worth worrying about for this assignment.

Go up to about $T = 3.0$ for $N = 1$, $T = 5.0$ for $N = 2$ and 3, and $T = 7.0$ for $N = 4$.

2. Are the Fermi energies you found in your computations consistent with what you had expected? Explain.
3. Explain the occupation number you found in your computations at low temperatures for a state with energy $\epsilon = 6A$ for $N = 2$ and $N = 3$.

PROBLEM 29.2
Fermi–Dirac Quantum Gas Problem 2: Specific heat

Using the programs that you've written for quantum gases, carry out the following calculations for Fermi–Dirac statistics.

The Fermi temperature is given by the equation

$$k_B T_F = A \left(\frac{3N}{\pi} \right)^{2/3}.$$

It is discussed in the book and will be discussed in class. For this assignment, all you need to know is that it is a characteristic temperature for an ideal gas of fermions.

For $N = 1000$ particles, make three plots of the specific heat as a function of temperature. Let the lowest temperature be $T = 0.01$ for all three plots. Let the highest temperatures be $T = T_F/100$, $T = T_F/10$ and $T = 2T_F$.

1. For $T < 0.1$, the specific heat is seen to be very small. Why is this true?
2. In a temperature range of about $T = 3.0$ to $T = 0.1T_F$, how does the specific heat behave. Explain why this is to be expected.
3. For T much larger than T_F, the specific heat seems to be going to a constant value. Derive the value of this constant. Is the value you derived consistent with your plots?

PROBLEM 29.3
Fermions in a two-level system (with degeneracy)

Consider a system of N-independent fermions.

Assume that the single-particle Hamiltonians have only two energy levels, with energies 0 and ϵ. However, the two levels have degeneracies n_0 and n_1, which are, of course, both integers.

1. First take the simple case of $n_0 = n_1 = 1$, with $N = 1$. Find the chemical potential, μ, as a function of temperature. What is the Fermi energy, $\epsilon_F = \mu(T = 0)$?
2. Now make it more interesting by taking arbitrary values of n_0 and n_1, but specifying that $N = n_0$. Again find the chemical potential, μ, as a function of temperature *for low temperatures*. That is, assume that $\beta\epsilon \gg 1$. What is the Fermi energy?
3. Keep arbitrary values of n_0 and n_1, but consider the case of $N < n_0$. Again find the chemical potential, μ, as a function of temperature *for low temperatures*. That is, assume that $\beta\epsilon \gg 1$. What is the Fermi energy?

4. Keep arbitrary values of n_0 and n_1, but consider the case of $N > n_0$. Again find the chemical potential, μ, as a function of temperature *for low temperatures*. That is, assume that $\beta\epsilon \gg 1$. What is the Fermi energy?

PROBLEM 29.4
Fermions in a *three*-level system (with degeneracy)

Consider a system of N-independent fermions.

Assume that the single-particle Hamiltonians have three energy levels, with energies ϵ_1, ϵ_2, and ϵ_3, where $\epsilon_1 < \epsilon_2 < \epsilon_3$. The three energy levels have degeneracies n_1, n_2, and n_3, which are, of course, integers. The values of the n_js are to be left arbitrary.

1. First take the case of $N = n_1$. Find the chemical potential, μ, for low temperatures. What is the Fermi energy, $\epsilon_F = \mu(T = 0)$?
2. Now take the case of $N = n_1 + n_2$. Find the chemical potential, μ, for low temperatures. What is the Fermi energy, $\epsilon_F = \mu(T = 0)$?
3. Next, consider the situation with $n_2 \geq 2$ and $N = n_1 + 1$. Find the chemical potential, μ, for low temperatures. What is the Fermi energy, $\epsilon_F = \mu(T = 0)$?

PROBLEM 29.5
Ideal Fermi gas in two dimensions

Consider an ideal Fermi gas in two dimensions. It is contained in an area of dimensions $L \times L$. The particle mass is m.

1. Calculate the density of states.
2. Using your result for the density of states, calculate the number of particles as a function of the chemical potential at zero temperature. ($\mu(T = 0) = \epsilon_F$, the Fermi energy.)
3. Calculate the Fermi energy as a function of the number of particles.
4. Again using your result for the density of states, calculate the total energy of the system at zero temperature as a function of the Fermi energy, ϵ_F .
5. Calculate the energy per particle as a function of the Fermi energy ϵ_F.

PROBLEM 29.6
More fun with the ideal Fermi gas in two dimensions

Consider the same two-dimensional, ideal Fermi gas that you dealt with in the previous assignment. You will need the result of that assignment to do this one.

1. Calculate the average number of particles as a function of μ *exactly*. (This is one of the few problems for which this can be done.)
2. Calculate μ as a function of N and T. Then find the high- and low-temperature behavior of μ.

PROBLEM 29.7
The specific heat of the Maxwell–Boltzmann quantum gas

Using the programs that you've written for quantum gases, carry out the following calculations for Maxwell–Boltzmann statistics.

1. For both $N = 1$ and $N = 10$ particles, plot c_N as a function of temperature for T between $T = 0.1$ and $T = 8.0$.
   What differences do you see between the two plots? Explain.
2. For a number of particles of your choice, plot c_N vs. T between $T = 2.0$ and $T = 50.0$.
3. What was the basis for your choice of N in the previous problem?
4. For large values of T, the specific heat seems to be going to a constant. Find the value of the constant, and compare it to the corresponding constant for Fermi–Dirac statistics.

PROBLEM 29.8
An artificial model of a density of states with a gap in the energy spectrum

Consider a system with the following density of states

$$\mathscr{D}(\epsilon) = \begin{cases} 0 & 0 > \epsilon \\ A(\epsilon_1 - \epsilon) & \epsilon_1 > \epsilon > 0 \\ 0 & \epsilon_2 > \epsilon > \epsilon_1 \\ A(\epsilon - \epsilon_2) & \epsilon > \epsilon_2 \end{cases}$$

where A is a constant, and $\epsilon_2 > \epsilon_1 > 0$.

1. Find the Fermi energy $\epsilon_F = \mu(T = 0)$ for the following three values of the total number of particles, N.
   (a) $N = A\epsilon_1^2/4$
   (b) $N = A\epsilon_1^2/2$
   (c) $N = 3A\epsilon_1^2/4$

2. For each of the three cases, find the specific heat at low temperatures from the Sommerfeld expansion.
3. For one of the cases studied above, the Sommerfeld expansion for the specific heat can be summed exactly to all orders. *And the answer is wrong!* Explain which case is being referred to, and why the Sommerfeld expansion has failed.

PROBLEM 29.9
An artificial model of a density of states with a gap in the energy spectrum— continued

Again consider a system with the following density of states:

$$\mathcal{D}(\epsilon) = \begin{cases} 0 & 0 > \epsilon \\ A(\epsilon_1 - \epsilon) & \epsilon_1 > \epsilon > 0 \\ 0 & \epsilon_2 > \epsilon > \epsilon_1 \\ A(\epsilon - \epsilon_2) & \epsilon > \epsilon_2 \end{cases}$$

where A is a constant, and $\epsilon_2 > \epsilon_1 > 0$. The number of particles is given by $N = A\epsilon_1^2/2$, so the Fermi energy is $\epsilon_F = (\epsilon_1 + \epsilon_2)/2$.

In the last assignment we found that the Sommerfeld expansion gives incorrect results for this model. In this assignment we will calculate the low-temperature behavior correctly.

1. Show that if $\epsilon_1 \gg k_B T$, that $\mu = \epsilon_F$ is a very good approxmation, even if $k_B T \approx \epsilon_2 - \epsilon_1$.
2. Calculate the energy of this model as a function of temperature for low temperatures. Assume that $\epsilon_1 \gg \epsilon_2 - \epsilon_1 \gg k_B T$.
3. Calculate the heat capacity of this model as a function of temperature for low temperatures from your answer to the previous question. Assume that $\epsilon_1 \gg \epsilon_2 - \epsilon_1 \gg k_B T$.

30

Insulators and Semiconductors

Mankind is a catalyzing enzyme for the transition from a carbon-based intelligence to a silicon-based intelligence.

Gérard Bricogne, French crystallographer

Although the ideal gas of fermions is a useful model for the behavior of metals, real systems deviate from this model in a very important way: real energy spectra contain gaps in which no energy levels occur. When the Fermi energy lies in one of those gaps, the behavior of the electrons is qualitatively different from that described in the previous chapter.[1] Such materials do not have the high electrical conductivity of metals. They are insulators or semiconductors, and their properties are the subject of the current chapter.

The origin of the strange band gaps that arise in the electron density of states in real materials is that the regular arrangement of the nuclei in a crystal lattice creates a periodic potential. This affects the energy levels of the electrons and gives rise to gaps in the energy spectrum. The mechanism by which the gaps arise can be understood from two perspectives. The first initially assumes that the atoms are very far apart, so that the electrons cannot easily jump from one atom to the next. Since an electron is tightly bound to a nucleus, this is known as the 'tight binding approximation'. The second way of looking at it starts from the opposite extreme, where the electrons act almost like an ideal gas, but are presumed to be subject to a very weak periodic potential. This is known as the 'nearly free electron approximation'. In both cases, gaps in the energy spectrum occur due to the periodicity of the lattice.

In the following sections we will explore both ways of seeing how gaps arise in the energy spectrum in a crystal. After that we will investigate the consequences of energy gaps, which will lead us to the properties of insulators and semiconductors.

30.1 Tight-Binding Approximation

Begin by considering a single, isolated atom. Since we know that electrons in isolated atoms have discrete energy levels, the energy spectrum of a single atom will look qualitatively like the diagram on the far left of Fig. 30.1.

---

[1] Be cautious in comparing our use of the Fermi energy with that of other books. We have defined the Fermi energy to be the $T \to 0$ limit of the chemical potential, as in eq. (29.6). This is often referred to in other books as the 'Fermi level'.

An Introduction to Statistical Mechanics and Thermodynamics. Robert H. Swendsen, Oxford University Press (2020).
© Robert H. Swendsen. DOI: 10.1093/oso/9780198853237.001.0001

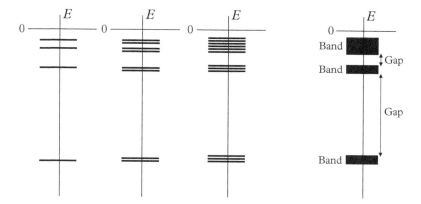

Fig. 30.1 *Schematic plot of the energy levels in a crystal. The diagram on the far left shows the energy spectrum of an isolated atom. The second diagram shows the effect of bringing two atoms close together, and the third diagram corresponds to three neighboring atoms. The diagram on the far right shows the band structure that arises when the lattice constant is of the order of nanometers, as in a real crystal. The 10^{20} or more energy levels make each energy band a quasi-continuum.*

If we bring a second atom to within a few tenths of a nanometer of the first, the wave functions will overlap, and the energy levels will be split according to quantum perturbation theory, as shown in the second diagram in Fig. 30.1.

If there are three atoms close together, each energy level will split into three levels, as shown in the third diagram in Fig. 30.1.

Now assume that we have a crystal; that is, a regular array of atoms in which the neighboring atoms are separated by a few tenths of a nanometer. If the crystal contains 10^{20} atoms, the energy levels will be split into 10^{20} closely spaced energy levels, which will form a quasi-continuum. Such a quasi-continuum of energy levels is called a 'band'. This is shown schematically in the diagram on the far right of Fig. 30.1.

In the schematic diagram in Fig. 30.1, the spreading of the energy levels has been large enough to merge the top two single-atom energy levels into a single band. However, the gap between the bottom two two single-atom energy levels remains, despite the spreading of the energy levels into bands. These gaps, or 'band gaps', represent forbidden energies, since there are no single-particle wave functions with energies in the gaps.

This qualitative description of how gaps arise in the single-particle excitation spectrum can be made quantitative by applying perturbation theory to the weak overlap of the wave functions of nearly isolated atoms. Since the electrons are tightly bound to individual atoms when the overlap is very weak, this is known as the 'tight-binding approximation'. We will not go into the details of the theory here, since our purpose is only to understand qualitatively how gaps in the excitation spectrum arise.

Before going on to discuss the opposite extreme, in which electrons experience only a weak periodic potential due to the crystal lattice, we need some mathematics. The following section introduces Bloch's theorem for a particle in a periodic potential (Felix

Bloch (1905–1983), Swiss physicist, Nobel Prize 1952). Once armed with this theorem, we will discuss the effects of the crystal lattice on the single-electron energy spectrum.

30.2 Bloch's Theorem

For simplicity, we will again only treat the one-dimensional case. The three-dimensional case is not really more difficult, but the necessary vector notation makes it look more complicated than it is.

Consider a single particle in a periodic potential, $V(x)$, so that the Hamiltonian is given by

$$H = \frac{p^2}{2m} + V(x). \tag{30.1}$$

Denoting the period by a, we have

$$V(x+a) = V(x). \tag{30.2}$$

Define translation operators, T_n, for any integer n. such that

$$T_n H(p,x) = H(p, x+na) = H(p,x). \tag{30.3}$$

Translation operators have the property that they can be combined. For an arbitrary function $f(x)$,

$$T_n T_m f(x) = T_n f(x+ma) = f(x+ma+na) = T_{m+n} f(x), \tag{30.4}$$

or

$$T_n T_m = T_{m+n}. \tag{30.5}$$

The translation operators also commute with each other

$$[T_n, T_m] = T_n T_m - T_m T_n = 0, \tag{30.6}$$

and every T_n commutes with the Hamiltonian.

$$[T_n, H] = T_n H - H T_n = 0 \tag{30.7}$$

Therefore, we can define wave functions that are simultaneously eigenfunctions of all the translation operators and the Hamiltonian.

Denote a single-particle eigenfunction by $\phi(x)$, and the eigenvalue of T_n by C_n.

$$T_n\phi(x) = \phi(x+na) = C_n\phi(x) \tag{30.8}$$

From eq. (30.5), we can see that

$$T_n T_m\phi(x) = C_n C_m\phi(x) = T_{m+n}\phi(x) = C_{n+m}\phi(x) \tag{30.9}$$

or

$$C_n C_m = C_{n+m}. \tag{30.10}$$

We can solve eq. (30.10) for C_n in terms of a constant k,

$$C_n = \exp(ikna), \tag{30.11}$$

where k is a real number. It is easy to confirm that this is a solution by inserting eq. (30.11) into eq. (30.10). The reason that k must be real is that $\phi(x)$ – and therefore C_n – must be bounded for very large (positive or negative) values of x.

The next step is to note that the eigenfunction can be written as

$$\phi_k(x) = \exp(ikx)u_k(x), \tag{30.12}$$

where $u_k(x)$ is periodic.

$$u_k(x+a) = u_k(x) \tag{30.13}$$

The eigenfunctions $\phi_k(x)$ also have an important periodicity with respect to the wave number k. If we define a vector

$$K = \frac{2\pi}{a}, \tag{30.14}$$

and consider the wave function

$$\phi_{k+K}(x) = \exp[i(k+K)x]\,u_{k+K}(x), \tag{30.15}$$

then

$$\begin{aligned}
\phi_k(x) &= \exp(ikx)u_k(x) \\
&= \exp[i(k+K)x]\exp(-iKx)\,u_k(x) \\
&= \exp[i(k+K)x]\,u_{k+K}(x) \\
&= \phi_{k+K}(x).
\end{aligned} \tag{30.16}$$

It is easy to confirm that $u_{k+K}(x) = \exp(-iKx)u_k(x)$ is a periodic function of x with period a.

By extending this argument to $\phi_{k+qK}(x)$, where q is any integer, we see that ϵ_k must be a periodic function of k with period $K = 2\pi/a$.

$$\epsilon_k = \epsilon_{k+K} \tag{30.17}$$

The following section uses Bloch's theorem to calculate the single-particle excitation spectrum for nearly free electrons. The periodicity of ϵ_k will be crucial to the discussion.

30.3   Nearly-Free Electrons

In this section we will look at the consequences of a weak periodic potential for a gas of non-interacting electrons. We will begin by assuming that the potential is exactly zero, and then investigate the effects of turning on a periodic potential, such as that due to a regular crystal lattice.

The discussion will consist of three stages:

1. Reminder of the energy spectrum of a non-interacting particle in a box.
2. The effects of Bloch's theorem on the representation of the energy spectrum of non-interacting particles in the limit that the amplitude of the periodic potential vanishes.
3. The effects of turning on a periodic potential with a small amplitude.

30.3.1   Free Electrons

We begin with the energy levels for free electrons, which are given in eq. (27.12) of Chapter 27.

$$\epsilon_k = \frac{\hbar^2}{2m}k^2 \tag{30.18}$$

Here

$$k = \frac{n\pi}{L}, \tag{30.19}$$

and n is a positive integer.

Fig. 30.2 shows a plot of ϵ_k as a function of the wave number k.

30.3.2   The 'Empty' Lattice

The next step is to look at the consequences of the periodicity of the lattice on the energy spectrum, while keeping the amplitude of the periodic potential equal to zero. This might

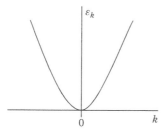

Fig. 30.2 *Free electron energy spectrum.*

seem rather strange, since a potential of amplitude zero cannot change the energy of a state. Nevertheless, the *representation* of the energy can change.

The representation of the energy spectrum in a periodic potential differs from that shown in Fig. 30.2 because of the periodicity in k-space imposed by eq. (30.17). This representation is known as the 'empty lattice', because even though we are assuming periodicity, there is nothing there to give rise to a non-zero potential.

The wave functions in the empty lattice representation can be written in the form given in eq. (30.12). The empty-lattice energy, ϵ_k, is periodic in k, as indicated in eq. (30.17). The empty-lattice energy spectrum is plotted in Fig. 30.3.

> The degeneracies seen in Fig. 30.3 all come from the symmetry that $\epsilon_k = \epsilon_{k+K}$. The value of the empty-lattice representation is that it highlights those points at which the splitting due to a periodic lattice removes the degeneracies.

Fig. 30.3 shows how the periodicity indicated by eq. (30.17) affects the energy spectrum. The parabolic spectrum shown in Fig. 30.2 is repeated at intervals of $K = 2\pi/a$

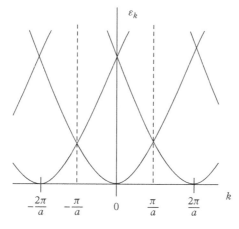

Fig. 30.3 *Empty lattice energy spectrum.*

along the entire k-axis. This repetition has the consequence that all information about the energy spectrum is contained in any region of length $K = 2\pi/a$ along the axis. It is conventional to choose the region from $-\pi/a$ to π/a. This region is called the Brillouin Zone, after Léon Brillouin. In Fig. 30.3, the limits of the one-dimensional Brillouin Zone are indicated by two vertical dotted lines.

In more than one dimension, a more general definition is used for the Brillouin Zone. The periodicity in $\vec{k}$-space is defined by a set of three vectors, $\{\vec{K}_1, \vec{K}_2, \vec{K}_3\}$. The set of all vectors of the form $n_1\vec{K}_1 + n_2\vec{K}_2 + n_3\vec{K}_3$, where the n_js are integers, forms a 'reciprocal lattice'. The Brillouin Zone is then defined as including those points closer to the origin in $\vec{k}$-space than to any other point in the reciprocal lattice. For the one-dimensional case, the reciprocal lattice consists of the points, $2n\pi/a$, where n is an integer. The points between $-\pi/a$ and π/a then satisfy this definition of the Brillouin Zone.

30.3.3   Non-Zero Periodic Potential

Now we are going to discuss the effects of making the amplitude of the periodic potential non-zero.

If we consider Fig. 30.3, we see that there are crossings of the energy levels at $k = n\pi/a$, where n is an integer. These crossings represent degeneracies, where two states have the same energy (in the original representation, they are traveling waves going in opposite directions). A non-zero periodic potential splits the degeneracy at these crossings, according to the usual result of quantum perturbation theory. After the splitting, the energy spectrum looks like Fig. 30.4.

After the splitting of the crossing degeneracies, gaps in the energy spectrum open up. There are now ranges of energies for which there are no solutions to the wave-function equation.

30.4   Energy Bands and Energy Gaps

The consequences of both the tight-binding and the nearly-free electron analyses are qualitatively the same. In certain ranges of energies, there exist single-particle energy states. These energy ranges are called 'bands'. In other energy ranges, no such states are found. These energy ranges are called 'gaps'.

Both the tight-binding and the nearly-free electron analysis can be used as starting points for quantitative calculations of the energy spectrum and the density of states. However, that would take us beyond the scope of this book.

The existence of energy gaps is clearly an important feature of the density of states. As we will see in the next section, when the Fermi energy falls within a band gap, the

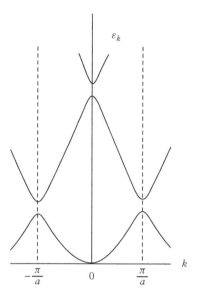

Fig. 30.4 *Band structure with energy gaps occurring at $k = \frac{\pi}{a}, 0,$ and $-\frac{\pi}{a}$.*

behavior of the system differs dramatically from what we found for the free-electron model.

30.5 Where is the Fermi Energy?

When the energy spectrum has band gaps, the behavior of the material will depend strongly on where the Fermi energy is – specifically, on whether the Fermi energy lies within a band or within a gap between bands. This, in turn, depends on whether the bands near the Fermi energy are empty, completely filled, or partially filled at zero temperature. If there is a partially filled band at $T = 0$, the Fermi energy will lie at the boundary between the filled and empty states. If there are only completely filled and completely empty bands at $T = 0$, the Fermi energy will lie halfway between the highest filled band and the lowest empty band.

To understand why the behavior of a material depends strongly on where the Fermi energy is, first recall that when a single-particle state is filled, no other electron can enter that state. If all single-particle states at a given energy are filled, nothing can happen at that energy. Neither can anything happen at an energy for which all states are empty. It is only when electrons can change states without changing their energy significantly that they can move from one area to another within the system.

The explicit dependence of the electrical properties of systems on their band structures gives rise to the differences between metals, insulators, and semiconductors, which will be discussed in the following sections.

30.6   Fermi Energy in a Band (Metals)

If the Fermi energy falls within a band, it requires only a tiny amount of thermal energy to raise an electron into an empty energy level with a slightly higher energy. The electron is then able to move to other states with the same energy in different parts of the system. Since electrons are charged, their motion is electric current. Systems with Fermi energies that fall within a band are metals.

In Chapter 29 we found from the Sommerfeld expansion that the specific heat at low temperatures was given by eq. (29.46),

$$C_V = \frac{\pi^2}{3} D_F(\epsilon_F) k_B^2 T. \tag{30.20}$$

For free electrons, this gave the particular result given in eq. (29.53).

If the Fermi energy lies in one of the energy bands, eq. (30.20) still holds. The only change from the free electron result is quantitative, due to the change in the density of single-electron states at the Fermi energy. For some important conductors, such as copper, silver, and gold, the free-electron specific heat is within 40% of the measured value of the specific heat at low temperatures. For other metals, the magnitude of the change can be quite large—up to an order of magnitude larger than the free-electron specific heat. The details of band structure that give rise to these differences can be quite interesting, but go beyond the scope of this book.

If the Fermi energy falls in an energy gap, the behavior of the system is qualitatively different from that of a metal, and we find insulators and semiconductors. Their properties are dramatically different from those of metals—and from each other! We will consider each case separately in the following sections, but first we will discuss the generic equations that govern both insulators and intrinsic semiconductors (defined below).

30.7   Fermi Energy in a Gap

When there is an energy gap between the highest filled single-particle state and the lowest empty one, the Fermi energy must be exactly in the middle of the gap. This is a very general principle, which was already implicit in the calculations for the cases of discrete sets of energy levels that were treated in Chapter 29. To see that this is so, introduce a specific model for which the density of states is given by

$$D(\epsilon) = \begin{cases} A(\epsilon - \epsilon_C)^a & \epsilon > \epsilon_C \\ 0 & \epsilon_C > \epsilon > \epsilon_V \\ B(\epsilon_V - \epsilon)^b & \epsilon_V > \epsilon, \end{cases} \tag{30.21}$$

where A, B, a, and b are positive constants. The usual values of the exponents are $a = b = 1/2$—the demonstration of which is left as an exercise for the reader—but for now we will keep the notation general.

The subscripts C and V in (30.21) refer to the 'conduction' (upper) and 'valence' (lower) bands. Conduction will occur only in so far as electrons are thermally excited from the valence band into the conduction band.

Since we are interested in very low temperatures when calculating the Fermi energy, states with energies much greater than ϵ_C are nearly empty, while those with energies much less than ϵ_V are nearly full. Therefore, the form of the density of states away from the gap is often unimportant.

It might seem extremely improbable that the filled single-particle state with the highest energy (at $T=0$) is exactly at the top of a band, while the lowest empty state is at the bottom of the next higher band. Actually, it is quite common. The reason is most easily seen from the tight-binding picture. A band is formed from the splitting of atomic energy levels. If these energy levels are filled, then the bands states will also be filled. If the energies in a band do not overlap the energies of any other band, it will be completely filled, while the higher bands will be empty.

As usual, we will use eq. (27.61), which gives the number of particles N as a function of μ, to find the chemical potential

$$N = \int_0^\infty D(\epsilon)\,(\exp[\beta(\epsilon-\mu)]+1)^{-1}\,d\epsilon. \tag{30.22}$$

Splitting the integral into two parts—one for each band—gives

$$N = \int_0^{\epsilon_V} D(\epsilon)\,(\exp[\beta(\epsilon-\mu)]+1)^{-1}\,d\epsilon + \int_{\epsilon_C}^\infty D(\epsilon)\,(\exp[\beta(\epsilon-\mu)]+1)^{-1}\,d\epsilon. \tag{30.23}$$

By our assumptions, the lower band is exactly filled at zero temperature,

$$N = \int_0^{\epsilon_V} D(\epsilon)\,d\epsilon. \tag{30.24}$$

Subtracting eq.(30.24) from eq.(30.23), we can eliminate N from the equation for the chemical potential

$$0 = \int_{\epsilon_C}^\infty D(\epsilon)\,(\exp[\beta(\epsilon-\mu)]+1)^{-1}\,d\epsilon + \int_0^{\epsilon_V} D(\epsilon)\left[(\exp[\beta(\epsilon-\mu)]+1)^{-1}-1\right]d\epsilon. \tag{30.25}$$

Using the identity from Section 29.3, we can rewrite the resulting equation in a form that shows the equality between the number of particles taken out of the lower band and the number of particles added to the upper band

$$\int_{\epsilon_C}^{\infty} D(\epsilon)\left(\exp[\beta(\epsilon - \mu)] + 1\right)^{-1} d\epsilon = \int_{0}^{\epsilon_V} D(\epsilon)\left(\exp[-\beta(\epsilon - \mu)] + 1\right)^{-1} d\epsilon. \quad (30.26)$$

Anticipating that $\mu \approx \epsilon_F = (\epsilon_C + \epsilon_V)/2$ at low temperatures, we see that the exponentials in the denominators of both integrands are large. This allows us to approximate eq. (30.26) as

$$A \int_{\epsilon_C}^{\infty} (\epsilon - \epsilon_C)^a \exp[-\beta(\epsilon - \mu)] d\epsilon \approx B \int_{-\infty}^{\epsilon_V} (\epsilon_V - \epsilon)^b \exp[\beta(\epsilon - \mu)] d\epsilon, \quad (30.27)$$

where we have inserted the approximate form for the density of states in eq. (30.21). We have also changed the lower limit of the integral on the right to minus infinity, since the exponential in the integrand makes the lower limit unimportant at low temperatures.

We do not need to evaluate the integrals in eq. (30.27) explicitly at this point, but we do need to extract the β-dependence. We do this by defining new variables for the integral on the left of eq. (30.27),

$$x = \beta(\epsilon - \epsilon_C), \quad (30.28)$$

and for the integral on the right,

$$y = -\beta(\epsilon - \epsilon_V). \quad (30.29)$$

Eq. (30.27) then becomes

$$A \exp\left(\beta(\mu - \epsilon_C)\right) \beta^{-1-a} \int_{0}^{\infty} x^a \exp[-x] dx =$$

$$B \exp\left(\beta(\epsilon_V - \mu)\right) \beta^{-1-b} \int_{0}^{\infty} y^b \exp[-y] dy. \quad (30.30)$$

Now bring the exponentials with μ to the left side of the equation,

$$\exp\left(\beta(2\mu - \epsilon_C - \epsilon_V)\right) = \left(\frac{B \int_{0}^{\infty} y^b \exp[-y] dy}{A \int_{0}^{\infty} x^a \exp[-x] dx}\right) (k_B T)^{b-a}. \quad (30.31)$$

Denoting the dimensionless constant X by

$$X = \frac{B \int_{0}^{\infty} y^b \exp[-y] dy}{A \int_{0}^{\infty} x^a \exp[-x] dx}, \quad (30.32)$$

we find

$$\mu = \frac{1}{2}(\epsilon_C + \epsilon_V) + \frac{k_B T}{2}(b-a)\ln(k_B T) + \frac{k_B T}{2}\ln X. \tag{30.33}$$

Clearly, the Fermi energy is given by

$$\epsilon_F = \lim_{T \to 0} \mu = \frac{1}{2}(\epsilon_C + \epsilon_V), \tag{30.34}$$

for any values of the constants A, B, a, and b.

As mentioned at the beginning of this section, the most usual values of the exponents are $a = b = 1/2$. For this case, eq. (30.33) simplifies.

$$\mu = \frac{1}{2}(\epsilon_C + \epsilon_V) + \frac{k_B T}{2}\ln\left(\frac{B}{A}\right) \tag{30.35}$$

The calculation in this section also gives us an explicit estimate for the number of electrons in partially filled energy levels that can contribute to an electric current. The number of electrons that are thermally excited to the upper band is given by the first term on the right of eq. (30.23),

$$N_e = \int_{\epsilon_C}^{\infty} D(\epsilon)(\exp[\beta(\epsilon - \mu)] + 1)^{-1}\, d\epsilon \tag{30.36}$$

$$\approx \int_{\epsilon_C}^{\infty} D(\epsilon)\exp[-\beta(\epsilon - (\epsilon_C + \epsilon_V)/2]d\epsilon$$

$$\approx \exp(-\beta(\epsilon_C - \epsilon_V)/2)\int_{\epsilon_C}^{\infty} D(\epsilon)\exp[-\beta(\epsilon - \epsilon_C)]d\epsilon$$

$$\approx \exp(-\beta(\epsilon_C - \epsilon_V)/2)A\int_{\epsilon_C}^{\infty} (\epsilon - \epsilon_C)^{1/2}\exp[-\beta(\epsilon - \epsilon_C)]d\epsilon$$

$$= \exp(-\beta(\epsilon_C - \epsilon_V)/2)A\int_{0}^{\infty} (x/\beta)^{1/2}\exp[-x]\beta^{-1}dx$$

$$= \exp(-\beta(\epsilon_C - \epsilon_V)/2)A(k_B T)^{3/2}\int_{0}^{\infty} x^{1/2}\exp[-x]dx$$

$$= \frac{A\sqrt{\pi}}{2}(k_B T)^{3/2}\exp(-\beta(\epsilon_C - \epsilon_V)/2).$$

The factor $\exp(-\beta(\epsilon_C - \epsilon_V)/2)$ in the last line of eq. (30.36) shows that the number of electrons in the upper band will be exponentially small when the band gap $\epsilon_{gap} = \epsilon_C - \epsilon_V$ is much greater than $k_B T$.

For an insulator, the band gap, ϵ_{gap}, between the highest filled ('valence') and lowest empty ('conduction') band is large; that is, greater than roughly $3\,eV$. The reason for the choice of this (somewhat arbitrary) number is that a band gap of $3\,eV$ implies that at room temperature $(T = 300\,K)$, $\beta \epsilon_{gap}/2 \approx 58$, so that $\exp(-\beta \epsilon_{gap}/2) \approx 6.5 \times 10^{-26}$. This factor in eq. (30.36) is so small that essentially no electrons are found in the conduction band of an insulator with a band gap of $3\,eV$ or greater. Without conduction electrons, no current can flow, and the material is an insulator.

30.8 Intrinsic Semiconductors

A rough definition of a pure (or intrinsic) semiconductor is a material with a band structure like an insulator, but with a smaller band gap. Examples are given in Table 30.1.

For an intrinsic semiconductor like germanium (Ge), the factor of $\exp(-\beta \epsilon_{gap}/2)$ in eq. (30.36) is about $\exp(-\beta \epsilon_{gap}/2) \approx \exp(-13) \approx 2 \times 10^{-6}$ at room temperature $(300\,K)$. While this value is small, there are still a substantial number of electrons that are thermally excited. The conductivity of pure germanium is much smaller than that of the good insulator, but it is not negligible.

30.9 Extrinsic Semiconductors

The main reason why semiconductors are so interesting—and extraordinarily useful in modern electronics—is that their electrical properties are very strongly influenced by the presence of relatively small concentrations of impurities. Indeed, that property provides a better definition of a semiconductor than having a small band gap.

Since the word 'impurities' tends to have negative connotations that would not be appropriate in this context, impurities in semiconductors are generally called 'dopants', and impure semiconductors are called 'extrinsic' semiconductors. There are two basic classes of dopant: donors and acceptors. The distinction is based on whether the particular dopant has more (donor) or fewer (acceptor) electrons than the intrinsic semiconductor. In the following two subsections, the two kinds of dopant are discussed.

Table 30.1 *Band gaps of semiconductors. The third column expresses the band gap in units of $k_B T$ at room temperature.*

Material	ϵ_{gap} (eV)	$\epsilon_{gap}/k_B T$
Ge	0.67	26
Si	1.14	43
GaAs	1.43	55
GaSb	0.7	27
InAs	0.36	24
InN	0.7	27

To have a concrete example, we will use silicon (*Si*) in both cases as the intrinsic semiconductor that is being doped.

The presence of dopants is important because they introduce new energy levels into the band gap. These energy levels have the effect of moving the Fermi energy from the middle of the gap, where it would be for an intrinsic semiconductor, to either just under the conduction band (donor) or just above the valence band (acceptor).

30.9.1   Donor Impurities

Phosphorus (*P*) is an example of a donor impurity for silicon (*Si*). The phosphorus ion, P^+, goes into the lattice in place of a *Si* atom, keeping its net positive charge. The remaining electron from the *P* atom goes into the conduction band. A semiconductor that contains donor dopants is called an '*n*-type' semiconductor because the dopants provide extra carriers with *negative* charge.

Since there is a Coulomb attraction between the net positive charge on the P^+ ion in the lattice and the donated electron in the conduction band, the combination looks very much like a hydrogen atom. The main differences are that the effective mass of electrons in the conduction band differs from that of a free electron, and there is a screening effect due to the dielectric constant of *Si*. The combination of these two effects results in a bound state with an energy of about $-0.044\,eV$, relative to the bottom of the conduction band. This donor state is close to the bottom of the conduction band, because the binding energy is much smaller than the band gap of $1.14\,eV$. The extra electron is in the new donor-state energy level, which lies in the band gap, about $0.044\,eV$ below the bottom of the conduction band.

The new energy level is the highest occupied level in the system. The lowest unoccupied level at the bottom of the conduction band, which is only about $0.044\,eV$ higher. As a result, the Fermi energy is moved from the center of the gap to about $0.022\,eV$ below the conduction band. Since the binding energy is only about half of $k_B T$ at room temperature for the example of *P* impurities in *Si*, even without calculations we can see that many more electrons will be excited into the conduction band with dopants than without.

30.9.2   Acceptor Impurities

Boron (*B*) is an example of an acceptor impurity for *Si*. The *B* atom goes into the lattice in place of a *Si* atom and binds an electron from the valence band, creating a B^- ion. Since an electron was removed from the valence band, the band is no longer full. The state with the missing electron is known as a 'hole'. For reasons that go beyond the scope of this book, the missing electron acts just like a *positive* mobile charge. A semiconductor that contains acceptor dopants is called a '*p*'-type semiconductor, because the dopants provide extra carriers with *positive* charge.

In analogy to the situation for donors, the mobile positive hole in the valence band is attracted to the negative B^- ion fixed in the lattice. Again, this creates a new localized state with an energy in the band gap. However, for '*p*'-type semiconductors the state is located just above the top of the valence band (about $0.010\,eV$ above for B in Si). The equations for calculation of the energy are essentially the same as those for the bound state in 'n'-type semiconductors. The resultant Fermi energy is located about $0.005\,eV$ above the top of the valence band.

30.10   Semiconductor Statistics

Although electrons are fermions, so that we might expect the equations for Fermi statistics to apply to the impurity states, the true situation is rather more interesting. The occupation of the donor and acceptor states is determined by what we might call 'semiconductor statistics', which we have not seen before in this book.

The reason for the difference between semiconductor statistics and Fermi statistics arises from two features of the impurity energy levels:

1. The impurity levels in doped semiconductors are two-fold degenerate because there are both spin-up and spin-down states.
2. The two states in a given level cannot both be occupied, because of a strong Coulomb repulsion.

These features mean that the probabilities for occupying the two degenerate states that make up the impurity energy level cannot be regarded as independent: the possibility of the energy level containing two electrons is excluded. Because of the lack of independence of the two impurity states, the derivation of semiconductor statistics given in the next subsection calculates the occupation of the entire energy level.

30.10.1   Derivation of Semiconductor Statistics

To derive semiconductor statistics—that is, the occupation number for an impurity energy level—we return to eq. (27.31) in Chapter 27,

$$\mathcal{Z} = \prod_{\alpha} \sum_{n_\alpha} \exp[-\beta(\epsilon_\alpha - \mu)n_\alpha]. \tag{30.37}$$

For an donor energy level with energy ϵ_d, the sum over n_{ϵ_d} must only contain terms corresponding to (1) both states empty, (2) only the spin-up state occupied, and (3) only the spin-down state occupied. Double occupancy of the donor energy level is excluded due to the Coulomb repulsion energy,

$$\sum_{n_{\epsilon_d}} \exp[-\beta(\epsilon_d - \mu)n_{\epsilon_d}] = \exp[0] + \exp[-\beta(\epsilon_d - \mu)] + \exp[-\beta(\epsilon_d - \mu)] \tag{30.38}$$

$$= 1 + 2\exp[-\beta(\epsilon_d - \mu)].$$

Inserting eq. (30.38) in eq. (27.44) for the average occupation number, we find

$$\langle n_d \rangle = \frac{\sum_{n_\alpha} n_\alpha \exp[-\beta(\epsilon_\alpha - \mu)n_\alpha]}{\sum_{n_\alpha} \exp[-\beta(\epsilon_\alpha - \mu)n_\alpha]} \tag{30.39}$$

$$= \frac{2\exp[-\beta(\epsilon_d - \mu)]}{1 + 2\exp[-\beta(\epsilon_d - \mu)]}$$

$$= \left[\frac{1}{2}\exp[\beta(\epsilon_d - \mu)] + 1 \right]^{-1}.$$

The extra factor of $1/2$ in the denominator of the last line in eq. (30.39) distinguishes semiconductor statistics from fermion statistics. Naturally, the single-particle states in semiconductor electron bands still obey Fermi–Dirac statistics.

30.10.2   Simplified Illustration of Semiconductor Statistics

As an example of the consequences of semiconductor statistics, we will first consider a simplified model in which there are N_d donor impurities with a corresponding energy level ϵ_d, which lies slightly below the bottom of the conduction band (ϵ_C). We will assume that the valence band is filled, so that we can ignore it. To simplify the mathematics and the interpretation of the results, the conduction band will be approximated by assigning all N_C states the same energy, ϵ_C—not because it is realistic (it is not), but to show the effects of semiconductor statistics. A more realistic model will be presented in the next subsection.

As was the case for problems with either Fermi–Dirac or Bose–Einstein statistics, the first task is to find the chemical potential. The equation governing the chemical potential is, as usual, the equation for the total number of particles. Since each donor impurity contributes a single election, the total number of electrons is equal to N_d, the number of donor levels,

$$N = N_d = N_d \left[\frac{1}{2}\exp[\beta(\epsilon_d - \mu)] + 1 \right]^{-1} + N_C \left[\exp[\beta(\epsilon_C - \mu)] + 1\right]^{-1}. \tag{30.40}$$

Note that the electrons in the conduction band still obey Fermi–Dirac statistics.

We might have imagined that the Fermi energy coincides with the energy of the impurity level, ϵ_d, because of the two-fold degeneracy and single occupancy at $T = 0$. However, due to the exclusion of double occupancy of the impurity level, the Fermi energy turns out to be half way between the impurity-state energy and the bottom of the conduction band, $\epsilon_F = (\epsilon_C + \epsilon_d)/2$. We will assume that this is true in the following derivation of the Fermi energy, and confirm it by the results of the calculation.

To solve eq. (30.40), we can use a variation on the identity in eq. (29.7).

$$N_d \left[2\exp[-\beta(\epsilon_d - \mu)] + 1\right]^{-1} = N_C \left[\exp[\beta(\epsilon_C - \mu)] + 1\right]^{-1}. \tag{30.41}$$

At low temperatures, since we expect $\mu \approx (\epsilon_d + \epsilon_C)/2$, we have both $\beta(\mu - \epsilon_d) \gg 1$ and $\beta(\epsilon_C - \mu) \gg 1$. This means that the exponentials dominate in both denominators in eq. (30.41), and the neglect of the '+1' in both denominators is a good approximation,

$$\frac{N_d}{2} \exp[\beta(\epsilon_d - \mu)] \approx N_C \exp[-\beta(\epsilon_C - \mu)]. \tag{30.42}$$

We can solve this equation for the chemical potential.

$$\mu \approx \frac{\epsilon_C + \epsilon_d}{2} + \frac{1}{2} k_B T \ln\left(\frac{N_d}{2N_C}\right) \tag{30.43}$$

Since the Fermi energy is the zero temperature limit of the chemical potential, we can immediately see from eq. (30.43) that

$$\epsilon_F = \frac{\epsilon_C + \epsilon_d}{2} \tag{30.44}$$

On the other hand, because the difference in energy between ϵ_C and ϵ_d can be of the order of $k_B T$, the second term in eq. (30.43) is not always negligible. There are usually far fewer impurities than conduction states, and we can easily have $N_d \approx 10^{-5} N_C$. The second term in eq. (30.43) would be roughly $-12 k_B T$. To see how significant this effect is, we need to calculate the occupancy of the donor levels:

$$\langle n_d \rangle = N_d \left[\frac{1}{2} \exp\left[\beta \left(\epsilon_d - \mu\right)\right] + 1 \right]^{-1} \tag{30.45}$$

Inserting eq. (30.43) for the chemical potential, we find

$$\langle n_d \rangle \approx N_d \left[\frac{1}{2} \exp\left[\beta \left(\epsilon_d - \frac{\epsilon_C + \epsilon_d}{2} \right) - \frac{1}{2} \ln\left(\frac{N_d}{2N_C}\right) \right] + 1 \right]^{-1} \tag{30.46}$$

or

$$\langle n_d \rangle \approx N_d \left[\frac{1}{2} \left(\frac{2N_C}{N_d} \right)^{1/2} \exp\left[-\beta \left(\frac{\epsilon_C - \epsilon_d}{2} \right) \right] + 1 \right]^{-1}. \tag{30.47}$$

There are two competing factors in the first term in the denominator of eq. (30.47). While the factor of $\exp[-\beta(\epsilon_C - \epsilon_d)/2]$ can be small for very low temperatures, for *As* in *Ge* at room temperature it has a value of about 0.6. On the other hand, if $N_d \approx 10^{-5} N_C$, the factor of $\sqrt{2N_C/N_d} \approx 450$, so that it would dominate, making the occupation of the donor state quite small. Under such circumstances, almost all of the donor levels would be empty, and the electrons would be in the conduction band.

Note that the emptying of the donor levels is an entropic effect. There are so many more states in the conduction band that the probability of an electron being in a donor level is very small, even though it is energetically favorable. This feature is not an artifact of the simple model we are using for illustration in this section; it is a real feature of semiconductor physics.

In the next section we will consider a (slightly) more realistic model that removes the unphysical assumption of an infinitely thin conduction band.

30.10.3   A More Realistic Model of a Semiconductor

Although our neglect of the valence band in the model used in the previous subsection is often justifiable, giving all states in the conduction band the same energy is certainly not. In this subsection, we will introduce a model that is realistic enough to bear comparison with experiment – although we will leave such a comparison to a more specialized book.

We will assume the donor levels all have energy ϵ_d, as in Subsection 30.10.2. However, we will model the conduction band by a density of states as in eq. (30.21). The total density of states can then be written as

$$D(\epsilon) = \begin{cases} A(\epsilon - \epsilon_C)^{1/2} & \epsilon \geq \epsilon_C \\ N_d \delta(\epsilon - \epsilon_d) & \epsilon < \epsilon_C. \end{cases} \tag{30.48}$$

Since only single-particle states near the bottom of the conduction band will be occupied, the upper edge of the band is taken to be infinity.

The equation for the total number of donor electrons, which gives us the chemical potential, is then

$$N_d = N_d \left[\frac{1}{2} \exp[\beta(\epsilon_d - \mu)] + 1 \right]^{-1} \tag{30.49}$$

$$+ \int_{\epsilon_C}^{\infty} A(\epsilon - \epsilon_C)^{1/2} \left[(\exp[\beta(\epsilon - \mu)] + 1)^{-1} - 1 \right] d\epsilon.$$

As in the previous section, we will again use a variation on the identity in eq. (29.7) to simplify the equation. We will also introduce the same dimensionless variable defined in eq. (30.28),

$$N_d \left[2\exp[-\beta(\epsilon_d - \mu)] + 1 \right]^{-1} = A \exp\left(\beta(\mu - \epsilon_C)\right) \beta^{-1-a} \int_0^{\infty} x^{1/2} \exp[-x] dx. \tag{30.50}$$

Since we can evaluate the integral, this becomes

$$N_d \left[2\exp[-\beta(\epsilon_d - \mu)] + 1 \right]^{-1} = \frac{A\sqrt{\pi}}{2} (k_B T)^{3/2} \exp\left(\beta(\mu - \epsilon_C)\right). \tag{30.51}$$

The Fermi energy again turns out to be half way between the energy level of the donor state and the bottom of the conduction band,

$$\epsilon_F = \frac{\epsilon_d + \epsilon_C}{2}. \tag{30.52}$$

The remaining behavior of this more realistic model is qualitatively the same as that of the simplified illustration in the previous subsection. The details will be left to the reader.

30.11   Semiconductor Physics

This brings us to the end of our introduction to insulators and semiconductors. The field of semiconductors, in particular, is extremely rich. We have covered the bare basics, but we have not even mentioned how semiconductors can be used to create lasers or the transistors that are the basis of all modern electronics. However, this introduction should be sufficient preparation to read a book on semiconductors without too much difficulty. The field is both fascinating and extraordinarily important in today's world.

31

Phase Transitions and the Ising Model

Magnet, n. *Something acted upon by magnetism.*
Magnetism, n. *Something acting upon a magnet.*
The two definitions immediately foregoing are condensed from the works of one thousand eminent scientists, who have illuminated the subject with a great white light, to the inexpressible advancement of human knowledge.

Ambrose Bierce, in *The Devil's Dictionary*

The Ising model is deceptively simple. It can be defined in a few words, but it displays astonishingly rich behavior. It originated as a model of ferromagnetism in which the magnetic moments were localized on lattice sites and had only two allowed values. This corresponds, of course, to a spin-one-half model, but since it is tiresome to continually write factors of $\hbar/2$, it is traditional to take the values of a 'spin' σ_j, located on lattice site j to be

$$\sigma_j = +1 \text{ or } -1. \tag{31.1}$$

The energy of interaction with a magnetic field also requires a factor of the magnetic moment of the spin, but to make the notation more compact, everything will be combined into a single 'magnetic field' variable h, which has units of energy,

$$H_h = -h\sigma_j. \tag{31.2}$$

Eq. (31.2) implies that the low-energy state is for $\sigma_j = +1$ when $h > 0$.

There should not be any confusion of the 'h' in eq. (31.2) with Planck's constant, because the latter does not appear in this chapter.

The energy of interaction between spins on neighboring lattice sites, j and k, is given by the product of the spins,

An Introduction to Statistical Mechanics and Thermodynamics. Robert H. Swendsen, Oxford University Press (2020).
© Robert H. Swendsen. DOI: 10.1093/oso/9780198853237.001.0001

$$H_{\mathcal{J}} = -\mathcal{J}\sigma_j\sigma_k. \tag{31.3}$$

The constant $\mathcal{J}$ is taken to have units of energy. The low-energy states of eq. (31.3) are $\sigma_j = \sigma_k = +1$ and $\sigma_j = \sigma_k = -1$ for $\mathcal{J} > 0$. This form of interaction is also called an 'exchange' interaction, due to the role of exchanging electrons in its derivation from the quantum properties of neighboring atoms. That derivation is, of course, beyond the scope of this book.

If we put the two kinds of interactions together on a lattice, we can write the Hamiltonian as

$$H = -\mathcal{J}\sum_{\langle j,k \rangle}\sigma_j\sigma_k - h\sum_{j=1}^{N}\sigma_j \tag{31.4}$$

where the notation $\langle j, k \rangle$ denotes that the sum is over nearest-neighbor pairs of sites. In the rest of this chapter we will discuss the remarkable properties of the model described by eq. (31.4).

31.1   The Ising Chain

The first calculations on the Ising model were carried out by Ernst Ising (German physicist, 1900–1998), who was given the problem by his adviser, Wilhelm Lenz (German physicist, 1888–1957). In 1924 Ising solved a one-dimensional model with N spins, in which each spin interacted with its nearest neighbors and a magnetic field,

$$H = -\mathcal{J}\sum_{j}\sigma_j\sigma_{j+1} - h\sum_{j=1}^{N}\sigma_j. \tag{31.5}$$

This model is also referred to as an 'Ising chain'. The limits on the first sum have not been specified explicitly in eq. (31.5) to allow it to describe two kinds of boundary conditions.

1. Open boundary conditions
      The first sum goes from $j = 1$ to $j = N - 1$. This corresponds to a linear chain for which the first spin only interacts with the second spin, and the spin at $j = N$ spin only interacts with the spin at $j = N - 1$.
2. Periodic boundary conditions
      A spin $\sigma_{N+1} = \sigma_1$ is defined, and the first sum goes from $j = 1$ to $j = N$. This corresponds to a chain of interactions around a closed loop.

In both cases, the second sum runs from $j = 1$ to $j = N$ to include all spins.
      We will investigate various aspects of this model in the following sections before going on to discuss the (qualitatively different) behavior in more than one dimension. We begin by ignoring the interactions and setting $\mathcal{J} = 0$ in the next section.

31.2   The Ising Chain in a Magnetic Field ($J = 0$)

Begin by setting $\mathcal{J} = 0$ in eq. (31.5),

$$H = -h \sum_{j=1}^{N} \sigma_j. \tag{31.6}$$

The canonical partition function is then given by

$$Z = \sum_{\{\sigma\}} \exp[-\beta(-h \sum_{j=1}^{N} \sigma_j)], \tag{31.7}$$

where we have used the notation that $\{\sigma\} = \{\sigma_j | j = 1, \cdots, N\}$ is the set of all spins.

Eq. (31.7) can be simplified by using the best trick in statistical mechanics (factorization of the partition function: Sections 19.12 and 24.7),

$$Z = \sum_{\{\sigma\}} \prod_{j=1}^{N} \exp[\beta h \sigma_j] = \prod_{j=1}^{N} \sum_{\sigma_j} \exp[\beta h \sigma_j] = \prod_{j=1}^{N} Z_j, \tag{31.8}$$

where

$$Z_j = \sum_{\sigma_j} \exp[\beta h \sigma_j]. \tag{31.9}$$

This reduces the N-spin problem to N-identical single-spin problems, each of which involves only a sum over two states.

Since each $\sigma_j = \pm 1$, the sums can be carried out explicitly:

$$Z_j = \exp(\beta h) + \exp(-\beta h) = 2 \cosh(\beta h). \tag{31.10}$$

It is easy to show that the average magnetization is given by

$$m = m_j = \langle \sigma_j \rangle = \frac{\exp(\beta h) - \exp(-\beta h)}{\exp(\beta h) + \exp(-\beta h)} = \tanh(\beta h). \tag{31.11}$$

Since it will be important later to be familiar with the graphical form of eq. (31.11), it is shown in Fig. 31.1. The shape of the curve makes sense in that it shows that $m = 0$ when $h = 0$, while $m \to \pm 1$ when $h \to \pm\infty$.

The average energy of a single spin is then easily found,

$$U_j = \langle -h\sigma_j \rangle = -hm_j = -hm = -h \tanh(\beta h). \tag{31.12}$$

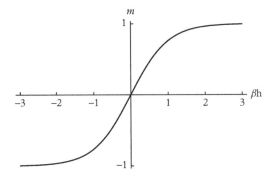

Fig. 31.1 *Plot of the magnetization of an isolated spin against the magnetic field,* $m = \tanh(\beta h)$.

The energy of the entire Ising chain is given by simply multiplying by N,

$$U = \langle H \rangle = -h \sum_{j=1}^{N} \langle \sigma_j \rangle = -hNm = -hN \tanh(\beta h). \tag{31.13}$$

The specific heat is found by differentiating with respect to the temperature,

$$c = \frac{1}{N} \frac{\partial U}{\partial T} = \frac{-1}{N k_B T^2} \frac{\partial U}{\partial \beta} = k_B \beta^2 h^2 \operatorname{sech}^2(\beta h). \tag{31.14}$$

The specific heat goes to zero at both $T = 0$ and $T = \infty$.

31.3 The Ising Chain with $h = 0$, but $J \neq 0$

Next, we will put the interactions between spins back into the Ising chain, but remove the magnetic field. To simplify the problem we will use open boundary conditions

$$H' = -\mathcal{J} \sum_{j=1}^{N-1} \sigma_j \sigma_{j+1}. \tag{31.15}$$

At first sight, finding the properties in this case looks considerably more difficult than what we did in the previous section. Each spin is linked to its neighbor, so the Hamiltonian cannot be split up in the same way to factorize the partition function. However, we can factorize the partition function if we introduce new variables.

Let $\tau_1 = \sigma_1$, and $\tau_j = \sigma_{j-1}\sigma_j$ for $2 \leq j \leq N$. Clearly, each τ_j takes on the values ± 1. Given the set of N values $\{\tau_j | j = 1, \ldots, N\}$, we can uniquely find the original values of the σ_j's.

The Hamiltonian in eq. (31.15) can now be rewritten as

$$H = -\mathcal{J} \sum_{j=2}^{N} \tau_j. \tag{31.16}$$

This makes the Hamiltonian look just like eq. (31.6), which we derived in the previous section for the case of independent spins in a magnetic field. Although the $N-1$ τ-variables all appear in the Hamiltonian, σ_1 does not. Summing over $\sigma_1 = \pm 1$ in the partition function is responsible for an overall factor of two in the equations below.

Because eq. (31.16) is a sum of independent terms, the partition function can again be calculated by factorization.

$$Z = \sum_{\{\tau\}} \prod_{j=1}^{N} \exp[\beta \mathcal{J} \tau_j] = 2 \prod_{j=2}^{N} \sum_{\tau_j} \exp[\beta \mathcal{J} \tau_j] = 2 \prod_{j=2}^{N} Z'_j \tag{31.17}$$

where

$$Z_j = \sum_{\tau_j = \pm 1} \exp[\beta \mathcal{J} \tau_j] = \exp[\beta \mathcal{J}] + \exp[-\beta \mathcal{J}] = 2 \cosh(\beta \mathcal{J}) \tag{31.18}$$

and

$$Z = 2(Z_j)^{N-1} = 2 \left(2 \cosh(\beta \mathcal{J})\right)^{N-1}. \tag{31.19}$$

The Massieu function, $\tilde{S}[\beta]$, for the Ising chain with $h = 0$ is given by the logarithm of the partition function in eq. (31.19)

$$\tilde{S}[\beta] = \ln Z = \ln 2 + (N-1) \ln Z_j = \ln 2 + (N-1) \ln \left(2 \cosh(\beta \mathcal{J})\right). \tag{31.20}$$

Relating the partition function to a Massieu function is preferable here because this model exhibits negative temperatures.

The energy of the system is easily found to be

$$U = -(N-1) \mathcal{J} \tanh(\beta \mathcal{J}), \tag{31.21}$$

and the specific heat per spin is

$$c = k_B \beta^2 \mathcal{J}^2 \operatorname{sech}^2(\beta \mathcal{J}) (1 - 1/N) \approx k_B \beta^2 \mathcal{J}^2 \operatorname{sech}^2(\beta \mathcal{J}). \tag{31.22}$$

Note that the form of eq. (31.22) is essentially the same as that of eq. (31.14), which we found in the previous section. The specific heat in this case also goes to zero at both $T = 0$ and $T = \infty$.

31.4   The Ising Chain with both $J \neq 0$ and $h \neq 0$

The full Hamiltonian in eq. (31.5) with both exchange interactions and a magnetic field
is more challenging. We cannot factorize the partition function in the same way that we
have done with other problems, so we will have to use a different approach, known as
the Transfer Matrix method. The basis of the method is that the partition function can
be expressed as a product of relatively small matrices—in this case, just 2×2 matrices—
which can then be diagonalized and their determinants evaluated.

31.4.1   Transfer Matrix

We will consider the case of periodic boundary conditions, so that $\sigma_{N+1} = \sigma_1$, as
discussed earlier. To make the matrices more symmetric, we will write the Hamiltonian,
eq. (31.5), in the form

$$H = -J \sum_{j=1}^{N} \sigma_j \sigma_{j+1} - \frac{h}{2} \sum_{j=1}^{N} (\sigma_j + \sigma_{j+1}).$$
(31.23)

The upper limits on the both summations is N, so that the last terms in the sums are
$-J\sigma_N\sigma_1$ and $(h/2)(\sigma_N + \sigma_1)$, due to the periodic boundary conditions.

Using eq. (31.23), we can transform the partition function into a product of 2×2
matrices,

$$Z = \sum_{\{\sigma\}} \exp\left(-\beta\left(-J\sum_j \sigma_j\sigma_{j+1} - \frac{1}{2}h\sum_j (\sigma_j + \sigma_{j+1})\right)\right)$$
(31.24)

$$= \sum_{\{\sigma\}} \prod_j \exp\left(\beta J\sigma_j\sigma_{j+1} + \frac{1}{2}\beta h(\sigma_j + \sigma_{j+1})\right)$$

$$= \sum_{\{\sigma\}} \prod_j T(\sigma_j, \sigma_{j+1}).$$

In this equation,

$$T(\sigma_j, \sigma_{j+1}) = \exp\left(\beta J\sigma_j\sigma_{j+1} + \frac{1}{2}\beta h(\sigma_j + \sigma_{j+1})\right),$$
(31.25)

is defined as the transfer matrix. Note that we have not exchanged the sums and products,
because we have not eliminated the coupling between variables. The sums remain and
can be interpreted as matrix multiplication. As the name implies, we can also write the
transfer matrix explicitly as a matrix,

$$T(\sigma_j, \sigma_{j+1}) = \begin{pmatrix} \exp(\beta\mathcal{J} + \beta h) & \exp(-\beta\mathcal{J}) \\ \exp(-\beta\mathcal{J}) & \exp(\beta\mathcal{J} - \beta h) \end{pmatrix}. \tag{31.26}$$

Eq. (31.24) is equivalent to representing the partition function as a product of matrices. To see this more clearly, consider the factors involving only three spins, somewhere in the middle of the chain, at locations $j - 1, j$, and $j + 1$,

$$\sum_{\sigma_j = \pm 1} T(\sigma_{j-1}, \sigma_j) T(\sigma_j, \sigma_{j+1}) = T^2(\sigma_{j-1}, \sigma_{j+1}) \tag{31.27}$$

The expression on the left of eq. (31.27) can also be written in terms of the explicit matrices from eq. (31.26),

$$\begin{pmatrix} \exp(\beta\mathcal{J} + \beta h) & \exp(-\beta\mathcal{J}) \\ \exp(-\beta\mathcal{J}) & \exp(\beta\mathcal{J} - \beta h) \end{pmatrix} \begin{pmatrix} \exp(\beta\mathcal{J} + \beta h) & \exp(-\beta\mathcal{J}) \\ \exp(-\beta\mathcal{J}) & \exp(\beta\mathcal{J} - \beta h) \end{pmatrix}. \tag{31.28}$$

Carrying out the matrix multiplication explicitly, we find

$$T^2(\sigma_{j-1}, \sigma_{j+1}) = \begin{pmatrix} \exp(2\beta\mathcal{J} + 2\beta h) + \exp(-2\beta\mathcal{J}) & \exp(\beta h) + \exp(-\beta h) \\ \exp(\beta h) + \exp(-\beta h) & \exp(2\beta\mathcal{J} - 2\beta h) + \exp(-2\beta\mathcal{J}) \end{pmatrix}. \tag{31.29}$$

If we carry out the sums over σ_2 through σ_N, we find the matrix $T^N(\sigma_1, \sigma_{N+1})$. Since $\sigma_{N+1} = \sigma_1$, the final sum over σ_1 yields the partition function

$$Z = \sum_{\sigma_1 = \pm 1} T^N(\sigma_1, \sigma_1). \tag{31.30}$$

Fortunately, we do not have to carry out all the matrix multiplications explicitly. That is the power of the Transfer Matrix method—as we will show in the next section.

31.4.2 Diagonalizing $T(\sigma_j, \sigma_{j+1})$

The next step is the one that makes this calculation relatively easy. The trace is invariant under a similarity transformation of the form

$$\tilde{T} = RTR^{-1}. \tag{31.31}$$

In this equation, R and its inverse R^{-1} are both 2×2 matrices. If we write out the product of T's we can see how it reduces to the original product of T's. The series of products brings the R and R^{-1} matrices together, so that their product gives unity. Using the compact notation, $T(\sigma_j, \sigma_{j+1}) \rightarrow T_{j,j+1}$, we see that

$$\tilde{T}_{1,2}\tilde{T}_{2,3}\tilde{T}_{3,4} = RT_{1,2}R^{-1}RT_{2,3}R^{-1}RT_{3,4}R^{-1} = RT_{1,2}T_{2,3}T_{3,4}R^{-1}. \tag{31.32}$$

This procedure can be extended to arbitrarily many T matrices. Since we have periodic boundary conditions, after applying the transformation eq. (31.31) N times and using $R^{-1}R = 1$ between the T-matrices, eq. (31.30) becomes

$$Z = \mathrm{Tr}R\tilde{T}^N R^{-1} = \mathrm{Tr}\tilde{T}^N. \tag{31.33}$$

For the last equality, we have used the fact that the trace of a product of matrices is invariant under a cyclic permutation.

Finding the trace is much easier if we capitalize on the freedom of making a similarity transformation to diagonalize the T matrix. The diagonal elements of T' will then be given by the eigenvalues λ_+ and λ_-,

$$\tilde{T}(\sigma_j, \sigma_{j+1}) = \begin{pmatrix} \lambda_+ & 0 \\ 0 & \lambda_- \end{pmatrix}. \tag{31.34}$$

The product of N diagonal matrices is trivial

$$(T')^N = \begin{pmatrix} \lambda_+^N & 0 \\ 0 & \lambda_-^N \end{pmatrix}, \tag{31.35}$$

and the partition function becomes

$$Z = \lambda_+^N + \lambda_-^N. \tag{31.36}$$

The only remaining task is to find the eigenvalues of the T matrix in eq. (31.25). This involves solving the eigenvalue equation

$$T\chi = \lambda\chi, \tag{31.37}$$

or

$$\begin{pmatrix} \exp(\beta\mathcal{J} + \beta h) & \exp(-\beta\mathcal{J}) \\ \exp(-\beta\mathcal{J}) & \exp(\beta\mathcal{J} - \beta h) \end{pmatrix} \begin{pmatrix} \chi_1 \\ \chi_2 \end{pmatrix} = \lambda \begin{pmatrix} \chi_1 \\ \chi_2 \end{pmatrix}. \tag{31.38}$$

This matrix equation is equivalent to two simultaneous equations for λ, and can be solved by setting the determinant equal to zero,

$$\begin{vmatrix} \exp(\beta\mathcal{J} + \beta h) - \lambda & \exp(-\beta\mathcal{J}) \\ \exp(-\beta\mathcal{J}) & \exp(\beta\mathcal{J} - \beta h) - \lambda \end{vmatrix} = 0. \tag{31.39}$$

Solving the resulting quadratic equation—and leaving the algebra to the reader—we find the required eigenvalues,

$$\lambda_\pm = e^{\beta \mathcal{J}} \cosh(\beta h) \pm \sqrt{e^{2\beta \mathcal{J}} \cosh^2(\beta h) - 2\sinh(2\beta \mathcal{J})}. \qquad (31.40)$$

The partition function of a system of N spins is given by inserting eq. (31.40) into eq. (31.36).

It is interesting to see how this form of the partition function behaves in the limit that $N \to \infty$ (the thermodynamic limit). Since $\lambda_+ > \lambda_-$, the ratio $\lambda_-/\lambda_+ < 1$. We can use this fact to rewrite eq. (31.36) in a convenient form

$$Z = \lambda_+^N \left(1 + \left(\frac{\lambda_-}{\lambda_+}\right)^N\right). \qquad (31.41)$$

The Massieu function then takes on a simple form

$$\tilde{S}[\beta] = \ln Z$$

$$= N\ln\lambda_+ + \ln\left(1 + \left(\frac{\lambda_-}{\lambda_+}\right)^N\right)$$

$$\approx N\ln\lambda_+ + \left(\frac{\lambda_-}{\lambda_+}\right)^N. \qquad (31.42)$$

The Massieu function per spin in the thermodynamic limit is just

$$\frac{\tilde{S}[\beta]}{N} \approx \ln\lambda_+ + \frac{1}{N}\left(\frac{\lambda_-}{\lambda_+}\right)^N \to \ln\lambda_+,$$

or, inserting λ_+ from eq. (31.40),

$$\lim_{N\to\infty} \frac{\tilde{S}[\beta]}{N} = \ln\left(e^{\beta \mathcal{J}} \cosh(\beta h) + \sqrt{e^{2\beta \mathcal{J}} \cosh^2(\beta h) - 2\sinh(2\beta \mathcal{J})}\right). \qquad (31.43)$$

Perhaps the most important thing to notice about eq. (31.43) is that the Massieu function is an analytic function of the temperature everywhere except at $\beta = \pm\infty$. Since a phase transition can be defined as a point at which the free energy is not analytic in the limit of an infinite system (thermodynamic limit), this means that the one-dimensional Ising model does not exhibit a phase transition at any non-zero temperature.

This lack of a phase transition was one of the main results that Ising found (by a method different from that which we have used). Based on this result, he speculated that the model would not show a phase transition in higher dimensions, either. In this speculation, he was wrong. However, it took from 1924, when Ising did his thesis work, until 1944, when Lars Onsager (Norwegian chemist and physicist, 1903–1976, who became an American citizen in 1945) derived an exact solution for the two-dimensional Ising model (without a magnetic field) that showed a phase transition.

Although Onsager's exact solution of the Ising model is beyond the scope of this book, we will discuss some approximate methods for calculating the behavior of the Ising model at its phase transition, beginning with the Mean-Field Approximation in the next section.

31.5   Mean-Field Approximation

The history of the Mean-Field Approximation (MFA) is interesting in that it precedes the Ising model by seventeen years. In 1907, Pierre-Ernest Weiss (French physicist, 1865–1940) formulated the idea of describing ferromagnetism with a model of small magnetic moments arranged on a crystal lattice. He assumed that each magnetic moment should be aligned by an average 'molecular field' due to the other magnetic moments, so that the idea was originally known as the 'Molecular Field Theory'. The justification of the model was rather vague, but comparison with experiment showed it to be remarkably good.

In the following subsections we will derive the Mean-Field Approximation for the Ising model in an arbitrary number of dimensions,

$$H = -\mathcal{J} \sum_{\langle j,k \rangle} \sigma_j \sigma_k - h \sum_{j=1}^{N} \sigma_j. \tag{31.44}$$

Recall that the notation $\langle j, k \rangle$ indicates that the summation in the first term is taken over the set of all nearest-neighbor pairs of sites in the lattice.

31.5.1   MFA Hamiltonian

The basic idea of MFA is to approximate the effects of interactions on a spin by its neighbors as an average (or 'mean') magnetic field. Specifically, we would like to calculate an approximation to the average magnetization of a spin at site j,

$$m_j = \langle \sigma_j \rangle. \tag{31.45}$$

For simplicity, we will assume periodic boundary conditions, so that the properties of the model are translationally invariant. Since all spins see the same average environment, they should all have the same average magnetization, m,

$$m = \frac{1}{N} \sum_k m_k = m_j. \tag{31.46}$$

For any given spin, σ_j, we can express the sum over the nearest-neighbor sites as a sum over the neighbors δ, and extract all terms in eq. (31.44) that contain σ_j,

$$H_j = -\mathcal{J}\sum_\delta \sigma_j\sigma_{j+\delta} - h\sigma_j = -\left(\mathcal{J}\sum_{j+\delta}\sigma_{j+\delta} + h\right)\sigma_j. \tag{31.47}$$

Although eq. (31.47) can be regarded as a reduced Hamiltonian for spin σ_j, we cannot interpret its average value as representing the average energy per spin, since $H \neq \sum_j H_j$ because of double counting the exchange interactions. This is a mental hazard that has caught many students and even a few textbook authors.

Note that eq. (31.47) has the same form as the Hamiltonian for a single spin in a magnetic field, $-h\sigma$, from eq. (31.2), except that the external magnetic field is replaced by an effective magnetic field,

$$h_{j,\text{eff}} = \mathcal{J}\sum_\delta \sigma_{j+\delta} + h. \tag{31.48}$$

The difficulty with the effective field found in eq. (31.48) is that it depends on the instantaneous values of the neighboring spins, which are subject to fluctuations. MFA consists of replacing the fluctuating spins in eq. (31.48) by their average (or mean) values, which must all be equal to the magnetization m, due to translational invariance,

$$h_o^{MFA} = \mathcal{J}\sum_\delta \langle\sigma_{j+\delta}\rangle + h = \mathcal{J}zm + h. \tag{31.49}$$

In the second equality in eq. (31.49) we have introduced the parameter z to denote the number of nearest-neighbor sites in the lattice of interest ($z = 4$ for a two-dimensional square lattice). The MFA Hamiltonian,

$$H_o^{MFA} = -h_o^{MFA}\sigma_o, \tag{31.50}$$

has exactly the same form as eq. (31.2). An equation for the magnetization can be written down immediately in analogy to eq. (31.11),

$$m = \tanh(\beta h_o^{MFA}) = \tanh(\beta\mathcal{J}zm + \beta h). \tag{31.51}$$

Unfortunately, there is no closed-form solution for eq. (31.51). However, there is a nice graphical solution that makes it relatively easy to understand the essential behavior of the solution, as well as guiding us to good approximations.

31.5.2   Graphical Solution of eq. (31.51) for $h = 0$

First consider the case of a vanishing magnetic field, $h = 0$, for which eq. (31.51) simplifies,

$$m = \tanh(\beta \mathcal{J} zm). \tag{31.52}$$

It might seem strange, but we can further simplify the analysis of eq. (31.52) by expressing it as two equations,

$$m = \tanh(x) \tag{31.53}$$

$$x = \beta \mathcal{J} zm. \tag{31.54}$$

The second equation can then be rewritten as

$$m = \left(\frac{k_B T}{z \mathcal{J}} \right) x. \tag{31.55}$$

Since eqs. (31.53) and (31.55) are both equations for m as a function of x, it is natural to plot them on the same graph, as shown in Fig. 31.2. Intersections of the two functions then represent solutions of eq. (31.52).

When the slope of the line described by eq. (31.55) equals 1, it provides a dividing line between two qualitatively different kinds of behavior. The temperature at which this happens is known as the critical temperature, T_c, or the Curie temperature, in honor of Pierre Curie (French physicist, 1859–1906, awarded the 1903 Nobel Prize, together with his wife, Maria Sklodowska-Curie, 1867–1934)

$$k_B T_c = z \mathcal{J}. \tag{31.56}$$

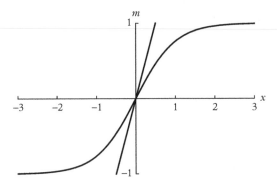

Fig. 31.2 *Plot of* $m = \tanh x$ *and* $m = (T/T_c)x$ *for* $T > T_c$.

This expression for the mean-field value of T_c means that eq. (31.55) can be rewritten as

$$m = \left(\frac{T}{T_c}\right) x. \tag{31.57}$$

If $T > T_c$, the slope of the straight line described by eq. (31.57) is greater than 1, and the two curves in Fig. 31.2 will only intersect at $m = 0$ and $x = 0$. This gives the reasonable result that the magnetization above T_c is zero.

For temperatures below T_c there are three intersections of the two curves, as shown in Fig. 31.3. The intersection at $m = 0$ is not stable, which will be left to the reader to show. The other two intersections provide symmetric solutions $m = \pm m(T)$, which are both stable.

The two solutions found in Fig. 31.3 are expected by the symmetry of the Ising Hamiltonian under the transformation $\sigma_j \rightarrow -\sigma_j$ for all spins.

31.5.3 Graphical Solution of eq. (31.51) for $h \neq 0$

The situation is qualitatively different when the magnetic field, h, is non-zero. The Ising Hamiltonian, eq. (31.4), is no longer invariant under the transformation $\sigma_j \rightarrow -\sigma_j$ for all spins, because the second term changes sign. Eq. (31.51) can again be simplified by writing it in terms of two equations,

$$m = \tanh(x) \tag{31.58}$$
$$x = \beta \mathcal{J} z m + \beta h. \tag{31.59}$$

The second equation can then be rewritten as

$$m = \left(\frac{k_B T}{z \mathcal{J}}\right) x - \frac{h}{z \mathcal{J}}. \tag{31.60}$$

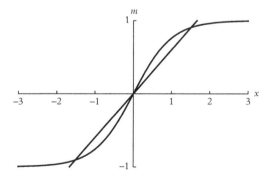

Fig. 31.3 *Plot of $m = \tanh x$ and $m = (T_c/T)x$ for $T < T_c$.*

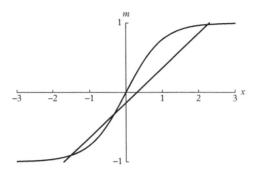

Fig. 31.4 *Plot of $m = x$ and $m = (T_c/T)x - h/k_BT$ for $T < T_c$.*

Fig. 31.4 shows a plot of eqs. (31.58) and (31.59) on the same graph. In analogy to the previous section, the intersections of the two functions represent solutions to eq. (31.11).

In Fig. 31.4 , $T < T_c$ and $h > 0$. Of the three points of intersection that mark solutions to eq. (31.11), only the one at the far right, with $m > 0$, is stable. The intersection to the far left is metastable, and the intersection nearest the origin is unstable. (Proofs will be left to the reader.)

It can be seen by experimenting with various parameters that there is always a stable solution with $m > 0$ when $h > 0$. Similarly, when $h < 0$ there is always a stable solution with $m < 0$.

As long as $h \neq 0$, the MFA prediction for the magnetization is an analytic function of the temperature, and there is no phase transition. This property is also true for the exact solution; a phase transition appears only in the absence of an applied magnetic field.

31.6   Critical Exponents

A curious aspect of phase transitions is that many properties turn out to have some sort of power-law behavior. For example, the magnetization near the critical temperature can be approximated by the expression

$$m \approx A|T - T_c|^\beta , \tag{31.61}$$

where A is a constant.

There is a serious source of potential confusion in eq. (31.61). The use of the symbol β as a dimensionless critical exponent is in direct conflict with the use of the same symbol as $\beta = 1/k_BT$. Unfortunately, both uses of β are so firmly rooted in tradition that an alternative notation is impossible. I can only council patience and a heightened degree of alertness when you encounter the symbol β anywhere in statistical mechanics or thermodynamics.

There is a full Greek alphabet soup of critical exponents defined in a similar manner. For example, for the specific heat,

$$c \propto |T - T_c|^{-\alpha}, \tag{31.62}$$

and for the magnetic susceptibility χ,

$$\chi = \left(\frac{\partial m}{\partial h} \right)_{h=0} \propto |T - T_c|^{-\gamma}, \tag{31.63}$$

$$\chi = \left(\frac{\partial m}{\partial h} \right)_{T=T_c} \propto |h|^{1/\delta}. \tag{31.64}$$

Correlations between spins can be described by a correlation function, which gives rise to additional critical exponents. The correlation function has the following asymptotic form for large separation $r \equiv |\vec{r}_j - \vec{r}_k|$ between spins,

$$f(r) \equiv \langle \sigma_j \sigma_k \rangle - \langle \sigma_j \rangle \langle \sigma_k \rangle \propto r^{-d-2+\eta} \exp[-r/\xi]. \tag{31.65}$$

An exponent η describes the power-law fall-off of the correlation function. The correlation length ξ, in the exponent in eq. (31.65), is the characteristic size of fluctuations near T_c. ξ diverges with an exponent ν,

$$\xi \propto |T - T_c|^{-\nu}. \tag{31.66}$$

All of these critical exponents can be evaluated in MFA, as illustrated in the following subsection.

31.7 Mean-Field Exponents

We can derive the value of β *within the mean-field approximation* by making use of the expansion of the hyperbolic tangent in eq. (31.67)

$$\tanh x = x - \frac{1}{3} x^3 + \cdots . \tag{31.67}$$

Near the critical point, the magnetization m is small, so that the argument of the hyperbolic tangent in eq. (31.52) is also small. We can therefore use eq. (31.67) to approximate eq. (31.52)

$$m = \tanh(\beta \mathcal{J} z m) \approx \beta \mathcal{J} z m - \frac{1}{3} (\beta \mathcal{J} z m)^3 + \cdots . \tag{31.68}$$

One solution of eq. (31.68) is obviously $m = 0$, which is the only solution above T_c. Below T_c, the non-zero solutions for the magnetization are stable, and to find them we can divide eq. (31.68) by $m \neq 0$. Solving the resulting equation for the magnetization, we find

$$m^2 = 3 \left(\frac{T}{T_c} \right)^2 \left(1 - \frac{T}{T_c} \right) + \cdots . \tag{31.69}$$

Close to T_c, the magnetization behaves as

$$m = \sqrt{3} \left(\frac{T}{T_c} \right) \left(1 - \frac{T}{T_c} \right)^{1/2} + \cdots , \tag{31.70}$$

so that we find

$$\beta_{MFA} = \frac{1}{2}. \tag{31.71}$$

The calculation of each of the critical exponents can be carried out within the mean-field approximation. However, it is only fair to let the reader have some of the fun, so we will leave the evaluation of the other critical exponents as exercises.

31.8   Analogy with the van der Waals Approximation

In Chapter 17 we discussed the van der Waals fluid and showed that it predicted a transition between liquid and gas phases. Although we did not emphasize it there, the van der Waals fluid is also an example of a mean-field theory. The interactions between particles are approximated by an average (mean) attraction strength (represented by the parameter a) and an average repulsion length (represented by the parameter b).

In the van der Waals model, the difference in density between the liquid and gas phases was found to be given by the expression

$$\rho_{liq} - \rho_{gas} \propto |T - T_c|^{\beta_{vdW}} \tag{31.72}$$

close to the critical point.

Eq. (31.72) for the discontinuity in density in the vdW fluid is directly analogous to eq. (31.61) for the discontinuity in magnetization in the Ising model from $+|m|$ to $-|m|$ as the applied magnetic field changes sign. Indeed, the analogy between these two models can be extended to all critical properties. Just as the density discontinuity is analogous to the magnetization, the compressibility is analogous to the magnetic susceptibility.

The analogy between the vdW fluid and the MFA Ising model even extends to both models having the same value of the critical exponent with a value of $\beta_{vdW} = 1/2 = \beta_{MFA}$. Indeed, all critical exponents are the same in both models, which will be discussed in the next section.

31.9 Landau Theory

The great Russian physicist, Lev Davidovich Landau (1908–1968) made a remarkable generalization of the mean-field approximation. His theory included not only the van der Waals and Ising models that we have discussed, but any extension of them with more distant or more complicated interactions.

Landau first introduced the concept of an 'order parameter'; that is, an observable quantity that describes a particular phase transition. For the Ising model, the order parameter is the magnetization, which measures the degree of ordering of the spins. For a fluid, the order parameter is given by the discontinuity of the density. In both cases, the order parameter vanishes above the critical temperature, T_c.

To indicate the generality of the theory, we will denote the order parameter by ψ. Landau postulated that the free energy F should be an analytic function of ψ, as indeed it is for both the vdW fluid and the MFA model of ferromagnetism. Since we are interested in the behavior near the critical temperature, we expect ψ to be small, so Landau expanded the free energy in powers of ψ

$$F = h\psi + r\psi^2 + u\psi^4 + \cdots. \tag{31.73}$$

In eq. (31.73), the first term indicates the interaction with an external magnetic field, which is why h multiplies the lowest odd power of ψ. There will, of course, also be higher-order odd terms, but we will ignore them in this introduction to the theory. For a magnetic model, all odd terms vanish when $h = 0$ because of the symmetry of the Hamiltonian.

The thermodynamic equilibrium state will be given by the value of the order parameter that corresponds to the minimum of the free energy. This leads us to assume that $u > 0$, because otherwise the minimum free energy would be found at $\psi = \pm\infty$. With this assumption we can find the minimum by the usual procedure of setting the first derivative equal to zero. Ignoring higher-order terms, we find

$$\frac{\partial F}{\partial \psi} = h + 2r\psi + 3u\psi^3 = 0. \tag{31.74}$$

First look at the solutions of eq. (31.74) for $h = 0$. There are two solutions. The simplest is $\psi = 0$, which is what we expect for $T > T_c$. The other solution is

$$\psi = \pm\sqrt{\frac{-2r}{3u}}. \tag{31.75}$$

Clearly, there is only a real solution to eq. (31.74) for $r < 0$. This led Landau to assume that r was related to the temperature, and to lowest order,

$$r = r_o(T - T_c). \tag{31.76}$$

Eq. (31.75) now becomes

$$\psi = \pm\sqrt{\frac{2r_o\,(T_c - T)}{3u}}.$$

(31.77)

Since the definition of the critical exponent β in Landau theory is that for very small ψ,

$$\psi \propto (T_c - T)^\beta,$$

(31.78)

we see that quite generally, $\beta = 1/2$.

All other critical exponents can also be calculated within Landau theory, and their values are identical to those obtain from MFA or the vdW fluid. The derivations will be left to the reader.

The calculation of the critical exponents is remarkable for two reasons. The first is that the same set of values are obtained from Landau theory for *all* models. This result is known as 'universality'. It is a very powerful statement. It says that the particular interactions in any model of interest are completely unimportant in determining the power laws governing critical behavior. The range, strength, or functional form of the interactions make no difference at all.

With some important limitations, universality is even true!

The second reason for the importance of Landau theory is that the particular values he found for the critical exponents are wrong! They disagree with both experiment and some exact solutions of special models.

The discovery that the Landau theory is wrong was a disaster. A great strength of Landau theory is its generality; tweaking parameters has no effect on the values of the critical exponents. Unfortunately, that means that if those values are wrong, there is no easy way to modify the theory to obtain the right answer.

31.10 Beyond Landau Theory

The resolution of the crisis created by the disagreement of Landau theory with experiment was not achieved until 1971, when Kenneth G. Wilson (American physicist, 1936–2013) applied renormalization-group theory to the problem. The key to the solution turned out to be that at a critical point, fluctuations at all wavelengths interact with each other; approximating the behavior of the system without including all interactions between different length scales always leads to the (incorrect) Landau values of the critical exponents.

The story of how physicists came to understand phase transitions is fascinating, but unfortunately would take us beyond the scope of this book. On the other hand, the intention of this book is only to be a starting point; it is by no means a complete survey.

We come to the end of our introduction to thermal physics, which provides the foundations for the broad field of condensed matter physics. It is my hope that you are now well prepared for further study of this field, which is both intellectually challenging and essential to modern technology.

31.11 Problems

....

PROBLEM 31.1
An Ising chain

We have solved the one-dimensional Ising chain *with free boundary conditions* for either $\mathcal{J} = 0$ or $h = 0$. We also solved the one-dimensional Ising chain *with periodic boundary conditions* with both $\mathcal{J} \neq 0$ and $h \neq 0$.

  For this assignment we will return to the one-dimensional Ising chain with $\mathcal{J} \neq 0$, but $h = 0$.

1. Use the same matrix methods we used to solve the general one-dimensional chain to find the free energy with $\mathcal{J} \neq 0$, but $h = 0$. (Do not simply copy the solution and set h equal to zero! Repeat the derivation.)
2. Find the free energy per site F_N/N in the limit of an infinite system for both free boundary conditions and periodic boundary conditions. Compare the results.
3. The following problem is rather challenging, but it is worth the effort.

    A curious problem related to these calculations caused some controversy in the late 1990s. Consider the following facts.

    (a) Free boundary conditions.

        The lowest energy level above the ground state corresponds to having all spins equal to $+1$ to the left of some point and all spins equal to -1 to the right. Or vice versa. The energy of these states is clearly $2\mathcal{J}$, so that the thermal probability of the lowest state should be proportional to $\exp(-2\beta\mathcal{J})$, and the lowest-order term in an expansion of the free energy should be proportional to this factor.

    (b) Periodic boundary conditions.

        The ground state still has all spins equal to 1 (por -1), but the deviation from perfect ordering must now occur in *two* places. One of them will look like

        and the other will look like

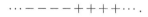

        The energy of these states is clearly $4\mathcal{J}$, so the thermal probability of the lowest state will be $\exp(-4\beta\mathcal{J})$ *for all N*. In this case, the lowest-order term in an expansion of the free energy should be proportional to $\exp(-4\beta\mathcal{J})$.

In the limit of an infinite system, this difference is not seen in the free energies per site. Can you find the source of this apparent contradiction?

PROBLEM 31.2
Mean-Field Approximation (MFA) for the Ising model

We derived the basic MFA equation for the magnetization in class,

$$m = \langle \sigma_j \rangle = \tanh(\beta z \mathcal{J} m + \beta h)$$

z is the number of nearest neighbors. For a two-dimensional square lattice, $z = 4$. We found the critical temperature to be given by

$$k_B T_c = z \mathcal{J}.$$

1. Consider the case of zero magnetic field ($h = 0$). For temperatures below the critical temperature, the magnetization is non-zero. As $T \to T_c$, the magnetization goes to zero as a power law

$$m \propto (T_c - T)^\beta .$$

   Find the value of β.
2. For temperatures above the critical temperature, find an exact solution for the magnetic susceptibility

$$\chi \equiv \frac{\partial m}{\partial h}$$

   at zero field.

   The magnetic susceptibility diverges at the critical temperature with an exponent called γ.

$$\chi \propto (T - T_c)^{-\gamma}$$

   What is the mean-field value of γ?

Suggestions for writing programs:
The next few problems deal with Monte Carlo simulations of the two-dimensional Ising model. In writing the programs, use a sequential selection of spins for the MC steps. Even though it is not exactly correct, the differences are very small and a sequential selection is much faster. You can implement the algorithm any way you like, but I recommend using a loop of the form:

```
while n < N_MC:
    nxm = L - 2
    nx  = L - 1
    nxp = 0
```

```
while nxp < L:
    nym = L - 2
    ny  = L - 1
    nyp = 0
    while nyp < L:
        STATEMENTS INSERTED HERE FOR MC UPDATE
        nym = ny
        ny  = nyp
        nyp += 1
    nxm = nx
    nx = nxp
    nxp += 1
STATEMENTS TO RECORD MC DATA INSERTED HERE
n += 1
```

PROBLEM 31.3
Monte Carlo simulation of a $d = 2$ Ising model

Write a computer program to perform a Monte Carlo computer simulation of the two-dimensional Ising model in zero magnetic field,

$$H = -\mathcal{J} \sum_{\langle i,j \rangle} \sigma_i \sigma_j$$

where the sum is over nearest-neighbor pairs of sites on a square lattice.

1. Write a program to perform a Monte Carlo simulation of the two-dimensional Ising model. You should be able to simulate any size $L \times L$ lattice, with L determined at the beginning of the program. You should be able to set the temperature T at the beginning of the program. You can also write the program for zero magnetic field. (We will use non-zero fields in later problems.)
2. Beginning with a random configuration on a fairly large lattice ($L = 32$), perform a single run of 20 MC sweeps at $T = 5.0$. and look at the printout of the configuration.
3. Beginning with a random configuration on a fairly large lattice ($L = 32$), perform a single run of 20 MC sweeps at $T = 3.0$. and look at the print-out of the configuration.
4. Beginning with a random configuration on a fairly large lattice ($L = 32$), perform a single run of 20 MC sweeps at $T = 2.0$, and look at the print-out of the configuration.
5. Compare the pictures of the configurations that you have generated at $T = 5.0, 2.0, 2.0$. What do you see?
6. Beginning with a random configuration on a fairly large lattice ($L = 16$ or $L = 32$), perform several runs with about 50 or 100 MC sweeps at $T = 0.5$. Keep the length of the runs short enough so that you can do it several times without wasting your time. Look at the print-outs of several configurations. If you see something that seems unusual, print it out. Explain why you think it is unusual.

PROBLEM 31.4
Monte Carlo simulation of a $d = 2$ Ising model (revisited)

For this problem we will modify the MC program to make a series of runs with temperatures separated by a value ΔT.

1. Use the Python function definition feature to create a function that does the entire MC simulation. Then put the function call inside a loop that first equilibrates the spins for some number of sweeps and then generates the data you want from a longer number of sweeps. The loop should change the temperature by an amount ΔT for every iteration of equilibration and data taking.
2. Calculate the average energy, specific heat, magnetization, and magnetic suscept-abililty for temperatures of $T = 2.0, 4.0$, and 6.0 for an 8×8 lattice. Use a large enough numbers of MC updates to obtain reasonable results, but not so many that you waste your time sitting in front of a blank computer screen.
3. What do you notice about the data from your MC simulation?

   This is a fairly open-ended question. The main purpose is to think about the results of the simulation.

PROBLEM 31.5
More Monte Carlo simulations of a two-dimensional Ising model

The Hamiltonian of a two-dimensional Ising model is: use periodic boundary conditions,

$$\mathcal{H} = -\mathcal{J} \sum_{<i,j>} \sigma_i \sigma_j - h \sum_j \sigma_j.$$

Recall that the notation $< i,j >$ denotes the nearest-neighbor pair i and j. For this entire assignment, take $h = 0$.

   Use or modify the program you wrote for an earlier problem to do the following simulations.

1. Simulate a 16×16 lattice with $\mathcal{J} = 1.0$. Begin with giving each spin a random value of $+1$ or -1. Scan the temperature from 5.0 to 0.5 at intervals of $\Delta T = -0.5$. Use about 100 or 1000 MC sweeps for each temperature, depending on the speed of your computer.

   Make a table of the results and plots of m vs. T, E vs. T, c vs. T, and χ vs. T.

   Does it look as if the critical temperature is at the mean-field value of $k_B T_c = 4\mathcal{J}$?
2. Repeat the sweeps, but this time choose the temperature range to concentrate on the peaks in the specific heat and the magnetic susceptibility. Estimate the location and height of the peak in specific heat and magnetic susceptibility. These will be estimates of the critical temperature.
3. And now for something completely different.

Do a simulation of a 32×32 lattice with *negative* temperature, and print out the spins at the end of the run. Begin with the spins in a random state $(T = \infty)$, and simulate for 100 sweeps at $T = -1.0$.

What do you see?

PROBLEM 31.6
A one-dimensional, spin-one model of magnetism

Consider a one-dimensional macroscopic model with spin-one quantum mechanical magnetic moments located on each of N sites. The Hamiltonian is

$$H = -\mathscr{J} \sum_{j=1}^{N-1} \sigma_j \sigma_{j+1} - h \sum_{j=1}^{N} \sigma_j - D \sum_{j=1}^{N} \sigma_j^2$$

where each σ_j takes on the values $-1, 0,$ or $+1$, $\mathscr{J}$ is an exchange interaction parameter, h is a magnetic field, and D is a parameter representing a 'crystal field'. The entire system is in contact with a thermal reservoir at temperature T.

1. Write down a formal expression for the partition function of this system. You do not have to evaluate any sums or integrals that appear in your expression for this part of the problem.
2. For general values of the parameters, derive a formal expression for the quantity

$$Q = \left\langle \sum_{n=1}^{N} \sigma_n^2 \right\rangle$$

in terms of a derivative of the partition function with respect to an appropriate parameter.
3. For $\mathscr{J} = 0$, but arbitrary values of h and D, calculate the partition function for this system.
4. For $\mathscr{J} = 0$, but arbitrary values of h and D, calculate the free energy of this system.
5. For $\mathscr{J} = 0$ and $h = 0$, calculate the entropy of this system.

Appendix
Computer Calculations and Python

There are a number of problems in this book that require computer calculations. These problems are not supplementary material; they are an essential part of the book. They are designed to provide a deeper insight into the physics of many-particle systems than would be possible through analytic calculations alone. This Appendix provides sample programs to support your own programming.

There are definite advantages to writing computer programs yourself. Writing a program requires focussing on details of the mathematical structure of a problem and, I hope, understand the physics more deeply.

The choice of programming language is up to you. Over the life of this book, programming languages will certainly change. However, the basic ideas of computer programming will continue to be useful and applicable to new improvements and developments.

If you are new to computation, I have changed my recommendation for the choice of programing language from that given in the first edition. I have found that MatPlotLib provides easier and more flexible commands for making plots, and the plots are clearer than with VPython. The Python programming itself is virtually unchanged. All VPython programs should run as regular Python programs, if you remove the VPython import commands at the top of the program, and replace the VPython plotting commands.

A.1 MatPlotLib

MatPlotLib is easily available on the Web. The tutorial is excellent. It was used to make the plots in figures 11.1 and 11.2.

The plotting commands are simpler than those for VPython, which are described later in this Appendix. The simplest way to describe them is just to give you an example program.

```
import random
import sys
import numpy as np
import math
import matplotlib.pyplot as plt
import scipy
from types import *
from time import clock, time

t0 = clock()

plt.close("all")
WIDE = 1              # 3
WIDE_MARK = 5.0       # 5.0
```

```
x_0 = 2.0
dx = 0.4
x_1 = 3.0
y_0 = 0.5
N = 10

bar_width =   0.5 * dx

x_plot = []
x_plot_bar = []
y_plot = []
n = 0
while n < N:
    x = x_0 + n * dx
    y = ( x - x_1 )**2  + y_0
    x_plot += [ x ]
    x_plot_bar += [ x - 0.5 * bar_width ]
    y_plot += [ y ]
    n += 1

plt.figure( 1 )

plt.plot( x_plot, y_plot, "-",  label = "curve", markersize = WIDE_MARK,
linewidth = WIDE )
plt.plot( x_plot, y_plot, "ko",  label = "dots", markersize = WIDE_MARK,
linewidth = WIDE )
plt.xlabel( "x" , fontsize = "xx-large" )
plt.ylabel( "y" , fontsize = "xx-large" )
plt.title ( 'First_example' )
plt.legend(loc = "lower_right", fontsize = "large" )

plt.figure( 2 )

plt.bar( x_plot_bar, y_plot, bar_width, color='r', label = "bar" )
plt.plot( x_plot, y_plot, "ko",  label = "dots", markersize = WIDE_MARK,
linewidth = WIDE )
plt.xlabel( "$x$" , fontsize = "xx-large" )
plt.ylabel( "$y$" , fontsize = "xx-large" )
plt.title ( 'Second_example' )
plt.legend(loc = "upper_left", fontsize = "large" )

plt.show()

print "\n_Plotting_time_=_" , '%_10.4f' % (clock() - t0)
```

Try running this program (it should only take about .25 seconds), and then change various commands to see what happpens.

A.2 Python

The examples presented here use the Python programming language, which I strongly recommend to newcomers to computation. It runs on any computer using any operating system. It is easy to learn, even if you've never programmed before. And the price is right: you can download it for free!

I've provided some information about Python programs to help you get started. The first program I've given you is almost—but not quite—sufficient to carry out the first assignment. However, if you know what every command means and what you want the program to do, the necessary modifications should be fairly straightforward. If there is anything about programming you do not understand, do not be shy about asking the instructor.

A.3 Histograms

Since an important part of the first program is a print-out of a histogram, it is important to understand how histograms work.

A histogram simply records a count of the number of times some event occurs. In our first program, 'OneDie.py' (given in the next section), an event is the occurrence of a randomly generated integer between '0' and '5'. In the program, the number of times the integer n occurs is recorded in the memory location '*histogram*[n]'. For example, if the number '4' occurs a total of 12 times during the course of the simulation, then the value stored in *histogram*[4] should be 12. The command

```
print( histogram[4] )
```

should contain the number '12'.

Histograms can be represented in various ways in a Python program. I have chosen to represent them by an array. The command

```
histogram = zeros(sides ,int)
```

creates an array named 'histogram' with 'sides=6' memory locations, and sets the value stored in every memory location equal to zero.

The command

```
histogram [r] =  histogram [r] + 1
```

increments the number stored in the r-th location. It could also be written as

```
histogram [r]  += 1
```

to produce the same effect. Commands of the form +=, -=, ^=, and /= probably seem strange the first time you see them, but they are very useful.

It is quite common to confuse a histogram with a list of numbers generated by the program. They are completely different concepts. The length of the histogram is given by the number of possibilities—six in the case of a single die. The length of the list of random numbers is stored in the memory location 'trials'; it can easily be of the order of a million. Printing out a histogram should not take very long. Printing out a list of random numbers can exceed both your paper supply and your patience, without producing anything useful. When writing a program, be careful that you do not make this error.

A.4 The First Python Program

The following is a sample Python program to simulate rolling a single die. Note that either single quotes or double quotes will work in Python, but they must match. Double single quotes or back quotes will give error messages.

```python
import random
import sys
from types import *
from time import clock, time

trials = 100
print( "_Number_of_trials_=_", trials)

sides = 6

histogram = zeros(sides, int)
print( histogram )

sum = 0.0
j=0
r=0

while j < trials:
    r = int(random.random() * sides )
    histogram[r] = histogram[r] + 1
    j=j+1

j=0
while j < sides:
    print(histogram[j], histogram[j] - trials/sides)
    j=j+1
```

The first four lines in the program given previously are called the 'header' and should simply be copied at the beginning of any program you write. Blank lines are ignored, so for clarity they can be put in anywhere you like. Indentations are meaningful: in the *while* command, every indented command will be carried out the required number of times before going on the next unindented command.

The function

```python
random.random()
```

produces a (pseudo-)random number between 0.0 and 1.0. Multiplying it by 6 produces a random number between 0.0 and 6.0. The command

```python
r = int(random.random() * sides )
```

then truncates the random number to the next lowest integer and stores the result in r, so that r is an integer between 0 and 5.

The program given previously should run exactly as written. Copy the program to your computer and try it! If it does not work, check very carefully for typographical errors.

Warning: Computers have no imagination. They do exactly what the program tells them to do—which is not necessarily what you want them to do. Sometimes a computer can recognize an error and tell you about it; but life being what it is, the computer will probably not recognize the particular error that is preventing your program from running properly. This might be another consequence of the Second Law of Thermodynamics, as discussed in Chapter 11.

Make a permanent copy of the program given previously, and then make your modifications in a new file. This is a good policy in general; base new programs on old ones, but do not modify a working program unless you have a back-up copy.

Experiment with making modifications. You cannot break the computer, so do not worry about it. If the program just keeps running after a few seconds, you can always stop it by closing the Python shell window.

If there is anything about the program that you do not understand, ask someone. If you understand everything in this program, the computational work for all of the rest of the problems will be much easier.

A.5   Python Functions

The Python language allows you to define functions in a very natural way. They are really quite easy to write, and can simplify your program enormously.

The following code is an example for defining a function to calculate the energy of a simple harmonic oscillator:

```
def ENERGY( x , K ):
    energy = 0.5 * K * x * x
    return energy
```

The energy E can then be calculated for any value of position x and spring constant K by the command:

```
E =  ENERGY( x , K )
```

One peculiarity of Python is that you must define a function *before* you use it in the program.

There is another interesting feature of Python that can affect how you use functions. If you are using 'global' variables—that is, variables defined in the main program—*and not changing their values in the program,* you do not have to include them in the list of arguments. For example, the code could be written as:

```
def ENERGY( ):
    energy = 0.5 * K * x * x
    return energy
```

This would give the same answer as before, as long as both 'K' and 'x' had been defined before 'ENERGY()' was called.

However, be careful if you want to change a global variable in a program. As soon as you write 'x=...', 'x' becomes a local variable, and Python forgets the global value. This can be a source of much annoyance. If you want to change a global variable inside a function, include a statement such as 'global x' in the function for each global variable.

Curiously enough, if you are using a global array, you *can* change its values in a function without declaring it to be global. I am sure that there is a very good reason for this discrepancy, but I do not have any idea of what it might be.

A.6   Graphs

The following section is from the first edition. As I mentioned earlier, I now recommend MatPlotLib instead

VPython has very nice programs to produce graphs. However, they can be rather daunting the first time you use them. The following VPython program provides examples for plotting multiple functions in a single graph. Three different ways to plot functions are demonstrated. A second window is also opened at a different location on the screen with yet another function.

```
from visual import *
from visual.graph import *
import random
import sys
from types import *
from time import clock, time

# Program to demonstrate various plotting options

MAX = 30

# The next command opens a plotting window

graph1 = gdisplay(x=0, y=0, width=400, height=400,
            title='Trig_functions',
                  xtitle ='x_',
                  ytitle ='y_',
          #xmax = 1.0 * MAX ,
          #xmin = 0.0 ,
            #ymax = 100.0,
              #ymin =0.0,
              foreground=color.black, background=color.white)

k=1
Apoints = []
while k < MAX:
    Apoints += [(k * pi / MAX, cos( k * pi / MAX ) )]
    k=k+1

data1 = gvbars(delta =2./MAX, color=color.red)
data1.plot( pos=Apoints)

k=1
Bpoints = []
```

```
while k < MAX:
    Bpoints += [(k * pi / MAX,  -sin( k * pi / MAX )    )]
    k=k+1

data2 = gdots(color=color.green)
data2.plot(pos = Bpoints)


k=1
Cpoints = []
while k < MAX:
    Cpoints += [(k * pi / MAX,  sin( k * pi / MAX ))]
    k=k+1

data3 = gcurve(color=color.blue)
data3.plot(pos = Cpoints)

# The next command opens a second plotting window

graph2 = gdisplay(x=401, y=0, width=400, height=400,
          title='Second_plot',
                 xtitle='X_',
                 ytitle='Y_',
          xmax = 2.0 ,
          #xmin = 0.0,
            #ymax = 1.10,
            #ymin =0.0,
          foreground=color.black , background=color.white)

k=0
Dpoints = []
while k < MAX:
    Dpoints += [(2.0 * k / MAX,  exp( - 2.0 * k / MAX ) )]
    k=k+1

data4 = gcurve(color=color.blue)
data4.plot( pos=Dpoints )
```

A.7   Reporting Python Results

Results of computational problems, including numbers, graphs, program listing, and comments, should be printed out neatly. Here are some hints for doing it more easily in LaTeX.

A.7.1   Listing a Program

LaTeX has a special environment that was used to produce the program listings in this Appendix. To use it, you need to include the following statement in the header:

```
\usepackage{listings}
\lstloadlanguages{Python}
\lstset{language=Python,commentstyle=\scriptsize}
```

The command for the listing environment is:

```
\begin{lstlisting}
      Program goes here
\end{lstlisting}
```

A.7.2   Printing—Pretty and Otherwise

The printed output of a Python program is always found in the Python shell window. The simplest procedure is just to copy it into your favorite word processor. However, you might want to clean up the output of your Python program first by specifying the width of a column and the number of significant digits printed. Python and Python have formatting commands that are inserted into a 'print' statement. I think the commands are most easily explained by giving an example.

Example: Setting Column Width

As an illustration of the default output, consider the following statements:

```
print( "spin", "hist", "hist-average", "freq", "freq-1/sides")
```

and, indented inside a loop,

```
    print( j, h, hDiff, hOverTrials, hOSdiff)
```

produce the following output:

```
spin hist hist-average freq freq-1/sides
2 0.333333333333 0.2 0.0333333333333
```

The more highly formatted statement (on multiple lines in this example, with commas indicating the continuations)

```
print( '%_6s' % "spin", '%_7s' % "hist",
         '%_10s' %"deviation", '%_8s' %"freq",
         '%_12s' %"deviation")
```

```
print( '%_6d' % j, '%_6.2f' % h, '%_10.4f' % hDiff,
          '%_8.4f' %h OverTrials, '%_12.4f' % hOSd)
```

produce the following output:

```
spin    hist   deviation    freq    deviation
  0   165.00    -1.3333    0.1650    -0.0017
```

In the expression '6s', the '6' refers to the column width, and 's' means that a string (letters) is to be printed. In the expression '6d', the '6' again refers to the column width, and 'd' means that an integer is to be printed. In the expression '10.4f', the '10' refers to the column width, the '.4' refers to the number of digits after the decimal point, and 'f' means that a floating point number is to be printed.

A.7.3   Printing Graphs in LaTeX

Under any operating system, there are efficient utility programs to capture an image of the screen, a window, or a selected area. Use them to make a file containing the output of the graphics window and copy it into your word processor. If you are using LaTeX, as I am, you can make a pdf file and then put it into your LaTeX file as follows:

```
\begin{figure}[htb]
\begin{center}
 \includegraphics[width=4in]{name_of_pdf_file}
  \caption{Caption goes here}
  \label{your choice of label here}
\end{center}
\end{figure}
```

You will also need to include

```
\usepackage{graphicx}
```

in your header.

A.8   Timing Your Program

If your program includes the header command

```
from time import clock, time
```

timing your program, or parts of your program, is easy. Placing the command

```
t0 = clock()
```

at the beginning of your program and something like

```
ProgramTime = clock() - t0
print( "_Program_time_=_" , ProgramTime )
```

at the end of the program, produces a printout of how long the program took to run. This information can be very useful in checking the efficiency of you program and planning how many iterations you should use in doing the assignments.

There is also a function 'time()' that can be used in the same way as 'clock()'. The difference is that 'clock()' will calculate how much CPU time was used by your program, while 'time()' will calculate how much real time has passed. Both are useful, but 'clock()' gives a better measure of the efficiency of the program.

A.9   Molecular Dynamics

The key loop in a molecular dynamics (MD) program is illustrated next. The function 'FORCE' is assumed to have been defined earlier, along with the parameter K.

```
# the velocity is assumed to have been initialized
# before the beginning of the while-loop
n=0
while n < N_MD:
    n += 1
    x        += dt * velocity
    velocity += dt * FORCE( x, K)
```

The algorithm given previously is suggested to simplify the programming. A better iteration method is the 'leap-frog' algorithm, in which the positions are updated at even multiples of $\delta t/2$, while velocities are updated at odd multiples. If you are feeling ambitious, try it out. You can check on the improvement by looking at the deviation from energy conservation.

A.10   Courage

The most important thing you need to run computer programs is courage. For example, when you are asked to run the first program for different numbers of trials, do not be afraid of using a large number of trials. Do not just use five trials vs. eight trials. Start with a small number such as 10, but then try 100, 1000, or higher. As long as the computer produces the answer quickly, keep trying longer simulations. However, if the computer takes more than about one second to finish, do not use longer simulations—your time is also valuable.

Index

Printed and bound by CPI Group (UK) Ltd, Croydon, CR0 4YY